BUILDING ROBOT DRIVE TRAINS

Other Titles in the TAB Electronics Robot DNA Series

Programming Robot Controllers by Myke Predko

Constructing Robot Bases by Gordon McComb

Other Great Books for Robot Builders

The Robot Builder's Bonanza, Second Edition, by Gordon McComb

Robot Builder's Sourcebook by Gordon McComb

Robots, Androids, and Animatrons, Second Edition, by John Iovine

Build a Remote-Controlled Robot by David R. Shircliff

Combat Robots Complete by Chris Hannold

Lego Mindstorms Interfacing by Don Wilcher

Building Robot Drive Trains

Dennis Clark
Michael Owings

McGraw-Hill
New York Chicago San Francisco Lisbon
London Madrid Mexico City Milan
New Delhi San Juan Seoul Singapore
Sydney Toronto

The McGraw·Hill Companies

Cataloging-in-Publication Data is on file with the Library of Congress

1 2 3 4 5 6 7 8 9 0 DOC/DOC 0 8 7 6 5 4 3 2

ISBN 0-07-140850-9

The sponsoring editor for this book was Scott L. Grillo and the production supervisor was Pamela A. Pelton. It was set in Century Schoolbook by MacAllister Publishing Services, LLC.

Printed and bound by RR Donnelley.

This book is printed on recycled, acid-free paper containing a minimum of 50 percent recycled, de-inked fiber.

CONTENTS

Dedication

For Shannon and Brendan, who, although they tried not to be, were quite a distraction—*DC*

For Glenn, Silas, and Avery—*MO*

Acknowledgments

A book of this type can't be written without a lot of help. We'd like to acknowledge the wisdom, experience, and help we've received from

Brian Petty of Servo City, for being a "techy" hobby servo kind of guy, and sharing.

The Seattle Robotics Society, The Encoder, and all the hobbyists who write for it: Wow.

The folks at Allegro Micro, for having an outstanding web site, and everyone else out there whose web sites gave us insight and inspiration.

Ray Kruse (www.johngalt.biz), for his crotchety-yet-helpful evaluation of our take on PID control loops, as well as all of those who helped with material and technical review.

Tony Ellis (www.conceptioneering.co.uk), for the reed-relay current sensing idea.

US Digital (www.usdigital.com), ActivMedia Robotics (www.activmedia.com), and Carnegie Mellon University (www.cmu.edu), for providing photographic material.

Glenn, for her patient (practically undying) support for this little literary adventure.

And finally,

Gordan McComb, our series editor, for his help in getting us through this.

Scott Grillo, our publisher, for being patient with a couple of first-time book authors.

About the Authors

Dennis Clark has degrees in electrical engineering technology and computer science and is a graduate student in behavioral robotics at Colorado State University. Mr. Clark has authored a series of articles on behavioral robotics for European hobbyist magazine *Elektor*.

Michael Owings is a freelance software developer. He lives and works in southern Louisiana.

INTRODUCTION: THE BASICS OF LOCOMOTION

Making Your Robot Move

Wherever amateur roboticists gather—online or at local club meetings—there's always a chance that at some point, that old question will arise:

What makes a robot a robot? Inevitably, the ensuing discussion becomes a heated debate or the newsgroup thread begins to grow and threaten to stretch on into perpetuity. Does a robot have to interact with its environment to be a real robot? Must it carry some kind of computer on board? Are sensors required? Are those Robot Wars robots really true robots or just armored remote control cars? And what about BEAM?[1] Everyone has an opinion, and so do we, but in the interest of avoiding controversy, we'll keep it to ourselves except to say this:

If it doesn't move, it isn't a robot.[2] You may build yourself the finest scale replica of R2-D2 seen outside of a Hollywood design studio. Without the capability to move, however, your creation is little more than statuary. Add in a microcontroller and a few sensors, and you'll finally have, well, a statuary with a computer inside. Although we certainly may concede the point that not all things that move are robots, we maintain that any true robot must at least be capable of *motion*.

Which is why we humbly suggest that you consider refraining from purchasing any of the other books in this series. Let's face it—they're unnecessary. Instead, invest the money we've saved you in six or seven copies of *this* book. That way you'll have extras in case you misplace one or spill diet soda on it, or accidentally set a copy alight with the lit

---

[1]BEAM robots essentially have no microcontroller for a brain, but they do move, so that's okay.

[2]Yes, we realize that so-called software robots or agents aren't capable of motion strictly speaking, but then, at the risk of sounding bigoted, software "robots" aren't really robots at all.

cigarette that might slip from your mouth as you doze on the couch.[3] Robotics is all about redundancy—and, of course, *motion*.

In this, the second book in the Robot DNA series, we will try to cover one specific type of motion, *locomotion*, the capability of a robot to move from place to place. We will primarily be focusing on the use of wheels, treads, and legs powered by variants of the standard electric motor—DC motors, stepper motors, and modified as well as unmodified hobby servos. We will *not* be discussing other means of motivation such as hydraulics, pneumatics, Nitinol wire, or air muscles; those topics range far beyond the scope of this book. Although certainly a worthwhile topic in its own right, robotic appendages, such as arms, grippers, and the like, will not be discussed here at all. Hopefully, the creative reader can find ways to adapt the methods, tips, and tricks presented here to other areas.

Prerequisites: Tools and Supplies

Most of this book has been written with the assumption that you own and are reasonably comfortable with standard hand tools. These include the following:

- Screwdrivers
- Pliers
- An assortment of files
- Hacksaws
- Clamps
- Vises
- Hammers
- Transdimensional subquantum protonic manipulator—and not one of those cheap plastic knock-offs.

We'll try to provide gentle introductions when it becomes desirable to bring more exotic items into the mix, such as tubing benders, taps, and dies.

[3]Hah! I knew you'd look here; as if we'd say "Just kidding!" We're not going to say that; you have to decide for yourself.

You should also own a few common power tools. We consider the following essential:

- A good hand drill is essential and a drill press is recommended. Suitable units can be had new for $100 or less if you shop around or buy used.
- A jigsaw with an assortment of blades.
- A Dremel or other rotary tool. This is essential for drilling circuit boards, but it has dozens of other uses.

Clearly, access to higher-end tools such as lathes, mills, and metal shear/bender/roller machines gives the lucky amateur a leg up, but again, they are not required.

All the materials we'll use in this book are available from either a well-stocked hobby shop or most large, local hardware depots. If you are unfortunate enough to live in an area where the former is unavailable, several reliable mail-order sources are included in Appendix C, "Resources." If you live in an area where the *latter* is unavailable, you're probably Amish and bought this book by mistake.

In some cases, we will be using components that you may not be able to find locally. It would be pretty unusual if your local Radio Shack carried the TI 754410 motor driver or IRFZ44 MOSFETs on the wall in back of the shop. You can, however, easily order these parts over the Internet. The components we choose will be those that are commonly available and not very expensive—so you can now start breathing again!

Skills You Need to Bring to the Party

We assume you understand how to operate these tools in a sane and safe manner. *Always* use goggles and gloves, and remember to secure your work piece. Naturally, the authors disclaim all responsibility for the loss of fingers, toes, or other appendages during the course of your robotic endeavors.

We also assume that you know which end of the soldering iron to hold because you will be creating some electronic circuits on your lab bench that in many cases you won't be able to buy. Besides, you can build some circuits more cheaply than buying them, and it's more fun to build it yourself, right? We further assume that you know and understand Ohm's

Law and can read a resistor value from the color bands. You should know the difference between ground and Vcc, and understand that all grounds in your system need to be connected together.

The Last Word

Finally, three elements must be considered when building a robot. They are related and cannot be separated or done independently: the environment your robot will inhabit, the task your robot is to perform, and the sensors on the robot itself. If you change any one of these elements in isolation from the other two, your robot may not function optimally or indeed may not function at all. A robot that does not have a specific task to do or is not designed properly for its environment is a robot that will never leave the workbench. You don't need to be exotic in your specifications and design, but you do need to consider all three elements if your robot is to survive and prosper. It is our goal that this book will help you get your robot running properly in your environment and speed you towards your robotic goals.

Most of the techniques presented here require nothing more than a bit of common sense, a steady hand, and a little patience. We assume no particular mechanical skills on your part, just a willingness to try, learn, make the occasional mistake, and—above all—to have fun.

Dennis Clark
Michael Owings

CHAPTER 1

The Basics of Robot Locomotion

This book deals primarily with locomotion via legs, wheels, or treads. Although the complexity inherent in multipod construction may seem obvious, you should also bear in mind that the humble wheel also presents its own unique challenges. We'll try to provide a quick overview of those challenges in this chapter. Subsequent chapters will delve into these subjects in much greater detail. You are encouraged to skip ahead to the more "meaty" material as the mood strikes you—we won't mind.

Your Robot's Niche

In recent years there has been an evolving trend in academic robotics toward looking to nature for solutions to the complex problems encountered in robot design. In particular, much has been drawn from biology and related subfields, such as entomology, ichthyology, and others. One area of great interest to roboticists has been the field of ethology, the study of the behavior of animals in their natural setting. An ethologist, for the most part, studies his or her subject in the wild, as opposed to a laboratory. For an ethologist, an organism and its environment are virtually inseparable.

We would urge you, when considering the design of your robot, to approach your subject in much the same way as an ethologist or other field biologist might—as a thing bound to a particular place. Think about the environment in which your robot is expected to operate as well as—broadly, at least—what kinds of behaviors you expect it to exhibit. By constraining your design in this way, you are far more likely to end up with a successful project. Attempt to design a general-purpose, indoor/outdoor, soccer-playing vacuum cleaner/BattleBot, and your robot is not likely to get far past the design stage.

Indoor Environments

Most robots built by amateurs are designed for use indoors if for no other reason than because the design challenges are substantially fewer. Plenty of exceptions exist, but because indoor robots tend to be the most common, we'll cover indoor robots first. We can divide indoor robot environments broadly into two categories:

- **Controlled environments** Controlled environments include mazes, sumo rings, the model houses used in firefighting robot contests, and others. These environments tend to be highly constrained with few surprises that might affect drive train design. Such robots tend to be designed with a specific task in mind; drive trains will tend to be correspondingly simple, often just wheels connected directly to a couple of drive motors.

- **Uncontrolled environments** Robots designed to be able to roam unconstrained or only moderately constrained within indoor spaces often face more terrain challenges than robots built to operate in controlled areas, such as contest arenas. Furniture and people or pets are among the more obvious obstacle types. Depending on the size of your robot, you'll also need to design for obstacles such as raised door thresholds, deep carpet, loose carpet, areas where rugs meet the floor, and more. You'll need to be able to handle ground clutter like clothing, shoes, and, if children are around, toys left on the floor. Of course, stairs or other "step downs" will also be things that you will at least want to be able to sense and avoid.

Outdoor Environments

Outdoor environments tend to complicate design by orders of magnitude. In addition to a wider variety of unexpected obstacles than might be encountered indoors, you will also need to take the following factors into account when you plan your robot:

- **Protection from dirt and moisture** You will want to be sure that dirt and grime stay out of motor housings and bearings, as well as, in the case of open gearboxes, the gears themselves. If you'll be puddle hopping, you'll likely want a means to keep water away from drive train mechanisms.
- **Weatherizing** Rain is an obvious weather hazard, but you'll also need to take into account possible problems associated with temperature extremes and excess humidity. You should also be aware that temperature extremes can have particularly nasty effects on microprocessor timing, as most timing circuits are susceptible to some degree to temperature-related errors.
- **Vibration damping** Vibration is an issue for almost all robots designed to work outdoors, even for those that will be confined to sidewalks, driveways, or roads. You may want to use vibration-damping techniques to mount motors or wheels to mitigate these effects. Electrical connections will need to be sturdier, and socketed

boards may require extra bracing so that they do not work loose over time.

Coping with Terrain Challenges

Although exceptions exist, most robots are likely to encounter the occasional bump or hole. You'll need to decide in advance what constitutes a bump or an obstacle for your robot. The operational difference between the two is simply this: bumps your robot goes over; obstacles it goes around. Will it be okay to run over a toy left carelessly in the robot's path? What about that new puppy?

The Role of Size and Weight Smaller and lighter robots will invariably have more difficulty in uncontrolled environments than larger robots will. A raised door threshold can become an insurmountable wall; an otherwise barely noticeable bump in the carpet can cause drive wheels to become raised unexpectedly, "hanging" the robot until someone comes along and fixes the problem.

Larger robots are generally affected less by small obstacles. With their (usually) larger wheel size and more powerful motors, they can often negotiate small environmental hazards reasonably well. At first glance, a larger robot may seem a better choice for unconstrained indoor environments.

Unfortunately, the design of larger robots presents its own set of challenges:

- As a robot grows larger, you will need larger (and more expensive) motors.
- Batteries will need to be larger as well, and since batteries usually constitute a disproportionate share of the robot's overall weight, the robot will become much heavier, especially if you have to move to the larger capacity but less power-dense lead-acid batteries.
- Heavier robots draw more current. If you expect to encounter continuous current draws of more than about 3 amps from a motor, you will be unable to use any of the less expensive single-chip H-bridge solutions on the market. Remember that a stalled motor can draw large amounts of current, and you'll want to be prepared for worst-case scenarios. Higher-current controllers will need to be bought off the shelf or built from discrete parts.

- Drive train components, such as shaft couplers and bearings, will need to be able to handle the significantly higher mechanical loads produced by large drive motors, which can complicate mechanical design. For example, you may need to consider indirectly coupling the drive wheels to the motors (through timing belts, chains, or other mechanisms) in order to keep the motor output shaft from being too heavily loaded.

A couple of other issues do not pertain directly to drive trains, but are still worth a mention regarding the construction of larger robots:

- Larger robots tend to require sturdier body construction than smaller robots.
- Sensor coverage for obstacle detection must be greatly enlarged for larger robots. Certain sensor types become somewhat more difficult or more expensive to use.

Wheels Versus Treads Treads often seem to be the obvious choice for outdoor robots that need to negotiate tough terrain. Moreover, treaded robots just look good. However, treads add a good bit of complexity to mechanical design. Many choose to purchase treaded platforms rather than construct one from scratch. For those of you who just can't resist, we'll explore the design of treaded robots in detail in Chapter 10, "Wheels and Tank Tracks."

The coolness factor of treads aside, wheels remain a central feature of most robot designs. For robots that need to negotiate tough terrain, many commercial and research robots continue to take the "monster truck" approach: four or more large wheels and plenty of ground clearance. Figures 1-1 and 1-2 show commercial and research robots that illustrate this concept.

The primary drawback to using large drive wheels lies in the fact that the maximum available torque will tend to decrease as the wheel diameter increases. We'll discuss torque in greater detail in Chapter 2, "Motor Types: An Overview," and Chapter 3, "Using DC Motors."

Legs Although mechanically far more complex than either wheels or treads, few robots impress a crowd more than those that use legs for locomotion, perhaps because legged robots above all others seem to be poised to cross the line between nature and artifice.

In theory, robots with legs should have the greatest capability to negotiate tricky terrain. In practice, most practical designs, from commercial

Figure 1-1 The Hyperion sun-seeking robot from Carnegie Mellon University (Source: Carnegie Mellon University)

Figure 1-2 A pair of ActivMedia Robotics model PT-AT robots in the snow (Source: ActivMedia Robotics)

research platforms to planetary rovers, continue to use wheels. Artificial legs with the same *degrees of freedom* (DOF) as natural legs have proven difficult (though certainly not impossible) to design in practice, and with the current technology available to amateurs, they require a substantial investment in servos (and batteries) even for legs with just a few DOF.

Nevertheless, legged robots are certainly doable for the amateur—we'll present a number of tips and designs in Chapter 11, "Locomotion for Multipods."

DC Motors—A Short History and Explanation

DC motors are a well-established technology, combining efficiency and widespread availability with ease of control.

When we say "well-established," we mean it—the electric motor goes back to 1821 when the first one was built by the noted British scientist and inventor Michael Faraday (see Figure 1-3). Faraday placed a magnet in a shallow dish of mercury. Above this, he suspended a wire. One end of the wire was connected to one pole of a battery. The other was allowed to dangle in the dish of mercury, which was itself connected to the other pole of the battery. Being conductive, the mercury acted as a sort of liquid commutator brush, and current flowed through the wire. As it did so, the resultant magnetic field created by the flowing current pushed the wire out and around the permanent magnet in the dish. Although this first single-pole motor could produce steady circular motion in the wire,

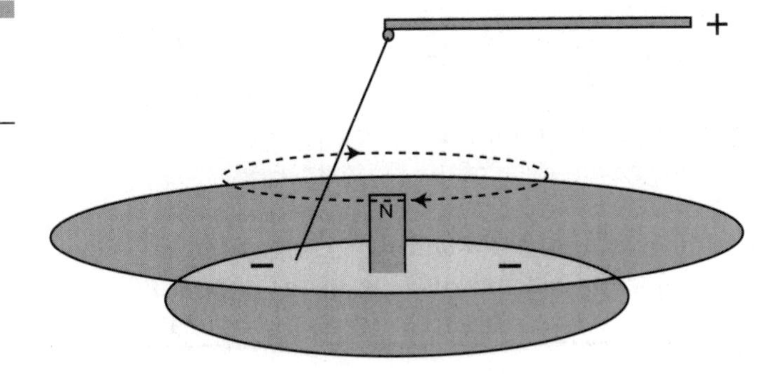

Figure 1-3 Faraday's motor

it wasn't capable of doing much real work. Faraday himself moved on to new endeavors, and it was left to others to refine the original concept into something more utilitarian.

By 1832, William Sturgeon—the British inventor of the electromagnet —had built a four-pole motor that drove a roasting spit, thus in essence creating the first modern kitchen appliance. He is better remembered, however, for inventing the brush commutator, which continues to be an integral part of most modern (nonbrushless) DC motors. Across the Atlantic in the United States, Thomas Davenport used his prototype motors to power a small model electric train and sundry shop equipment. He was granted a patent for his work in 1837.

The utility of the electric motor, however, was hobbled by the relative scarcity of electricity—at the time, acid-based batteries were almost the exclusive source of electrical power. The widespread use of electric motors in homes and industry would not occur until after the establishment of commercial generating stations and power grids. To this day, off-grid power sources for electric motors, such as batteries, remain problematic.

How an Electric Motor Works

Although the details of the electric motor have changed since the days of Davenport and Sturgeon, the fundamental elements have remained the same. The motion created by electric motors is a result of the so-called Lorenz Force. The Lorenz Force Law states that when a current-carrying conductor is placed in a magnetic field, a force (the Lorenz Force) is created orthogonally (at a right angle) to both the magnetic field flux and current flow. Figure 1-4 illustrates this phenomenon. As current flows through the conductor placed in a magnetic field, a downward force is created on the conductor.

The Lorenz Force Law is also sometimes termed *the Left-Hand Rule* (see Figure 1-5). With your left hand, point your fingers forward and your thumb up. If your fingers represent the direction of current flow, and your thumb the direction of magnetic flux (north to south), your palm faces in the direction of the Lorenz Force.

Now let's consider what happens when we place a simple one-turn coil in a magnetic field. In Figure 1-6, as segment B is forced down, segment A, in which current is running in the opposite direction in relation to segment B, is forced up. This gives us rotational motion until our coil lies

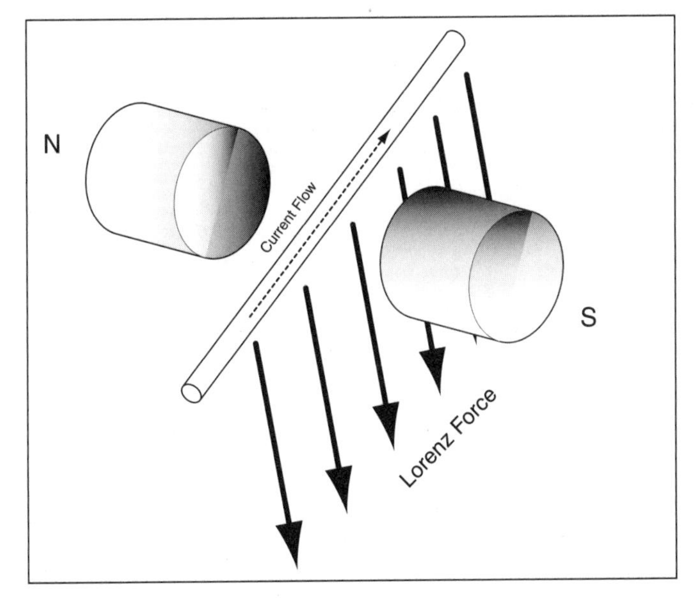

Figure 1-4 The Lorenz Force acting on a conductor

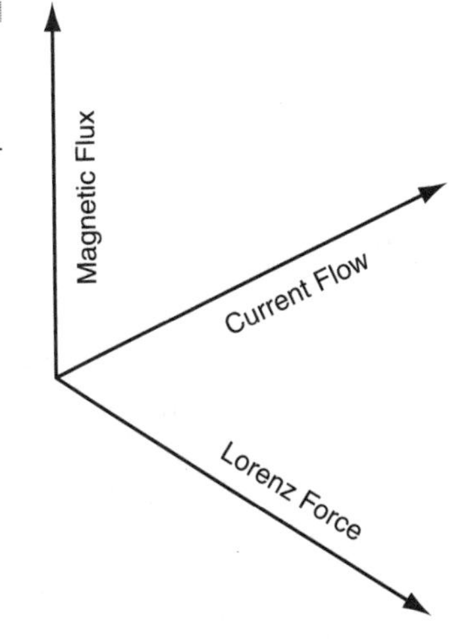

Figure 1-5 The Left-Hand Rule

in the vertical plane, at which point it will stop, since segments A and B are at their highest and lowest points, respectively.

If, on the other hand, we were to reverse the polarity of the current through our coil right at the moment when the vertical plane is reached, we would reverse the direction of the Lorenz Forces on the coil segments,

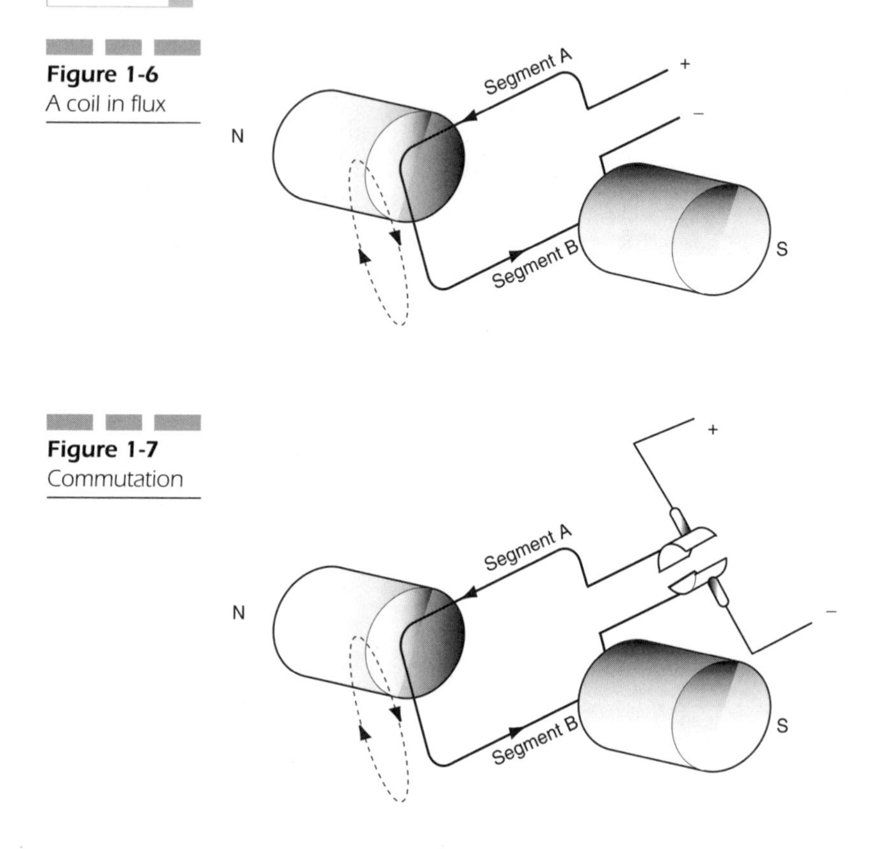

Figure 1-6
A coil in flux

Figure 1-7
Commutation

and our coil would continue to rotate through a full circle. The process of alternating current through the coil in this manner is termed *commutation* and is accomplished in our simple motor (and many DC motors) through a brush arrangement as in Figure 1-7. Each end of our coil is attached to a half-cylinder plate. Brushes attached to the power supply provide power to these split plates as they rotate with the motion of the coil. However, once the coil reaches the vertical plane, the plates have swapped positions, causing the polarity through the coil to be reversed. In this way, the coil rotates continuously.

A DC motor consists of a few key parts:

- **The armature** Consists of one or more coils mounted on a central shaft. Current is switched through these coils via the commutator assembly. As the Lorenz Force propels the coils, rotational force, or torque, is transmitted through the shaft.
- **The commutator** This consists of two split plates arranged on the motor shaft. These provide power to the armature coils. Since they

spin on the motor shaft, they are connected to the power supply via a pair of brushes, termed *commutator brushes.*

- **The stator** In the case of the motors we'll be discussing, the stator consists of a permanent magnet surrounding the shaft/armature assembly. It is this component that provides the magnetic field (flux). Some motors, especially larger motors, use an electromagnet as the stator. Prior to the widespread availability of powerful permanent magnets, stators were commonly implemented as electromagnets.

It should be noted that our simplified motor uses only a single coil comprising a single turn. Clearly, a real motor uses coils of multiple turns, which increases overall torque. Additionally, most real motors we'll be dealing with use at least three coils. This design ensures that the armature never becomes accidentally stuck in magnetic equilibrium with the stator magnetic field. It is also more efficient than the single-coil approach.

Of course, other electric motor types are available. Motors designed to run using AC current have no commutator/brush assemblies, because no need exists for a separate mechanism to reverse the polarity of the coil current. Brushless DC motors use a sensor to determine the armature position. Input from the sensor is used to trigger external circuitry that reverses the feed current polarity appropriately. Brushless motors have a number of advantages, including longer service life, less noisy (from an electrical standpoint) operation, and, in some cases, greater efficiency. Due to the difficulty of providing AC power to a mobile robot's motors, and because brushless DC motors appropriate for mobile robot use tend to be difficult to find and expensive when they *can* be found, we will not be discussing them in this book.

Stepper Motors

One type of DC brushless motor, however, will be covered here: the *stepper*. Stepper motors work somewhat differently than the standard DC motor described in the previous section. Figure 1-8 shows a simplified diagram of a 90-degree stepper.

In the center of the stepper motor is the *rotor*, which is the part of the motor that spins. The rotor is comprised of at least one permanent magnet (but often more than one). The rotor is mounted on the motor output shaft. Surrounding the rotor is the *stator*. Unlike typical brush DC

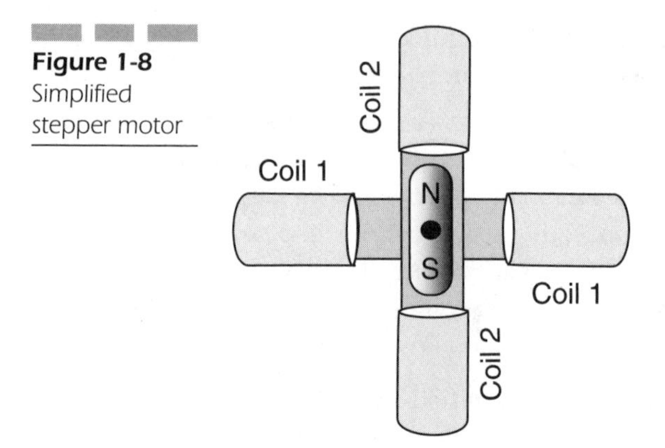

Figure 1-8 Simplified stepper motor

motors, the stator in a stepper is not a permanent magnet. Instead, the stator is a series of coils. As each coil is energized, the magnets on the stator cause it to swing into alignment with the field. If the coil is left energized, the stator will stay in place, resisting any further movement. This property is termed *holding torque* and is one reason steppers are often chosen over DC motors for certain types of applications.

To cause the rotor to move to its next position, or *step*, the first coil is switched off, and the next coil is energized. This causes the stator in our simple stepper to swing 90 degrees, as illustrated in Figure 1-9.

To get the motor to swing another 90 degrees, coil 2 is switched off, and coil 1 is reenergized, but this time with the polarity reversed. This forces the stator to swing another 90 degrees. To get another 90 degrees to the last position (4), we shut off coil 1 and reenergize coil 2 with an S-N polarity (see Figure 1-10).

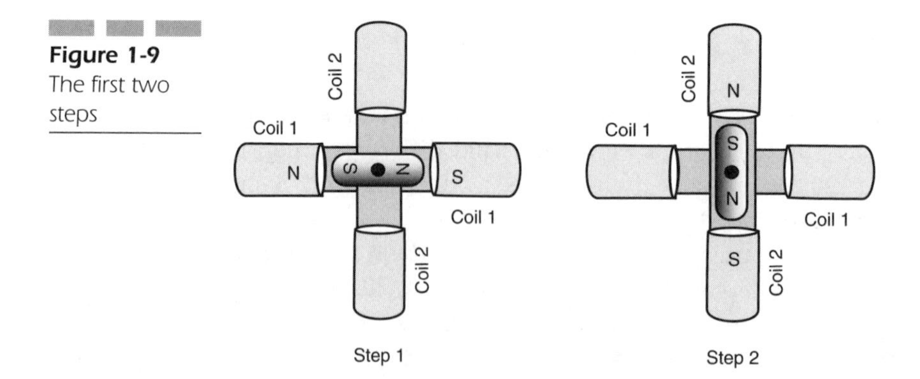

Figure 1-9 The first two steps

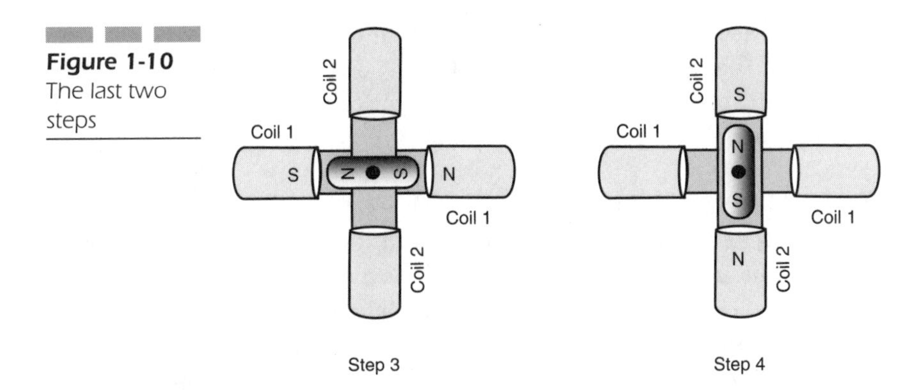

Figure 1-10 The last two steps

You can see that if we want to cause our stepper motor to rotate continuously, we must apply current to the coils in this repeating sequence. In essence, the user of a stepper motor is responsible for providing commutation. It should be obvious that if you choose to use stepper motors in your robot, you will need to deal with this additional control complexity. Moreover, steppers tend to provide less torque and consume more current than conventional DC motors.

Steppers have a few advantages over conventional motors, however. The first is the *holding torque* alluded to earlier. As long as power is provided to a stepper, the stator will resist any attempt to dislodge it from its current position. This property can come in quite handy for sumo robots or robots that need to "park" on the occasional incline.

The user commutation aspect of stepper motor operation implies built-in speed and position control. Assuming the motor does not slip or stall, accurate speed control can be accomplished without the use of a feedback mechanism, such as optical or magnetic encoders. This is a feat impossible with conventional motors.

Steppers also offer a high degree of positional control. Although our example stepper has a rather large 90-degree step, motors with angular resolutions of less than 1 degree are common. Steppers will be covered in detail in Chapter 5, "Using Stepper Motors."

Controlling DC Motors

Electric motors are remarkably easy devices to control. This, above all things, makes them suitable for use in a variety of devices. To start a DC motor, you need only apply power. Compare this with the typical ignition cycle of a gasoline engine. To reverse a DC motor, you merely reverse the polarity of the current feed—no need for a separate transmission mechanism. The speed and torque of a DC motor can be controlled by simply varying the voltage applied to the motor.

Pulse Width Modulation (PWM) Electric motor speed is controlled by raising or lowering the effective voltage of the current passing through the armature coils. In the old days, this was done through a large variable resistor, or *rheostat*, as in shown Figure 1-11.

Of course, in the old days, people were hanged for practicing witchcraft, and wooden undergarments were not uncommon. Please *never* use a resistor to lower motor voltage. As you can see, the rheostat works by dissipating excess power into heat. Mobile robots (and other battery-powered devices) cannot afford to waste precious battery power to heat. The modern solution is *Pulse Width Modulation* (PWM). PWM has a couple of variants, which will be discussed in greater detail along with other aspects of motor control in Chapter 7, "Motor Control 101, The Basics," and Chapter 8, "Motor Control 201—Closing the Loop with Feedback," but the fundamentals of PWM are pretty much the same for all varieties.

When PWM is used to control a motor, power is not applied to the motor continuously, but is instead applied as a square wave at a specific frequency, often around 20 kHz, but sometimes as low as 60 Hz or as high as 50 kHz. The power to the motor is controlled by changing the *duty cycle*, or pulse width, of the power signal, as in Figure 1-12.

Figure 1-11 Rheostat motor control

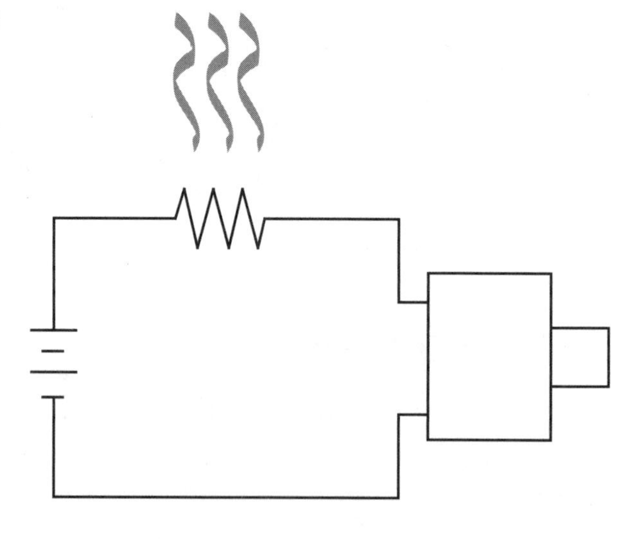

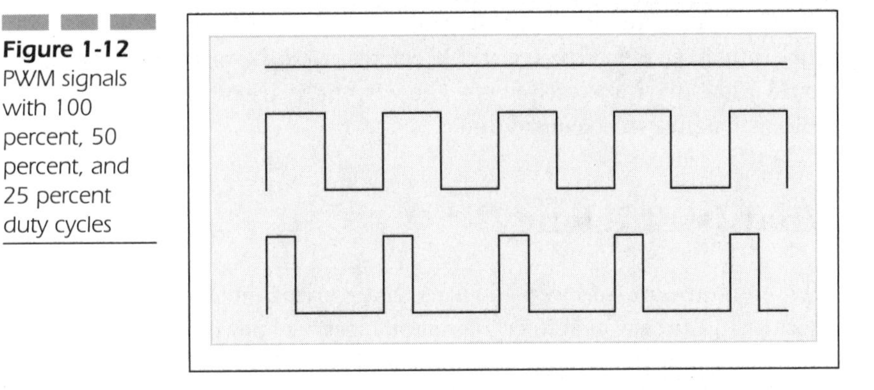

Figure 1-12 PWM signals with 100 percent, 50 percent, and 25 percent duty cycles

When the duty cycle is at 0 percent, the motor is completely off. When it is at 100 percent, the motor is completely on. At 50 percent, the motor is considered at half-power. This works because the motor, being a highly inductive load, tends to average out the rapidly changing input signal. In theory, a 12-volt PWM signal with a 50 percent duty cycle would create a voltage in the motor coils of 6 volts, a duty cycle of 25 percent would create a coil voltage of 3 volts, a duty cycle of 10 percent would create a voltage of 1.2 volts, and so on. That is, of course, in theory. In real life, the actual voltage in the coils will not have such a tidy, linear relationship with the power duty cycle. The exact voltage induced depends upon motor and coil construction as well as a host of other factors.

You should also note that the frequency of the power signal *never* changes, only the duty cycle. The exact choice of frequency remains a topic of some controversy in amateur robotic circles; we'll throw ourselves into the fray in Chapter 7 with some advice on how to choose an optimal PWM frequency.

PWM signals are easy to generate with most microprocessors. In fact, many have single- or dual-channel PWM hardware already built in, making logical implementation a trivial task.

DC Electric Motors and Efficiency

Another compelling reason electric motors are so widely used is that they are among the most efficient motor types available. Some of the larger motors used in electric vehicles can have efficiencies in the 90 percent range. By comparison, typical internal combustion engines scarcely reach much above 20 percent. Of course, if the story ended here, we would all be driving electric cars. Clearly, there are other factors besides motor efficiency to be considered.

What Is Efficiency?

All motors are *transducers*—devices that convert one form of power into another. In the case of an electric motor, electrical power is converted into mechanical power, usually in the form of the motor shaft rotation. The efficiency of a motor is simply the ratio of electrical power in to mechanical power out:

$$E = P_m/P_e$$

Both P_m and P_e are expressed in units of watts. In the case of a perfect motor, E would be equal to 1. In practice, E will generally be far lower than 1, and its exact value will vary over the range of motor output and speed.

Efficiency and the Roboticist We'll discuss efficiency issues in detail in Chapters 2 and 3. In the meantime, here are a couple of things to keep in mind when thinking about motors and efficiency:

- Efficiency tends to be highest at the low end of the torque range. Maximum efficiency often occurs when a motor is run at 10 to 20 percent of its maximum torque rating. This implies that you'll need to oversize your motors substantially if you want to run with high efficiency. The larger your motors, the more mass your robot will have to move around. Worse yet, higher-powered motors generally cost more than lower-powered motors.
- With the possible exception of Robot Wars/BattleBot robot designs, most robots require relatively low rotational velocity and high torque from their motors. Thus, motor output will usually be geared down. Unfortunately, most gearboxes—at least those accessible to hobbyists—introduce significant efficiency losses. Typically, the greater the gear ratio, the greater the losses you should expect. For example, one spec sheet for a high-end gearhead motor shows an average efficiency of 81 percent with a 9:1 gear ratio. This efficiency drops to 66 percent with a 100:1 ratio and all the way down to

50 percent with a 1000:1 ratio—and this for a *very* high quality motor.

- Your robot will likely rely on batteries as its primary energy source. Although this has no bearing on the efficiency of the motors themselves, the relatively poor energy density of batteries—the amount of energy they can store versus their weight—lowers the energy efficiency of the robot when taken as an entire system. We'll discuss energy density in a bit more detail in the next section.

The Special Challenges of Motorization

DC motors are in some ways very nasty little devices. As used in a typical robot, they are dependent upon relatively bulky batteries for power. DC motors are noisy—*always* in the electromagnetic sense and sometimes in the acoustic sense—and play poorly with other devices on your system. This section will introduce you to a few of the challenges posed by DC motors. We'll provide specific strategies and techniques for dealing with these challenges later in the book, particularly in Chapter 9, "Electronics and Microcontroller Interfacing." The advanced reader is, as always, encouraged to skip ahead.

Moving Your Power Supply—Batteries

Unless you're planning a tethered robot, you'll almost certainly need to cope with moving your power supply about with your robot. This leaves most of us with no option but to go with battery power. Although the focus of this volume is not on battery use and selection, the topic is difficult to separate entirely from locomotion, so we feel somewhat obliged to dedicate at least a few pages to the subject.

You should be aware of a few important battery properties when selecting a battery for your robot:

- **Rechargability** This tends to be an important factor for batteries in a mobile robot. Ultimately, nonrechargeable batteries are both expensive and inconvenient to replace over time. For this reason, we will confine our discussions in this section primarily to rechargeable battery types. You should be aware that not all battery types enable the same number of charge/discharge cycles. On the low end, the new rechargeable alkaline battery types only enable 25 or so discharge cycles. *Nickel-cadmiums* (NiCads), on the other hand, can be recharged upwards of 1,000 times with proper maintenance.

- **Energy density** This is usually expressed in either watt/hours per kilogram (WH/KG), or alternatively joules per kilogram. Sealed Lead-Acid (SLA) batteries tend to have the lowest energy density (25–30 WH/KG), and NiCads are about double that. Some exotic battery types have densities up in 2,000 range. By comparison, a kilogram of gasoline packs a whopping 13,000 watt/hours. The relatively poor energy density of most battery types has been one of the biggest obstacles to the widespread development and adoption of electric vehicles.
- **Voltage** Batteries are rated to supply a fixed voltage. Depending on the battery type, the actual voltage from the battery may vary significantly above or below the rated voltage, depending on where along the discharge cycle the voltage is measured. For example, a 12-volt battery might put out 13.5 volts when freshly charged, with output falling to around 10.8 volts before the battery is considered officially discharged. Different battery chemistries tend to have distinctly different discharge curves.
- **Capacity** Manufacturers generally rate batteries in terms of Amp-hours (AH) or sometimes milliamp-hours (mah). The idea is that a 4AH battery could supply 4 amps at the rated voltage for 1 hour, or alternatively 1 amp for 4 hours. Of course, in the real world things aren't that simple. In general, manufacturers test the battery discharge into a moderately high load over a 20- to 25-hour period. This longer period is used to derive the AH rating. Most battery types tend to be better at providing less current for a longer period than more current over a shorter duration. Thus, you should *not* expect your 4AH 12-volt battery to be able to supply 4 amps at 12 volts for an entire hour, or even 1 amp for 4 hours.
- **Internal resistance** Batteries are far from perfect conductors. All batteries can be modeled as shown in Figure 1-13. The internal resistance R acts as a current limiter, limiting both maximum output current as well as the maximum discharge rate. NiCads and Lead-Acid batteries both have resistances in the milliohm range. This makes them particularly suitable for supplying large surges of

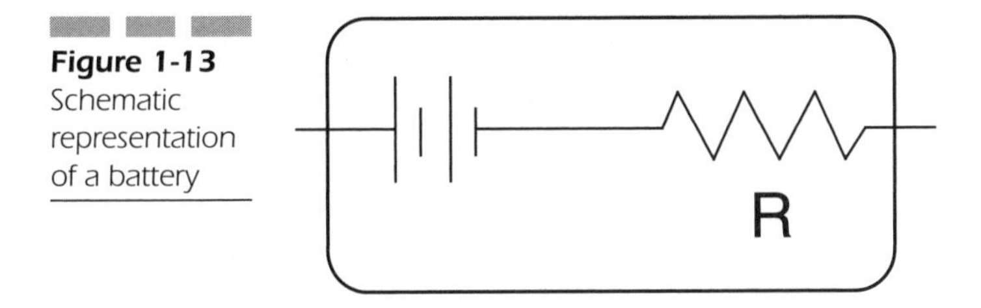

Figure 1-13 Schematic representation of a battery

current and explains the continuing popularity of NiCads with *radio controlled* (RC) car racing enthusiasts.

Battery Types As mentioned earlier, we'll be concentrating primarily on rechargeable batteries, and we'll limit ourselves even further to batteries that are commonly available to the amateur.

Nickel-Cadmium Cells (NiCads) NiCads are some of the most common rechargeable batteries around. They're available in a wide variety of shapes and sizes, including the standard consumer configurations (AAA, AA, C, D, 9-Volt) as well as packs of ganged cells popular with RC enthusiasts. These packs typically come in 6 to 7.2 volt versions, with current capacities ranging up to 4,000mah. NiCad packs are staples of the surplus market, some even coming bundled with their own chargers.

NiCads have traditionally not been known for their energy density (about 40 WH/KG). However, this is still an improvement over SLA cells. NiCads also provide relatively flat discharge curves—that is to say, they provide a fairly consistent voltage level over their discharge period, with a significant voltage fall-off occurring close to the point of depletion.

NiCads, because of their low internal resistance, are capable of providing high surge currents. Although this can be useful when required, it should also be kept in mind that this capability means that NiCads can pose a burn or even fire hazard in a short-circuit situation.

Charging NiCads properly can be a bit trickier than charging other battery types. To start with, NiCads are quite sensitive to overcharging, which can lead to *voltage depression*, where the afflicted cell fails to charge up to its proper voltage. If the cell is overcharged too much or too often the damage is permanent and you have a ruined pack. Investing in a decent charger can vastly improve the life span and performance of your NiCads. These chargers are available through RC equipment dealers in prices ranging from $40 on up.

Nickel-Metal-Hydride (NiMH) NiMH batteries share a number of characteristics with NiCads and are becoming increasingly popular with RC enthusiasts. This increased popularity has translated into increased availability and lower cost both for the batteries *and* chargers. NiMH batteries provide significantly greater energy density than NiCads, but at an increased cost. Furthermore, the number of charge cycles is more limited (around 400 maximum) than NiCads, making these batteries more expensive than NiCads on a per-W/H basis. Unlike NiCads, NiMH batteries contain no cadmium, making them significantly more environmentally friendly.

With a higher internal resistance, NiMH batteries cannot provide the same level of surge current as NiCads. Unless you expect your robot to require high surge currents—not usually the case for most designs—this is not necessarily a bad thing.

Today's modern electric radio controlled (RC) cars are using NiMH packs with up to 3,000mah ratings very successfully in this realm where the NiCd cell has ruled for years. As the technology has matured, the current carrying capacity has increased for NiMH cells to the point that there are fewer and fewer applications where this cell will be passed over for a NiCd.

Sealed Lead-Acid (SLA) SLA batteries provide the lowest energy density of all the rechargables, but have a number of advantages. They tend to cost much less than other battery types and are available in much higher power packages than NiCads—100AH SLAs are common; just look under the hood of your car. For larger robots requiring greater than 4AH batteries, SLAs are usually the way to go.

SLA charging regimens tend to be more forgiving than is the case with NiCads. Simple, constant voltage trickle charging with current limited to about a tenth of the AH rating works well. SLAs are somewhat more sensitive to discharge, however. Repeatedly discharging SLAs to depletion (typically below 20 percent of the rated voltage) will shorten the life of the battery significantly. When kept in long-term storage, SLAs need to be periodically recharged or else the battery will deteriorate and eventually no longer hold a charge. For this reason, it is best to be cautious when purchasing used SLA batteries.

Like NiCads, SLAs have a low internal resistance and can provide high surge currents when required. Clearly, you should take care to avoid shorts when using these batteries—the power produced through a short can easily melt insulation and cause painful burns or even fire.

Coping with Power Supply Noise

The amount of noise a motor can create on a power supply, especially when the current is switched on and off—as is the case when using PWM to control the motor speed—is truly astounding. At best, microprocessors connected to the same supply as the motor may periodically reboot or lock up. At worst, other circuitry sharing the power supply can be damaged. Power supply noise created by motors is rooted in a few causes:

- **Current demand** When a motor first starts or changes direction, it demands a great deal of current from the power supply, almost behaving as a short circuit. This causes significant dips in the power to the rest of the circuitry as the battery struggles to keep up with the motor's momentary demand.
- **Commutator brush noise** As the brushes make and break contact with the commutator plates, power to the armature coils is momentarily switched off and back on. Being inductors, the armature coils generate a brief high voltage as the current is switched off and the magnetic field around the coils collapses. These spikes, although brief, can reach into the hundreds of volts. Unless accounted for, the high voltages generated can cause significant damage to other circuitry.
- **PWM noise** Using PWM to control a motor necessarily entails switching the motor supply current on and off. As is the case during commutation, momentary removal of current from the armature coils causes a potentially damaging voltage surge (see Figure 1-14).

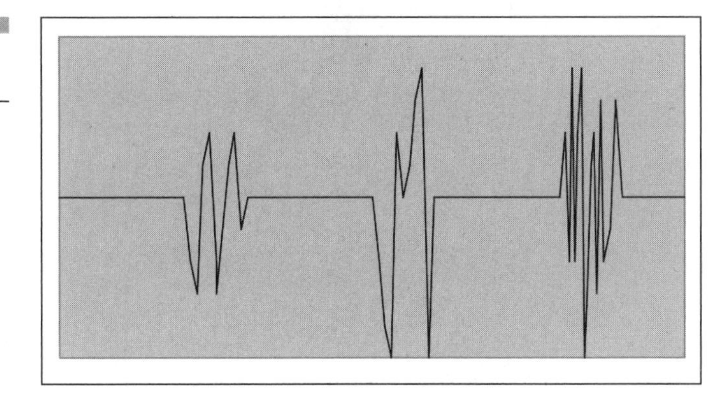

Figure 1-14 Motor noise

The electrical noise problem is sufficiently severe that some designers simply *punt*, choosing to power microcontrollers, sensors, and other digital logic from an entirely separate power source. Although this is certainly a valid approach, it can be inconvenient to carry two sets of batteries around, especially on smaller robots. This can also complicate recharging schemes.

Unless you are using unusually noisy motors, it is entirely possible to power motors and logic from the same supply. Chapter 9 will cover a number of techniques for sharing power supplies.

Electromagnetic Interference (EMI)

Electromagnetic noise is an inevitable byproduct of electric motor operation. It can be generated by a motor under PWM control as a result of the current pulsed through the motor coils. Electromagnetic noise can also be caused by arcing at the motor brushes as they make and break contact with the commutator segments. Once this noise begins interfering with other devices on the system, it formally becomes *electromagnetic interference* (EMI). Classically, an EMI problem is comprised of three elements, as shown in Figure 1-15:

- **A source** This could be a radio transmitter, power supply, or anything else capable of emitting electromagnetic energy. In our

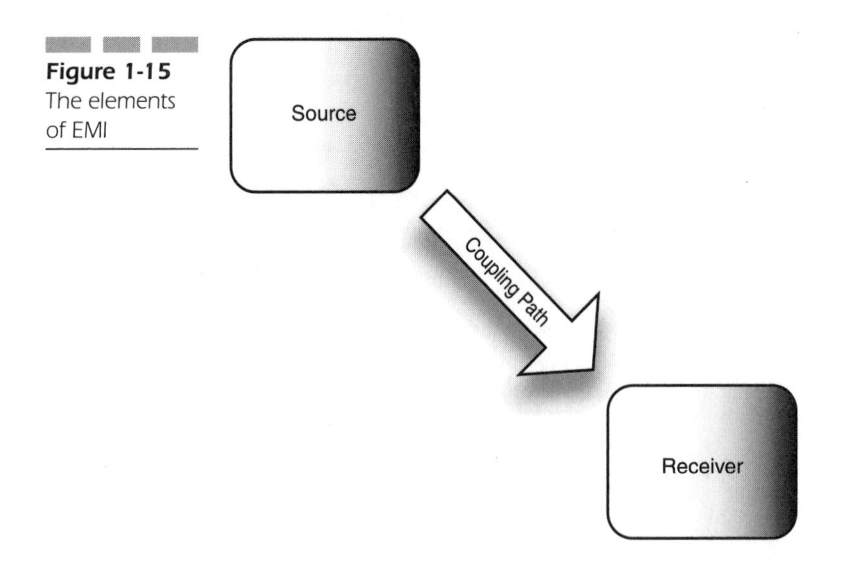

Figure 1-15 The elements of EMI

case, the most likely source of electromagnetic radiation is drive motors and servos.

- **A receiver** Sometimes termed *the victim,* the receiver is the device being adversely affected by the noise source. In the case of our robot, the victim might be the *Microcontroller Unit* (MCU), a radio-control unit, sensors, or other on-board electronics.
- **The coupling path** This is the path between the source and receiver over which interference travels. There can be multiple coupling paths in a system, and the paths can be inductive, capacitive, radiated, or conductive in nature.

EMI can be minimized in a variety of ways. In general, we can take steps to attenuate noise at the source, either by shielding or by filtering. Alternatively, we can shield the receiver. It may also be possible to identify and eliminate the coupling path, although dealing with coupling paths is notoriously difficult. Of course, EMI may never be a problem on your robot. Much will depend on your motors, cabling arrangement, robot construction, and other electronics installed. Chapter 3 will cover some common techniques for coping with EMI.

There are a few easy methods of avoiding EMI generated by DC motors:

- *Always* route motor supply lines as far away as possible from other electronics, especially antennae.
- *Avoid* large ungrounded metal sections on your robot. These tend to act as antennae.
- *Use capacitors* on your motor power leads. The recommended value is 0.1 uF with a maximum voltage rating at least 4x the expected motor power supply voltage. Be sure to attach each capacitor between the motor can and each respective motor power tab. These capacitors *must be installed at the motor*. You will have to use a small file to rough up the motor so the solder can adhere to it. Since motor cans are a very good heat sink, you will need at least a 40 watt soldering iron to solder to them.

Audible Noise

When compared to some motor types, electrically powered motors are reasonably quiet, but some configurations can produce a surprising amount of acoustic noise. Even a moderate amount of noise may be undesirable in some locations, such as office buildings or homes. Chapter 3 will discuss a few simple means you can use to attenuate DC motor and gearbox noise.

CHAPTER 2

Motor Types: An Overview

In robotics, few discussions are so rife with discord and confusion as the topic of which motor to use in order to move a robot. How powerful does the motor need to be? How big will it be? What kind of motor should I use: gearhead, servo, stepper, or whatever?

This chapter explains the differences between these motors and where we feel they are most useful and appropriate. We'll start with an overall survey and then dive into detailed discussions about what makes the motors go. Finally, at the end of this chapter we will show you how to determine the power you will need to have in order to move the robot that you are planning to build. This book is dedicated to helping the robot builder *use* motors to power his or her robots, not how to build a motor, so our descriptions of the motors will be brief and our discussions on how to design with motors will be more intense.

This chapter is full of math and formulae. If you don't know the specs for your motors, don't bother to read past the next couple of paragraphs. If you really want to design to your robot's capabilities, then read on; what you will need is found in this chapter.

Which Type of Motor Is Useful for What Kind of Robot?

Only a few reasonable ways exist for moving a robot: wheels, tracks, and legs. If we choose wheels, the easy one, we have the following options:

- Two driven wheels on a single motor and two steering wheels like a car
- Two driven wheels on a single motor and a single steering wheel
- Two driven wheels, each on their own motor with castors or skids that are driven like a tank (this is the most common choice by far)
- An actual tank-tread-style motivation (the hardest to achieve by far)

If we choose legs, then we need to decide how many legs: from two (the hardest) to six or more. After we choose our leg count, we need to decide upon the *degrees of freedom* (DOF) each leg will require. When we are choosing our means of motivation, we will need to factor in how heavy our robot will be and what kind of terrain it will need to traverse. To some extent, each of our choices will dictate the type of motor we will need to use and its power. The type of drive train we choose (wheels, tracks, or legs) will also determine our robot's speed and acceleration when used with a particular motor type (gearhead, stepper, and servo, as shown in Figures 2-1 through 2-3).

Motors: How They Compare

Before we deal with the nitty-gritty details of our choice of motors, let's rank the gearhead, stepper, and servo motors by their power and what task or style of locomotion each would perform best. In a nutshell, you

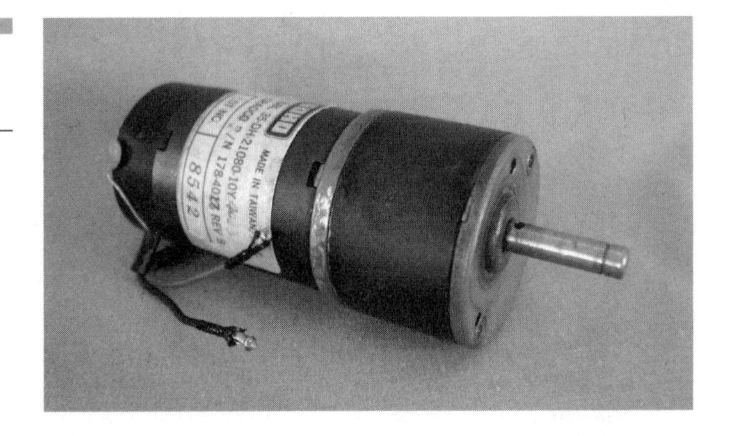

Figure 2-1 Gearhead motor

Figure 2-2 Stepper motor

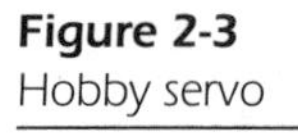

Figure 2-3
Hobby servo

Table 2-1
Relative power of robot motors

Relative Power	Motor Used	Weight Class
Most powerful	DC gearhead motor	For just about any robot
	Hobby servo	For robots up to 5 pounds[1]
Least powerful	Stepper motor	For light robots[2]

[1]Larger and heavier robots will need the larger, more powerful hobby servos.

[2]The most common steppers can't move over a pound or two.

can rank these motors from the most powerful to the least, as shown in Table 2-1.

Okay, that's fine, but you want to know what these motors are best at doing, right? Table 2-2 gives the high and low points of each motor type. In the succeeding sections of this chapter, we give more details about these motor types and in successive chapters we'll go into great detail on their use and control.

Every motor we discuss will have exceptions on the issue of cost. You can get a great deal on any of the motor types we've discussed. Conversely, you can pay as much as you want for a specific type of DC motor, hobby servo, or stepper motor to get exactly what you need. We are basing our estimates of cost on what is commonly available in the surplus market for DC motors and stepper motors. We buy hobby servos new, and our cost estimates are based upon the common hobby store or online hobby sources.

Table 2-2 Comparing features of robot motor types

Motor Type	Benefits	Detractors	Best For
DC motor	Commonly available Great variety Most powerful Easy to interface A must for large robots	Too fast, needs gearbox High current usually Harder to mount wheels More expensive Complex controls (PWM)	Large robots
Hobby servo	Gearbox included Great variety Good indoor robot speed Inexpensive Good small robot power Easy to mount Easy to mount wheels Easy to interface Medium power required	Low weight capability Little speed control	Small robots Legged robots
Stepper motor	Precise speed control Great variety Good indoor robot speed Easy to interface Inexpensive	Heavy for their power High current usually Bulky size Harder to mount wheels Low weight capability Not very powerful Complex controls needed	Line follower Maze solver

The DC Motor

The DC motor is the de facto standard robot platform motor. It is the most versatile motor with the greatest range of power, availability, and cost. But how does it work? Here we show you how the DC motor works and how to find a motor's power.

By far, the strongest motor is the DC motor. For decades, a great deal of research has gone into making DC motors faster, more powerful, and more efficient. Fine DC motors can reach 90 percent efficiency, but most of the motors that we will be able to afford and use do not reach such a lofty status; in general, our motors will be in the 40 to 70 percent range. Nevertheless, they are still powerful; they will just pull more current and drain our batteries faster.[3]

[3]"There ain't no such thing as a free lunch." Robert A. Heinlein.

How Does the DC Motor Work?

Quite a few books out there go into mind-shattering detail on the operation of and the physics behind the DC motor. This book isn't one of them. We're going to just touch on some of the basic traits as they relate to the circuits and the issues we solve here.

The permanent magnet DC motor has in its can two permanent magnets that provide a magnetic field in which the armature rotates. The armature, which is in the center of the motor, has an odd number of *poles*, each of which has a *winding*. The winding is connected to a contact pad on the center shaft, which is called the *commutator*. *Brushes* attached to the (+) and (−) wires of the motor provide power to the windings in such a fashion that one pole will be repelled from the permanent magnet nearest it and another winding will be attracted to another. As the armature rotates, the commutator changes which winding gets which polarity of the magnetic field. An armature always has an odd number of poles, and this ensures that the poles of the armature can never line up with their opposite magnet in the can, which would stop all motion.

Figure 2-4 illustrates the locations of the permanent magnets (North and South), the three armature poles (A, B, and C), and how the windings are powered by the brushes that contact the commutator. Near the center shaft of the armature are three plates attached to their respective windings (A, B, and C) around the poles. The brushes that feed power to the motor will be exactly opposite from each other, which enables the magnetic fields in the armature to forever *trail* the static magnetic fields of the magnets. This causes the motor to turn. The more current that

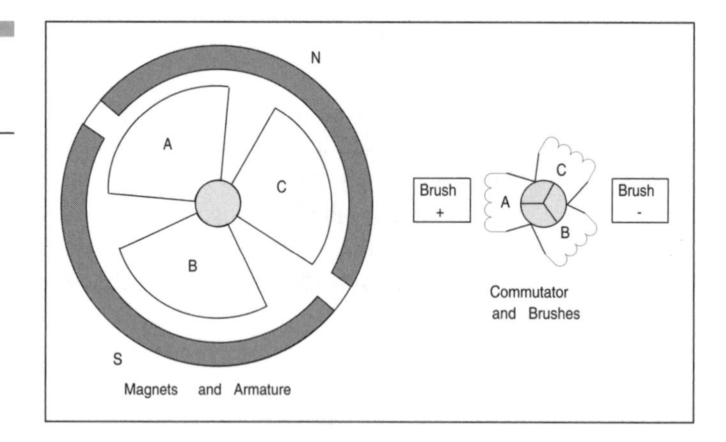

Figure 2-4 The basic DC motor

flows in the windings, the stronger the magnetic field in the armature and the faster the motor turns.

However, an interesting manifestation occurs as the motor spins. Even as the current flowing in the windings creates an electromagnetic field that causes the motor to turn, the act of the windings moving through the static magnetic field of the can causes a current in the windings. This current is opposite in polarity to the current the motor is drawing from the battery. The end result of this current and the *countercurrent* (CEMF) is that as the motor turns faster, it actually draws *less* current!

This is important because eventually the armature will reach a point where the CEMF and the drawn current balance out at the load placed on the motor, and the motor attains a *steady state*. If the motor has no load on it, then it is at this point that the motor is most efficient. It is also at this point that the motor is the weakest in its working range. The point where the motor is strongest is when there is no CEMF; all current flowing is causing the motor to try to move. This state is when the armature is not turning at all. This is called the stall or startup current, and it is when the motor's *torque* will be the strongest. This point is at the opposite end of the motor speed range from the steady state velocity.

Sizing a DC Motor

This is the part of the book that everyone hates, a part with lots of confusing math. You only need to read these next sections if you are seriously designing a robot with specific power needs and can obtain motors with full specification sheets before purchasing them. Of course, if you *really* want to know how an engineer would specify a motor for a project, then read on.

Torque is defined as the *angular force* the motor can deliver at some distance from the shaft. Think of this as a motor with a pulley of some diameter rotating to lift a weight hanging on a string. If a motor can lift one kilogram from a pulley with a radius of 1 meter, it would have a torque of 1 newton-meter. For those of you that don't speak metric, think in terms of pound-feet or ounce-inches instead. The relationship is the same. Okay, 1 newton equals 1 kilogram-meter/second^2, which is equal to 0.225 pounds, 1 inch equals 2.54 centimeters, and 100 centimeters are in a meter. You'll also want to know that there are 2π radians in one revolution.

The formula for mechanical power in watts is as follows:

$$P_m = T\omega$$

where T is torque measured in newton-meters and ω is angular velocity in radians/second. We use this formula to describe the power of a motor at any point in its working range.

A DC motor's maximum power is at half its maximum torque and half its maximum rotational velocity (also known as *no load velocity*), which gives us this formula:

$$P_{max} = 1/4 T_{max} \omega_{max}$$

This is simple to visualize from our discussion of the motor working range; where we have the maximum angular velocity (highest revolutions per minute [RPM]) we have the lowest torque, and where we have the highest torque, we have zero angular velocity. Figure 2-5 shows this relationship. Note that as motor velocity (RPM) increases, the torque decreases; at some point the power stops rising and starts to fall. This point is P_{max}.

When we specify motors for our robot, we want to have our motor running nearer its highest efficiency speed than its highest power in order to get the longest running time. In most DC motors, this will be at about 10 percent of its stall torque, which will be less than its torque at maximum power. To get the power we need from a DC motor, we then know

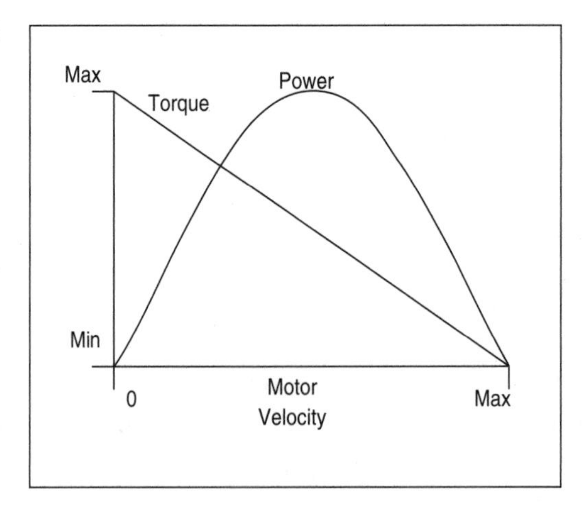

Figure 2-5 The relationship between torque, velocity, and power

that we need to find the power required to move our robot and then specify a motor whose P_{max} is higher than what we need. This is done so that we can operate the motor at the velocity we want and still have the power we want while minimizing the current required. By oversizing, we can run the motor nearer its highest efficiency speed. Obviously, we need to know the no-load velocity and the maximum torque of a motor to quantify its maximum power. We can measure both of these ourselves, but unfortunately if these specifications are not given by the seller, we have to buy the motor first.

To calculate P_{max}, you may need to convert units. Rather than make you look it up in other books, we'll show you how.

To convert RPM to radians/second for ω, do this:

$$\frac{rads}{\text{sec}} = \frac{revs}{\text{min}} \times \frac{2\pi rads}{rev} \times \frac{1\text{ min}}{60\text{ sec}}$$

For example, 100 RPM would be

$$100\frac{revs}{\text{min}} \times \frac{2\pi rads}{rev} \times \frac{1\text{ min}}{60\text{ sec}} = 10.5\,\frac{rads}{\text{sec}}$$

To convert ounce-inches to newton-meters for T, do this:

$$Nm = oz - in \times \frac{1lb}{16oz} \times \frac{1N}{0.225lb} \times \frac{2.54cm}{1in} \times \frac{1m}{100cm}$$

For example, 100 ounce-inches would be

$$100oz - in \times \frac{1lb}{16oz} \times \frac{1N}{0.225lb} \times \frac{2.54cm}{1in} \times \frac{1m}{100cm} = 0.71Nm$$

To calculate the power of a motor with this no-load RPM and this stall torque, we get a motor whose power is

$$P_{max} = \frac{1}{4}\,0.71Nm \times 10.5\,\frac{rads}{\text{sec}} = 1.86W$$

Since the no-load RPM is so low, we can assume our hypothetical motor is a gearhead DC motor.

Using and Finding the DC Motor

DC motors can run from 8,000 RPM to 20,000 RPM and more. Clearly, this is far too fast for our robots if we want to keep them under control. Fortunately, a solution exists to the motor speed problem: the gearbox.

The addition of a gear-down assembly benefits us in two ways: The final output is slower and the final output is more powerful. With a proper gearing choice, we will be running our motors at high RPM where the motor is most efficient and is using the least amount of current, thereby extending our battery life. Okay, that's a third benefit.

When you gear a motor down, you reduce the rotational velocity and At the same time increase the torque at the final output stage. This speed reduction and torque increase are linear and proportional. In other words, if we take a motor whose torque at 10,000 RPM is 1 newton-meter and gear it down by 4 times, its final speed will be 2,500 RPM and its final torque will be 4 newton-meter. This would be the case in a perfect world, but in reality small losses will occur due to friction of the gears in the gearbox that will cause some loss of speed and/or torque at the final stage.

We don't recommend that you build your own motor gearbox unless you have a well-stocked machine shop and are quite handy with a mill and a lathe (buy them with gearboxes). If you are fortunate, you will get the specifications on your motors that will allow you to make these calculations to find its power, or the motor may have its power in watts specified already and you won't have to calculate it. If neither of these options is available, you'll have to guess based on the size of the motor: bigger equals more powerful.

DC motors are available at almost every surplus store as well as a great many scientific stores and electronics suppliers. They range in cost from under a $1 to as much as you want to spend. Many of these suppliers will supply specification data to allow you to choose the proper motor for your job. For robots, you should get a gearhead motor so that your wheel RPM will be slow enough and you can get more power in a smaller package that uses less power. A (not exhaustive) list of motor suppliers is in Appendix C, "Resources."

In Chapter 3, "Using DC Motors," we will show you how to use your DC motor in greater detail (and with less math).

The Hobby Servo

It's a tossup as to whether the hobby servo is more powerful than the stepper motor. It is always possible to find a bigger, more powerful stepper, and hobby servo manufacturers are also always releasing more powerful (and more expensive) servos as well. Because we have to choose what comes next, we'll say that the hobby servo is the second most powerful motor for driving our robots. Several servo manufacturers make inexpensive and high-quality servos. The most common manufacturers are Hitec™, Futaba™, and Airtronics™, but several more are available to choose from. We include a list of online distributors of these and other hobby servos in Appendix C.

How Does the Hobby Servo Work?

The basic hobby servo is just a DC motor attached to a gearbox that gears the motor down to 180:1. This also raises the output torque 180 times as well. If you open the servo up, you'll see that the motor in there is pretty small, which is why you can't lift a house with one, even with such a massive increase in torque. Inside that little plastic case is also a controller board that converts the control signal sent to the servo into a movement of the output shaft. Hobby servos typically have a movement range of 60° specified but can often move 90° or more. A potentiometer (variable resistor) measures the position of the output shaft at all times so the controller board can accurately place and maintain the servo shaft at the desired setting.

This type of DC motor control is called *closed loop* motor control and is what the term *servo* means. An external controller tells the servo where to go with a signal known as *pulse proportional modulation* (PPM). This means that the width of the pulse is the encoded information needed by the controller. This is often mistakenly called *pulse width modulation* (PWM). PWM enables a duty cycle (the percentage of high versus low time) that is between 0 and 100 percent of the time period. PPM uses from 1 ms to 2 ms out of a possible 20 ms time period to encode its information.

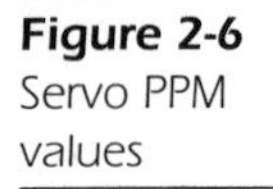

Figure 2-6 Servo PPM values

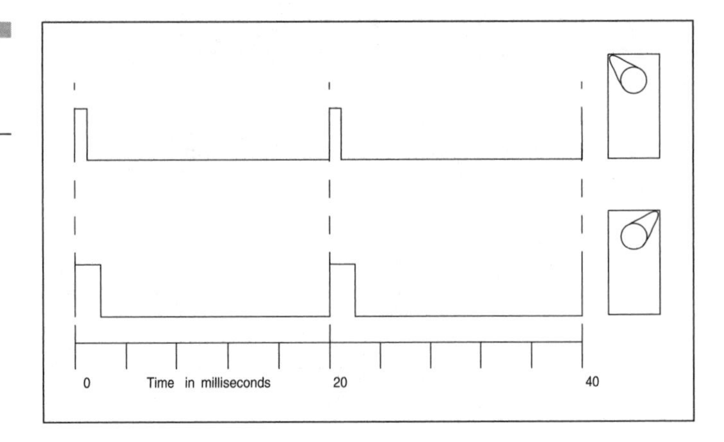

Figure 2-7 Differing PWM values

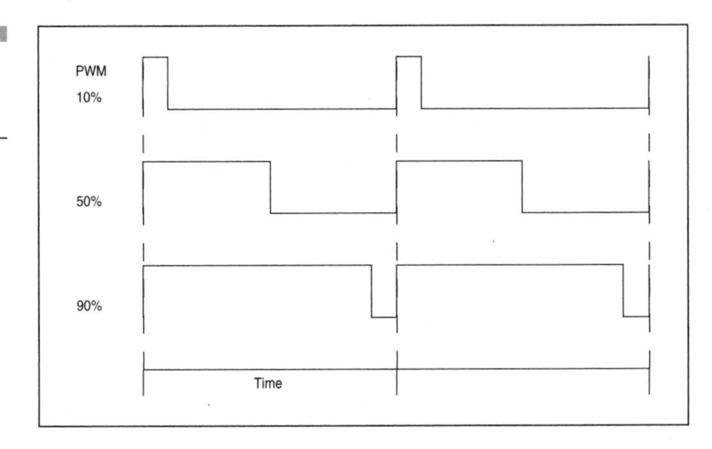

Figures 2-6 and 2-7 present a graphic depiction of the differences between PPM and PWM. In Figure 2-6, we show the servo control arm position at the two extremes of the typical PPM control signal. We will discuss how to generate and use servo PPM signals later on in Chapter 4, "Using RC Servo Motors." Figure 2-7 shows three different values of PWM and how those signals would look. In Chapter 7, "Motor Control 101, The Basics," we will show you how to generate PWM signals.

How Hobby Servos Are Rated

Hobby servos are rated by torque (ounce-inches/gram-cm) and speed. A servo rated 40 ounce-inches/.21 means that at 1 inch from the hub the servo will exert 40 ounces of force and move 60° in 0.21 seconds. Speeds are typically in the range of 0.11 seconds to 0.21 seconds. Hobby servos come in torque ranges from 17 ounce-inches to over 200 ounce-inches. Faster servos cost more, smaller servos cost more, and occasionally stronger servos cost more. Choose the servo that is strong enough to move your robot and is of the proper size and shape to fit your application.

As you can imagine, the smallest servos will be the weakest and the largest will be the strongest. Those servos that fall between these extremes can be a variety of sizes and strengths with some smaller servos being stronger than some larger ones. The given ratings are for 4.8 volts, and occasionally the ratings for 6 volts will be given. Usually, the 6-volt ratings are of higher power and faster speeds. Does this suggest something to you? It does to us: raise the voltage and you get more speed and power! Each manufacturer's servos will respond to our unauthorized increase of voltage differently. Some do get faster and stronger; some just get warmer and some cease to function at all. If you use a higher voltage, you void the warranty and risk the servo, so don't say we didn't warn you.

We've gone as high as 12 volts on Hitec™ and Futaba™ servos, and gotten higher speeds and greater power. They also got warmer, but because we were using them in the combat arena for competitions and for only five minutes at a time, it was a risk we were willing to take, and it paid off! Your mileage may vary.

Another criteria for choosing or rating a servo is whether or not the output shaft is supported by ball bearings. Ball bearings make a servo quieter, stronger, more durable, and longer lasting than plastic or oilite bushings. Bearings also make a servo more expensive. So if you want it to last longer and bear greater weight on the shaft, get a servo with ball bearings.

Sizing a Hobby Servo

We can use the hobby servo to motivate our robots in either of two ways: unmodified for leg articulation or modified for continuous rotation on wheeled robots. The hobby servo is a favorite of the small walking robot designer because it is small, it is strong, and it has a simple on-wire control interface. It's easy to connect to as well because a number of control horns attach to its output shaft in a secure manner that make linkages simple and reliable.

Sizing a Servo That Has Been Modified for Continuous Rotation A hobby servo is basically a DC motor with a built-in gearbox and built-in positional feedback electronics. If we modify the servo to remove the end-stop, which limits the servo's rotation, on its output gear, or *spline* gear, and remove its capability to sense location, then we can create a nicely compact gearhead motor that is quite powerful. We will discuss these modifications in great detail in Chapter 4, but for now let's assume we've made the modifications and get on with the (gasp!) math to size our servo to our robot's needs.

As we saw in the DC motor section earlier, this is the formula for power in a DC motor:

$$P_m = T\omega$$

This is the formula for finding the maximum power of a DC motor:

$$P_{max} = 1/4T_{max}\omega_{max}$$

Here are the specifications of the Hitec™ HS303 servo (Courtesy of Hitec RCD Inc.):

Operating Voltage		**4.8V**			**6.0V**	
	Oz-in		Kg-cm	Oz-in		Kg-cm
Torque	42		3.3	49		3.7
Speed at 60°		0.19			0.15	

Since we have modified our servo for continuous rotation, we can get the maximum no-load RPM and radians/second data from the speed specification:

$$\frac{60°}{0.19\text{sec}} \times \frac{2\pi rads}{360°} = \frac{5.5rads}{\text{sec}} \text{ and } \frac{60°}{0.19\text{ sec}} \times \frac{60\text{sec}}{1\text{ min}} \times \frac{1rev}{360°} = 52.7RPM$$

Our torque in newton-meters for this servo will then be

$$42oz - in \times \frac{1lb}{16oz} \times \frac{1N}{0.225lb} \times \frac{2.54cm}{1in} \times \frac{1m}{100cm} = 0.3\ Nm$$

or

$$3.3kg - cm \times \frac{9.8m/\sec^2}{1N} \times \frac{1m}{100cm} = 0.32Nm$$

Finally, our maximum power from this servo is

$$P_{max} = \frac{1}{4} 0.3Nm \times 5.5 \frac{rads}{\sec} = 0.42W$$

This is just for the servo when powered at 4.8 volts. If we power it at 6 volts, then

$$P_{max} = \frac{1}{4} 0.35Nm \times 7 \frac{rads}{\sec} = 0.61W$$

Quite a boost!

Sizing an Unmodified Servo for Leg Lifting Hobby servos have quite a bit of power built into them for their minute size. If we are to use a hobby servo to lift legs on a walking robot, we need to know how much the servo will need to lift. Unlike the rolling robots, we won't need to use the fancy formulae we've been using up to this point; we can just use the torque and speed values given in the servo specifications. Remember that the specification is based upon lifting or pushing a certain force at 1 inch or 1 centimeter from the shaft. If we make this lever arm longer, we reduce the force at the end of the lever; thus, the relationship is linear.

For example, if we have a servo with 42 ounce-inches of torque, this means that at 2 inches from the shaft that torque is still 42 ounce-inches, but it can only lift 21 ounces. Sizing an unmodified hobby servo is simple. If it has to lift 32 ounces 2 inches from the shaft of the servo, you need a hobby servo that has 64 ounce-inches of torque. This section was quite painless now, wasn't it?

If you need more torque to lift your robot than a servo can provide, you can magnify your servo's power by use of a lever arm. Essentially, the forces exerted are proportionally modified by the location of the fulcrum, or joint, in the lever arm. See Figure 2-8 to understand how torque and the movement arc are related when using a lever arm. F2 is proportionally more force than F1 by the ratio between R1 and R2. Note

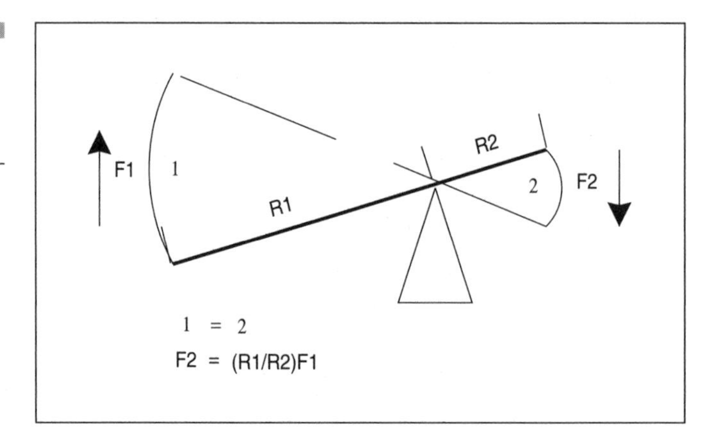

Figure 2-8 The lever arm and torque amplification

that the swing of the lever is also proportionally smaller for F2 than F1. Nothing comes for free.

At the end of this chapter, we'll discuss how to find the power needed in your servo by looking at how the robot will be walking.

Finding the Hobby Servo

Many servo manufacturers exist and each manufacturer has a great selection of servos of various sizes, shapes, powers, speeds, and, of course, costs. You can get all the information you need about each servo by visiting the manufacturers' web sites or simply by looking on the box where the specifications are printed. Brand-new, common servos can be as inexpensive as $10 a piece, or even less. You can spend over $100 on high-power, high-speed, or special servos as well. The vast majority of what the robot hobbyist will use for a hobby servo will cost in the neighborhood of $30 or less.

The Stepper Motor

The stepper motor is a completely different motor from those we've looked at so far. It is what is commonly called a *brushless* motor because the magnets are on the *rotor*, or shaft, and the windings are in the can of the motor. The shaft is free to rotate with no electrical contact to anything else. See Figure 2-9 for the basic details.

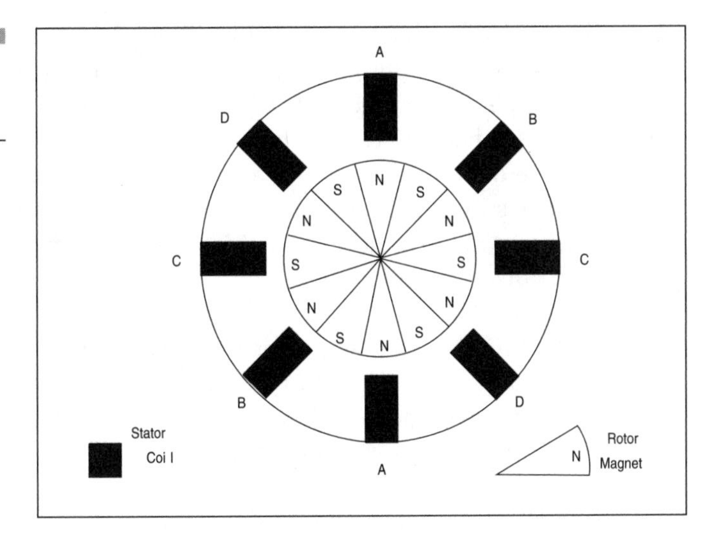

Figure 2-9 Stepper motor construction

Steppers are inherently slow-speed devices. They are designed for precision movement and to hold a position once they get there. This is fine for the hobby roboticist because we tend to want our motors to run slower. Rotational speeds in the 50 to 100 RPM range are well within the stepper motor's capabilities and what our robots will run best at. Steppers tend to be bulkier and heavier for a given power than a DC motor or hobby servo. A stepper is in general not a very strong motor, and because of this we don't recommend it for any robot over two pounds. Many stepper motors would have difficulty moving even that much weight.

How Does the Stepper Motor Work?

Whole books are dedicated to the discussion of the construction and inner workings of the stepper motor, so we won't go into grand detail here. However, we will cover the basic construction and operation of the stepper motor because a basic understanding will be useful to the hobby robot builder.

The stepper motor rotates by moving from one discrete position to another in turn. The number of steps needed to make a full rotation is one of the stepper motor's specifications and can be described in either the number of steps or the degrees per step. A number of windings in the can of a stepper motor are aligned in such a fashion as to enable the rotor to move precisely from one location to the next. The windings are located in the *stator* or can of the motor (labeled A, B, C, and D). As stator poles are energized, you can see how the rotor will be attracted to the opposite magnetic poles in the stator, causing the motor to turn. The rotor consists of a number of permanent magnets that will align with the stator coils as the coils are energized. Each magnet end facing out of the rotor is called a *tooth* and you will often see rotors described by the number of *teeth* they have. The stator coils are repeated, or stepped around the perimeter of the can, in an alternating fashion (refer to Figure 2-9). How the coils are arranged depends upon what type configuration of stepper motor you have.

Two basic kinds of stepper motor exist: *unipolar*, often called *four-phase,* and *bipolar*, often called *two-phase*. Unipolar stepper motors have four sets of windings alternating around the motor can, while the bipolar stepper motor has two sets of windings alternating positions. A unipolar stepper motor will energize its coils at a single polarity, hence the term *unipolar*. The bipolar stepper motor moves by energizing its coils first one polarity and then reversing the polarity as the rotor is turned, hence the designation *bipolar*. The number of steps in a full revolution of a stepper motor depends on the number of permanent magnets or teeth in the rotor. More teeth means more steps and finer resolution to each step.

Bipolar stepper motors are stronger and faster than unipolar stepper motors of the same size. Because a bipolar stepper motor has half the number of windings, it can use twice the amount of wire, or twice the size of the wire to generate a stronger electromagnetic field. However, a bipolar stepper motor requires that the driver be able to switch the polarity of the stator windings, and this makes the bipolar driver a more complex circuit. The unipolar stepper motor requires only that a driver switch one polarity of current to its windings. For this reason, a bipolar driver will require an entire H-bridge or a current-reversing circuit for each winding. Unipolar drivers need only use a simple transistor to energize a given winding. Since the driver circuitry is simpler for the unipolar stepper motor, it is the most commonly used stepper motor for robotic propulsion. In Chapter 5, "Using Stepper Motors," we'll go into great detail on the driver circuits for both types of steppers and how to use them.

How Can I Tell What Kind of Stepper Motor I Have?

Fortunately, determining what type of stepper motor you have is quite simple. A bipolar stepper motor has four leads, two for each coil/winding.

A unipolar stepper motor may have five, six, or eight leads. If the stepper motor has five leads, then one wire is the common and the other four are the ends of their respective windings. If the stepper motor has six wires (this is most common), then a pair of windings will have a common wire. This six-wire configuration can also function as a bipolar stepper motor if the manufacturer has designed it to function that way. If the stepper has eight wires, it could be what is called a *universal* stepper. This means that it could be used as either a unipolar with each pair of wires being a single coil, or it can be used as a bipolar stepper where you can put pairs of windings together either in a series or in parallel. In Chapter 5, we'll discuss how to sort out which wire is which on a stepper motor. If you are lucky, you'll get a wiring diagram with your stepper motor that shows the order of the windings. If not, you'll have to figure it out by trial and error.

With a random sample of three of our salvaged stepper motors, we found one of them on the web that we were able to get spec sheets for. That's not good odds, but it does suggest that you may be able to get specs for your stepper with a Google.com search.

Sizing a Stepper Motor

For a smoothly rotating wheel, you will want to use a stepper with at most 3.6° per step. More than that could seem to be jerky. You can find stepper motors with up to 30° per step, and we've seen stepper motors with resolutions down to 0.9° per step. A stepper motor is rated with a *detent torque*, *holding torque,* and *dynamic torque*. As you can probably guess, the detent torque is the torque of the motor when it's moving from step to step, and the holding torque is the torque of the motor when it is static and powered, holding its position. We will be interested in only the dynamic torque, as that is what will be moving our robot. This torque will be some maximum value and always depends on the pulse rate that the motor is stepping. The only other ratings used with steppers are the driver voltage and winding resistance.

There is no way to determine the power of a stepper with these values because steppers are built in a variety of ways, some stronger, some more accurate, some faster, and so on. The only way to determine the power of a stepper motor at a given RPM is with the manufacturer's torque/speed *derating* curves or charts. If the chart has a curve for both pull-in and pull-out, we are interested in the pull-out curve. The *pull-in* torque is defined as the maximum torque that may be applied to the motor before it starts skipping steps either when starting or stopping. The *pull-out* torque is defined as the maximum torque the motor can generate at a given pulse rate.

Using the stepper motor torque/speed chart, find the RPM you want the motor to spin at and cross-reference this with the torque at that RPM. The chart may be torque/PPS where PPS is *pulses per second.* In this case, divide PPS by the number of steps in a rotation to get the RPM and multiply this by 60. In other words

$$RPM = \frac{pulses}{\text{sec}} \times \frac{1rev}{steps} \times \frac{60\ sec}{1\text{ min}}$$

and

$$\frac{rads}{\text{sec}} = \frac{revs}{\text{min}} \times \frac{2\pi rads}{rev} \times \frac{1\text{ min}}{60\text{ sec}}$$

For example, if we have a stepper motor with 1.8° per step (200 steps per revolution) and we liked the torque of 0.08 newton-meter at 600 PPS (see Figure 2-10), then the RPM would be

$$RPM = \frac{600\ pulses}{\text{sec}} \times \frac{1rev}{200\ pulses} \times \frac{60\text{ sec}}{1\text{ min}} = 180\ RPM = \frac{18.8rads}{\text{sec}}$$

Then use our formula

$$P_m = T\omega$$

To find the power that this stepper motor will generate at that RPM, remember that torque (T) must be in newton-meters and that angular velocity (ω) must be in radians/second. In this case, the power of this stepper would be 1.5 watts.

Without a torque/speed-derating table, you are going to have to guess at the power of the stepper. In general, the harder it is to turn the shaft, the stronger the stepper is, but that's not much help without the specifications.

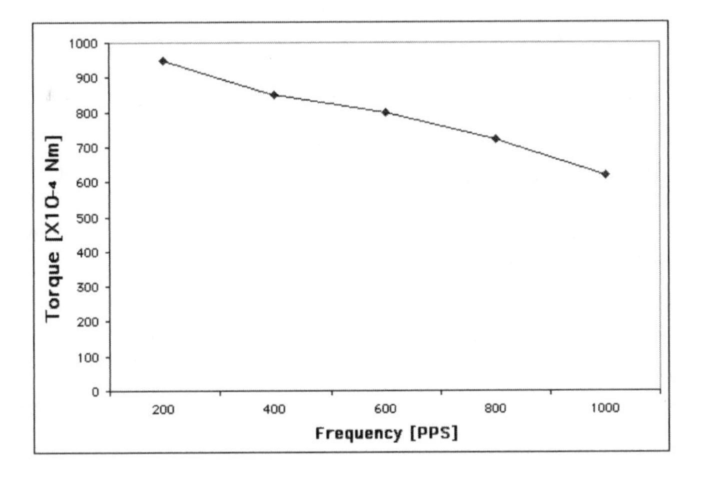

Figure 2-10 Stepper motor torque/speed curve

Finding and Using the Stepper Motor

Stepper motors are quite expensive if you buy them new. However, they are common on the surplus market for prices as low as $3 or less. Unfortunately, you will usually not get any kind of spec sheet with your stepper motor. Even worse, you may not even know what the voltage rating or steps per revolution is! This will make it rather difficult to properly choose your stepper.

Our advice is don't buy it unless you can get at least the step resolution and voltage rating. We've found that there are places that will give you the detent torque, step resolution, and current and voltage ratings—get your stepper motors there! One such place is www.jameco.com, which gives you most of the data you need. However, even they don't usually supply the derating curves for their stepper motors.

Two other cheap and common sources exist for stepper motors: old floppy disk drives and discarded printers. Disk drives will have very low power stepper motors; all they have to move is a low-mass read/write head. Printers will have some reasonably powerful stepper motors (for this type of motor) in them to move paper around inside the printer. We suggest you get your stepper motors from printers if you are on a budget. We could still find data sheets on the web for some stepper motors from discarded printers.

Determining the Power Needed to Move the Robot

We've discussed how to find the power of each of our motor types, and in this section we will calculate how much power our robot will need so we can choose our motors properly. Jones and Flynn[4] derive the origins of the formulae we will be using, and numerous other physics texts can also give you more background information on friction, force, and other factors that determine how to move a vehicle. If you want more information on these concepts, please consult those sources of knowledge. We will simply work with the "nuts and bolts" solutions that get us where we want to go.

Finding the Power Needed by a Wheeled Platform

We're going to be doing more physics again. Figure 2-11 graphically shows everything we'll talk about in this section. I promise I'll make this

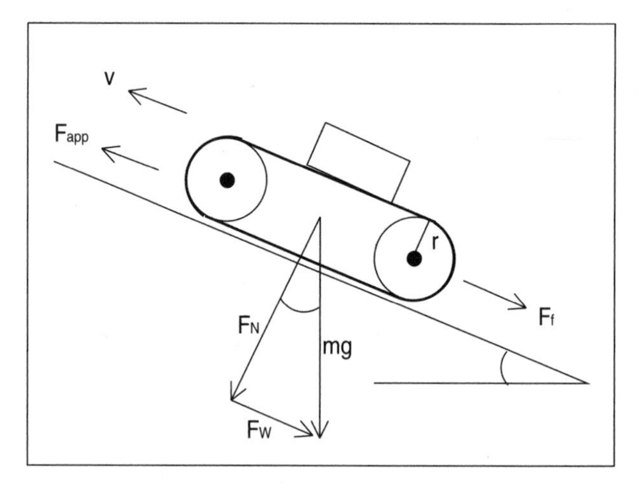

Figure 2-11 Factors of a robot's power requirements[5]

[4]Joseph L. Jones and Anita M. Flynn, *Mobile Robots: Inspiration to Implementation, 1st edition*. Wellesley, Mass: A. K. Peters, Ltd.

[5]Reprinted from Joseph L. Jones and Anita M. Flynn, *Mobile Robots: Inspiration to Implementation, 1st edition*, 1993, by permission from the publisher, A. K. Peters, Ltd.

as painless as possible. This is the easy part of the process after you have calculated the power of the motors.

Before we can choose motors to move our robot, we need to know how powerful those motors will need to be. To move our robot, we will be working against two related forces: friction and gravity. If we put these two forces together, we get this formula:

$$F_{app} = F_f + F_w$$

where F_{app} is the required force to move our robot, F_f is the frictional force resisting that movement, and F_w is the weight causing us to roll downhill. The F_w (weight) component is 0 on a flat surface because it can't fall down any more. Here is the formula for F_w:

$$F_w = mgsin\theta$$

where *mg* is the force of gravity where *m* is the mass and *g* is the acceleration due to gravity, 9.8m/sec^2, and sinθ is the angle of incline from the vertical. Sin(0) (straight down) is 0.00, and at any other angle this adds to the force required to move an object. To prove this to yourself, try pushing your car on the level ground. No problem, right? Now try to push it up a hill. It just got harder, didn't it? This is F_w at work.

Frictional force is always with us, no matter where we are, on level ground or hills. Frictional force, or F_f, is affected by the magnitude of the angle from the horizontal and by how much friction there is between the moving wheel and the surface on which it is moving, as shown by this formula:

$$F_f = \mu mgcos\theta$$

Here μ is the coefficient of friction, in our case the *static* coefficient of friction, *mg* is the force of gravity, and *cos*θ is the angle from the horizontal. A large μ means a large frictional resistance; think of trying to push a brick on a rough surface. A small value of μ is like a greased surface; things slide easily.

No easy way exists for measuring μ without actually testing your wheel material with a static scale and force scale; you're going to have to guess. A low value is not unreasonable if you are using wheels with a small contact patch[6] that is made of very hard rubber or plastic. A 0.3 or 0.4, for example, should suffice. If you are using tank treads, then you'll want a higher number, perhaps a 0.5 or 0.6 or even higher. You can't go wrong by guessing a bit high; you'll just be overpowering your robot, which, as we mentioned before, isn't always a bad idea. In fact, most of the time you will be overpowering your robot because F_f is only the startup or initial frictional force when the tire or track is essentially skidding. Once rolling, the rolling frictional force is a small fraction of that initial *skidding* friction.

If you want to be ultraconservative, the coefficient of the friction of rubber on dry concrete is between 0.9 and 1.0. The rolling friction of a rubber tire on dry concrete is about 0.01. Remember though, you won't get rolling if you don't have the power to overcome that initial skidding. If you really want to know what the coefficient of the friction of your wheel material is you can measure this directly. The definition of the coefficient of friction is

$$\mu_s = \frac{F_r}{F_n}$$

where μ_s is the static coefficient of friction, F_r is the resistive force of friction, and F_n is the normal force (perpendicular to the ground) acting to push the objects together. You can measure F_n easily with a scale; this is just mg, which is weight. It takes a bit of effort to measure F_r. The easiest way requires some kind of hanging scale that you can attach to your material. To find F_r, attach your force scale to your material and pull on it. The maximum weight registered is the force needed to get the objects to move relative to each other. Plug these values into the previous formula and you have your coefficient of friction.

By putting these formulae together to get the force we require for our robot, we get

$$F_{app} = mgsin\theta + \mu mgcos\theta$$

[6]The contact patch is the surface area of a tire that is in contact with the ground.

That was fun. Now let's make these formulae work for us. To figure out the power we need to move our robot, we have to use one more formula, the formula for power, which is

$$P = F_{app} v$$

Or power is force times velocity. We now know the power we need to move our robot over the terrain and slopes we predict it will face. But we need to know at what speed our motors need to rotate. Our final formula gets that rotational velocity and it is

$$\omega = v/r$$

Remember to keep your units consistent when using these formulae. Nothing messes up a project worse than using feet in one place and meters in another. If you are using two motors, divide your required power by 2. They are sharing the load and only need to be half as large as the total power needed. You might want to bump up your estimate just to move closer to the most efficient part of the speed curve regardless.

Finally, if at any time F_w is greater than F_f, or the gravity component acting on the slope is greater than the friction required to climb the slope, your robot will be slipping and all the power in the world will not help you. Increasing your traction, which is increasing your μ, will improve your hill-climbing capability, but it takes more power to move with a higher μ, so we see that we need to take our terrain into account when we design our robot.

Finding the Power Needed by a Walking Platform

Walking robots require a little more thought in their design. Walking, while far more natural and lifelike in a robot, is also far more complex than rolling. Several limitations and challenges must be overcome.

Moving Forward Walking robots that use hobby servos are easier to estimate power requirements for than rolling robots. We can use the torque ratings of the servos directly as our power ratings to lift a known weight. We can use the same formulae for legged platforms as we did for

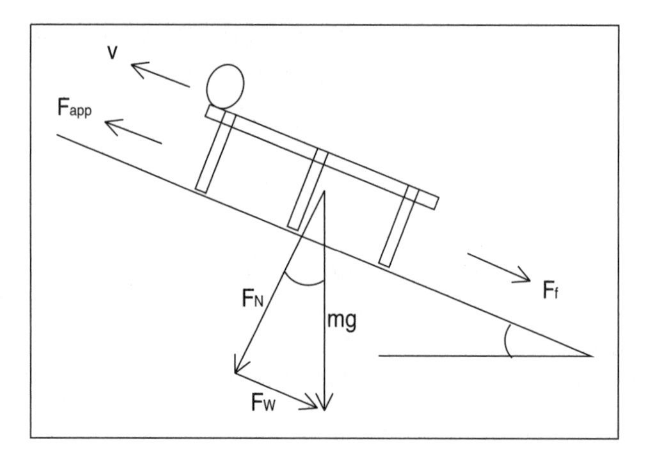

Figure 2-12 Factors calculated into a legged robot's power requirements

wheeled vehicles to estimate the power needed to move our robot with a small change. We'll just substitute a vaguely leggy robot for the vaguely wheeled robot in Figure 2-12 for a visual aid.

Because we are lifting our legs as they are moving forward, and we are counting on friction to force our robot to move forward as the legs move backward, F_f is *not* a force to overcome, rather, F_f is what assists us in enabling our robot to move. F_w (mgsinθ) and the bearing friction of the legs and servo motor are all we need to overcome to power our walking robot. Therefore $F_{app} = F_w$. While F_w is negligible on a flat surface, it becomes significant quickly when an incline is traversed, so take your expected terrain into account when you design your robot. Remember, if at any time F_w is greater than F_f or the gravity component acting on the slope is greater than the friction required to climb the slope, you will have a different problem, namely sliding backwards or toppling over.

Once you have calculated the power you need to move in the terrain (and slope) you have chosen, you simply use the servos that have the power to propel your robot up that slope. Again, remember, for instance, if you have two hobby servos pushing, each one needs only to be half the total required power to move your robot. If more than one leg is pushing, then each leg's servo can be smaller than the total required power.

Lifting a Leg Another issue must be resolved in legged robots that we don't have with wheels. We need to *lift* our robot as well as propel it forward or backward. We have a single concept to impart here; tilting a board whose pivot point is in the middle of the board is much easier than

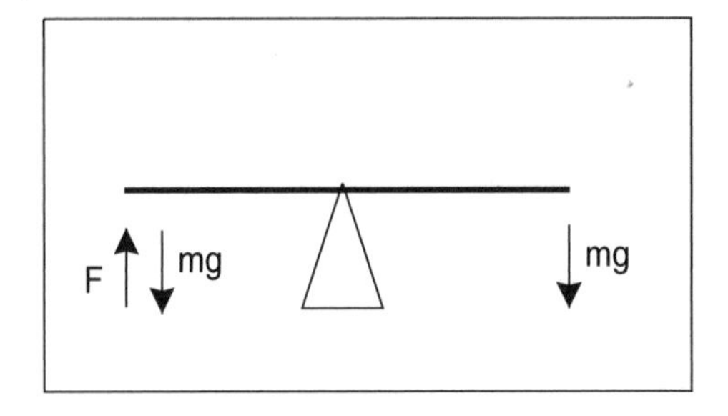

Figure 2-13 Lifting a board with a center pivot

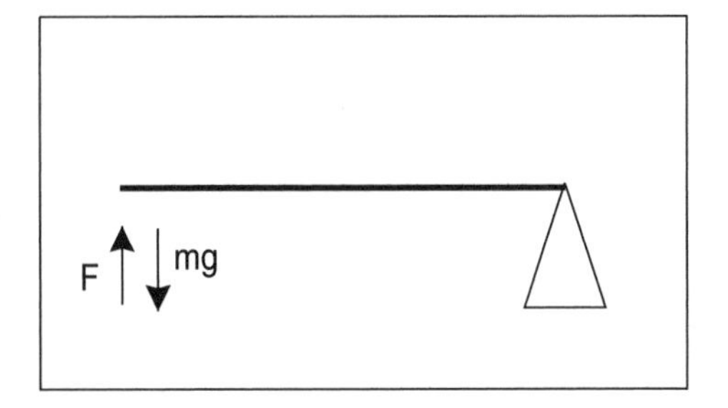

Figure 2-14 Lifting a board with an end pivot

tilting that same board whose pivot point is at one end. Figures 2-13 and 2-14 show what we mean.

In Figure 2-13 we are applying force at one end of the board whose pivot point is in the middle. This means that half the weight of the board (shown as mg) is *helping* to move the board. If you tilt the board with the pivot point at the other end of the board, as in Figure 2-14, you are moving the entire weight of the board with no help from gravity. When choosing hobby servos for lifting a legged robot, keep this concept in mind. In Chapter 11, "Locomotion for Multipods," we'll show a design for a three-servo walker that uses this concept to its advantage.

Walking Speed Obviously, we can't calculate the speed of your legged robot the way we calculated a wheeled vehicle's speed. We're walking, not rolling! Again, the hobby servo just gives us the characteristics we need to size and calculate our speeds. Is it any wonder these little workhorses are the robot hobbyist's best friend?

With a standard hobby servo, a walking robot will appear to waddle. This is because the servo will move the leg in an arc and not a straight line. The length of that arc depends upon the distance (in radians, of course) that the servo control horn turns and the length of the arm from

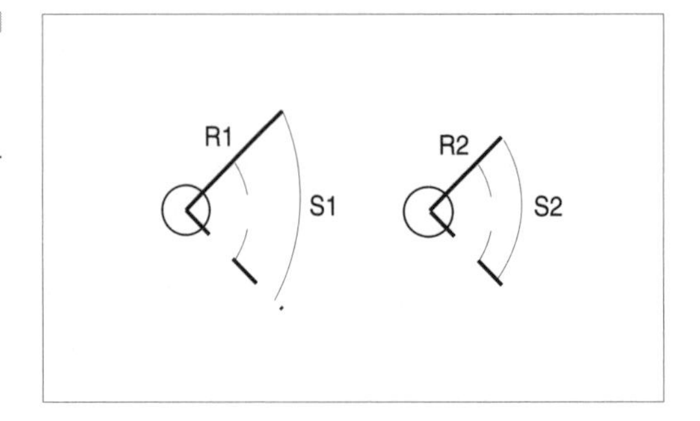

Figure 2-15 Servo arc and stride

the control horn to the upright part of the leg that contacts the ground. Figure 2-15 illustrates this relationship. The formula for calculating that arc is

$$\text{Stride} = 2\pi r\theta$$

where an easy way to calculate θ here is simply

$$\theta = \text{degrees}/360$$

In Figure 2-15, you see that although θ is the same, R1 is longer than R2, which makes S1 longer than S2.

Velocity is the distance traveled in a given length of time or

$$v = d/t$$

For instance, if our robot's upper leg length from the servo hub to where the leg turns downward to the ground is 2 inches (about 5 cm) and we allow a full 60° of arc, then our stride is

$$\text{Stride} = \frac{60}{360} \times 2\pi \times 2in = 2.1in(5.3cm)$$

So if we time our strides to be every 0.5 seconds, then our estimated velocity will be

$$v = \frac{2in}{0.5\text{ sec}} = \frac{4in}{\text{sec}}\left(\frac{10cm}{\text{sec}}\right)$$

This is, of course, only in a perfect world. The speed here would be a really fast walker. Remember that this motion is an arc, and the straight line distance is somewhat less. Also, legged robots tend to slip as they walk, so some straight line gain is lost when the other side of the robot takes a step. Although this calculation is not perfectly accurate, it is, however, a pretty good estimate of your walker's speed.

The Effects of Terrain and Debris

No special considerations are made for a rolling robot climbing a slope; those are covered in the previous power requirement calculations. However, if the robot will need to roll over some debris, that needs to enter into our calculations in a special way. In general, the rule of thumb is to size your wheel to be 10 times the diameter of the largest piece of debris it will encounter. If the tire is too small it will not roll over the debris; it will jam up against it and the debris then becomes an obstacle.

When a robot lifts a leg to move, not only must we consider how much power this will take, but we need to consider how high it needs to lift it. If we are expecting terrain height differences or debris, we need to make sure that the robot can lift its legs high enough to walk on or over debris. In the case of slopes, we need to make sure that the robot can lift its legs high enough that the slope will not affect the leg moving forward when the robot first encounters it.

This can be done in one of two ways: reduce the stride or raise the leg higher. Of these two choices, the former is best. Lifting legs too high may cause the walker to lose its balance and topple over. Taking shorter strides will allow quicker strides and have the high friction contact patch on the walking surface for a shorter period of time, which gives the robot a *scuttling* appearance.

Sometimes changing the gait of the walker will be the best way to handle a new slope. It will depend on whether or not your robot can actually detect the slope initially. The slope is only an issue to the robot when it first encounters it. When all the legs are on the slope, the strides can be essentially the same as when the robot is on a horizontal surface. When a walker is negotiating a highly variable surface, a shorter stride will minimize the changes in slope, and longer strides may cause our walker to topple. Slopes are very challenging to the walking robot. The taller the robot, the greater the challenge.

CHAPTER 3

Using DC Motors

If you aren't building a walker robot, odds are pretty good that you'll either be using modified servos, steppers, or vanilla DC motors for locomotion. This chapter will cover some of the basics of selecting and using regular DC motors, plain or gearhead.

Motor Selection

Before selecting a motor, you should have a good idea of your overall motor mechanical power requirements. Chapter 2, "Motor Types: An Overview," provides a set of techniques for these requirements based on robot weight and design. In addition to motor-running torque, you should keep in mind a number of factors when selecting a motor:

- **Motor speed** How fast do you need your robot to go? For an indoor robot, slower is generally better, but some kinds of robots may require high velocities. This is particularly true of competition robots, such as BattleBots or robots that compete in maze races. Remember that the final speed of your robot will be a product of both motor speed and wheel circumference.
- **Power requirements** As the mechanical power output of a motor rises, so does the required electrical power at its input. Consider the motor voltage requirements in addition to its current requirements. Unusual input voltages may complicate your battery and recharge requirements. A low-voltage motor, such as 6 volts, may cause problems if you intend to run other subsystems on your robot that need more than 6 volts from the same supply. Note that most linear 5-volt regulators such as the 7805 will not operate properly from a 6-volt battery.
- **Continuous runtime** How long will your robot need to operate between battery charges or changes? How long do you expect your motors to run without stopping? Some motors are meant for continuous operation, and some are not. There may be issues of gearbox wear and tear, as well as heating issues with some types of motors.
- **Output** Most motor output is via a round shaft. It will always be simpler to mate wheels and other hardware to a standard-sized flatted shaft. Many motors you might purchase surplus, however, may have odd-sized, odd-shaped, or even nonexistent shafts. Often, a great deal on a surplus or used motor may not be so great when you

consider the time you'll need to spend mating the shaft to a wheel. We'll cover some methods of shaft mating in detail in this chapter, as well as in Chapter 10.

- **Shaft load** Mounting wheels directly to motor output shafts is a common and simple strategy employed by amateurs. However, in the case of heavier robots, this practice may place more load on the motor shaft bearings or gearbox than either was designed for, shortening the life of the motor considerably. In this chapter, we'll attempt to develop some rules of thumb for determining maximum shaft loading and some simple methods of decoupling the shaft from the wheel.
- **Extra features** Some motors have built-in features that can really make your life easier. These include integral shaft encoders and associated logic, tachometers, and even on-board braking. Although these motors may cost more up front, you should consider the time they can save you down the road.
- **Noise** Some motors can be quite noisy. This may or may not be a problem. For an intermittent-duty robot confined to an office or workshop, noise may never be an issue. For a robot that roams a home 12 to 24 hours a day, noise can become an annoyance, especially to those insufficiently appreciative of your creation. Unfortunately, noise can be a difficult thing to predict and in some cases may only become apparent when a motor operates under load. In general, smaller, light-duty motors are not as likely to be noisy as larger motors. In this chapter, we'll cover a couple of techniques for coping with noisy motors.
- **Mountability** Motor mounting will be considered in detail in Chapter 6, "Mounting Motors," but as a rule almost all motors can be mounted given a little ingenuity and elbow grease. Nevertheless, you should be aware when choosing a motor, especially from a surplus outlet, that you may have to do more work than you planned in order to mount it to the chassis of your robot properly.

Speed Versus Torque—Plain Versus Geared Motors

Few robots need the kind of speed provided by plain, gearless motors, but there are, of course, exceptions. These include some BattleBot robots, maze-running robots, racing robots, and others. Small, plain permanent magnet DC motors, when fed their rated voltage, rarely run at no-load speeds lower than 5,000 *revolutions per minute* (RPM). Given a motor velocity in RPM, robot velocity can be computed with this simple formula:

$$V_r = (V_m D\pi)/60$$

Consider a robot with a 3-inch wheel diameter driven by a motor running at 5,000 RPM:

$$(5{,}000 \times 3 \times 3.141)/60 = 785.25 \text{ inches per second}$$

Naturally, this is only a theoretical maximum; the fact remains that gears allows us to do just that. Of course, we can slow down any electric motor by reducing the supply voltage. Unfortunately, any reduction in power to the motor will result in a proportional decrease in output torque. If we want to slow the motor down without reducing the output torque, we'll need to use a method that doesn't involve lowering the input voltage.

Gears—Fundamentals

At the most fundamental level, gears are a means of transmitting power. Power transmission via gears can take a number of useful forms:

- Gears can increase or decrease rotational velocity, with a corresponding decrease or increase in torque.
- Gears can change rotational direction or the angle of rotation.
- Gears can be used to convert rotational motion into linear motion.
- Gears can change the location of rotational motion.

We will be primarily interested in the first property of gears, the capability to modify torque and speed. Consider the two gears in Figure 3-1. Gear A, the smallest of the two gears, is termed the *pinion*. This gear has 16 teeth, half the number of teeth of the larger gear B (sometimes termed the *wheel*), which has 32 teeth. The *gear ratio*, the number of teeth on the wheel divided by the number of teeth on the pinion, is 2.

A gear ratio of 2 means that for each revolution of the pinion, the wheel will only spin half a turn. If we connect a motor to the pinion and an output shaft to the wheel, we will effectively cut the motor speed in

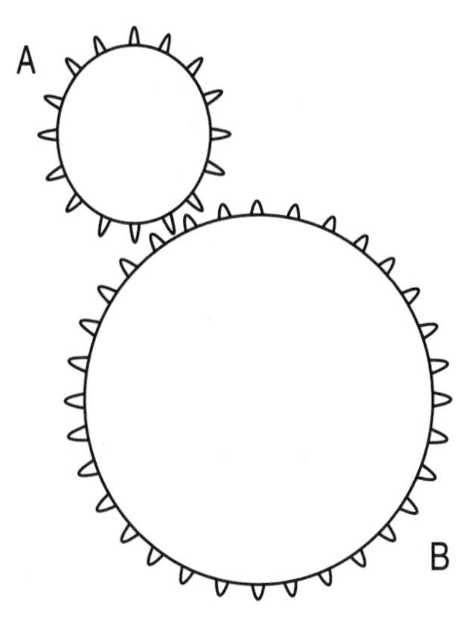

Figure 3-1 Gear pair with a ratio of 2

half at the output shaft. Happily, we will also double the output torque. Like their cousin, the lever, gears enable you to amplify force at the expense of the distance over which force is applied, or vice versa.

The equations for computing torque or the speed of the wheel (the larger gear) given the torque or speed of the pinion (the smaller gear) are as follows:

$$T_w = e(T_p R)$$

$$S_w = e(S_p / R)$$

where

T_w = wheel torque
T_p = pinion torque
R = gear ratio
e = gearbox efficiency constant between 0 and 1
S_w = wheel speed
S_p = pinion speed

Computing the torque and speed for the pinion given the torque or speed of the wheel is simply the inverse:

$$S_p = e(S_w R)$$

$$T_p = e(T_w / R)$$

For example, assume you have a pair of gears with a ratio of 2, where the pinion spins at 200 RPM with a torque of 10 inch-pounds. Assume also that the efficiency constant is .9 (90 percent efficiency). The output wheel will produce a torque of 18 inch-pounds (.9 × 10 × 2). The wheel will spin at 90 RPM (.9 × (200/2)).

It is important to note the efficiency constant, as it will tend to drop as more gears are added, implying that high gear reductions will incur a potentially steep efficiency penalty. These efficiency losses occur in the gears themselves as well as the associated bearings.

Achieving High Gear Reductions We can see that the reduction of any given gear is a function of the number of teeth on the input gear, called the *driver*, versus the number of teeth on the output gear, called the *follower*. As the follower grows larger to accommodate the larger number of teeth, however, its size may become impractical for use in a given application. The solution that usually comes to mind to beginners is to add intermediate gears, as in Figure 3-2.

With a little thought, though, it should be apparent that the gear ratio will remain unchanged with this arrangement. Because the gear ratio is solely a function of the ratio of the driver (A) gear to the follower gear (B), the intervening gears are referred to as *idlers*. Instead of the setup in Figure 3-2, gear reduction can be increased by attaching a pinion to the wheel shaft, and using this second pinion to drive another larger gear, as in Figure 3-3. This arrangement can be repeated many times to achieve high gear ratios and is commonly used in motor gearboxes.

Other means of getting large gear reductions in tight spaces include *worm gears*, often found in automotive electrical motors, and *planetary gearboxes*, frequently used in power tools. We'll discuss these in a bit more detail in the section "Types of Gears."

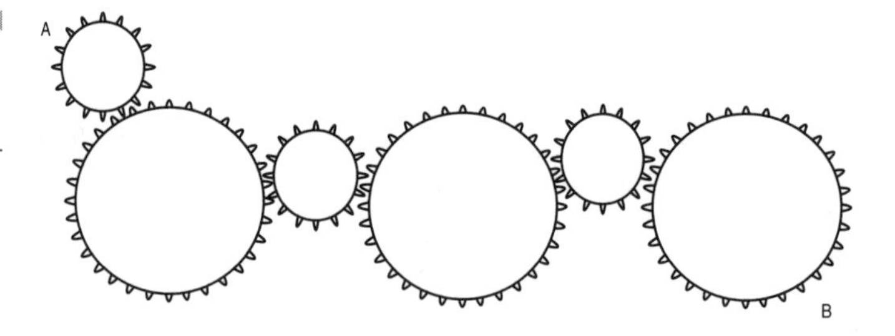

Figure 3-2 Gears and idlers

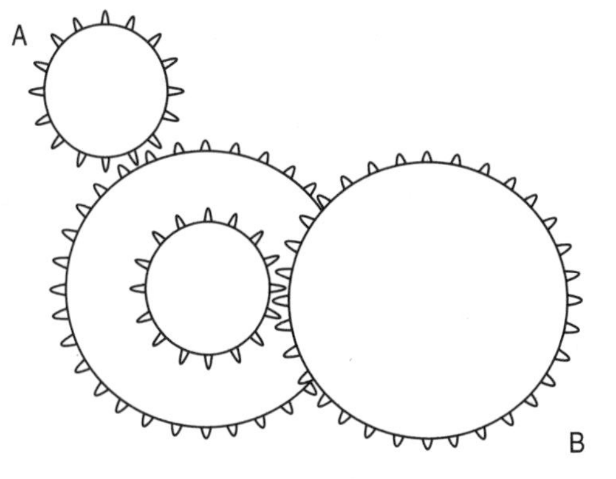

Figure 3-3 Spur gear reduction

Gear Vocabulary

You should understand a few key terms when talking about gears. Figure 3-4 illustrates a gear's anatomy and the key terms are listed here, in no particular order:

Pitch circle An imaginary circle upon which all other gear calculations and measurements tend to be based. It usually intersects the gear teeth at about their midsection.

Pitch diameter The diameter of the pitch circle.

Addendum circle An imaginary circle that forms the outer boundary of the gear. The addendum circle *just* touches the ends of the teeth.

Dedendum circle Another imaginary circle drawn in such a way as to pass through the bottom of the teeth and tooth spaces.

Addendum The distance between the addendum circle and the pitch circle.

Dedendum The distance between the dedendum circle and the pitch circle.

Clearance The distance between the top of an engaging gear tooth and the bottom of the engaged tooth space.

Tooth thickness The width of a gear tooth. This is measured along the arc of the pitch circle, not as a straight line.

Tooth space width The width of the space between teeth. Like tooth thickness, this value is measured along the arc of the pitch circle.

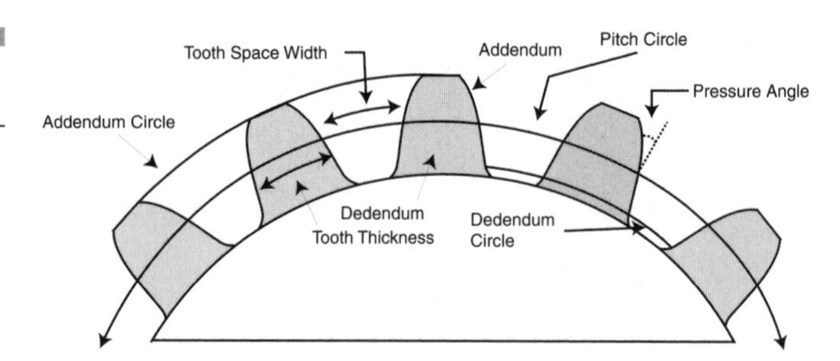

Figure 3-4
Gear anatomy

Backlash The difference between the tooth space width of an engaged gear and the tooth thickness of the engaging gear.

Pinion The smallest of a set of two mating gears.

Wheel The largest of a set of mating gears.

Diametral pitch Often just termed *pitch,* the diametral pitch of a gear indicates the number of teeth on the gear, usually in terms of teeth per inch of pitch circle diameter.

Driver In a set of two or more gears, the driver gear is the gear to which external force is applied. In a sense, this is the *input*.

Follower The follower gear is driven by the driver gear. In any set of gears, the follower can be considered the *output* gear.

Idler Idler gears are any intervening gears between the driver and a follower. Note that idler gears have no effect on the gear ratio—this is determined exclusively by pitches of the driver and follower.

Pressure angle The pressure angle is the term for the slope of the gear teeth. Since the sides of the gear teeth are usually curved, however, the pressure angle is more precisely defined as the angle between a tooth's inner edge (sometimes termed the *tooth profile*) and a line perpendicular to the pitch circle at the intersection of the tooth edge and pitch circle. Although odd pressure angle values are sometimes encountered, standard pressure angles are 20, 25, and 22.5 degrees. It is important that mating gears have matching pressure angles. When this is not the case, the gears will suffer excessive wear, emit noise, and will most likely exhibit larger losses through friction than would otherwise be the case.

Working depth The distance between the addendum circle and the clearance; the maximum depth of incursion of an engaging tooth.

Whole depth The distance from the addendum circle to the dedendum circle.

External gears Gears having their teeth along the outside surface. Spur gears are external gears.

Internal gears Gears having their teeth along an inner surface of a hollow cylinder.

Types of Gears

Several major classes of gears exist, and some of the most common you might encounter are as follows:

Spur gears Spur gears are one of the most frequently encountered gear types found in gearboxes (see Figure 3-5). These gears are cylindrical, with teeth evenly spaced along the outside surface of the cylinder, parallel with the direction of rotation. Spur gears are widely available on the surplus market and through specialty parts outlets.

Helical gears Like spur gears, helical gears are cylindrical in shape. However, the teeth on helical gears are set at an angle relative to the axis of rotation (see Figure 3-6). Helical gears tend to be much quieter than spur gears and are capable of carrying larger loads at higher speeds. For this reason, helical gears are widely used in automotive applications. Helical gears can be used in parallel, like spur gears, or at right angles.

Worm gears A worm gear is technically a pair of helical gears set at right angles to one another (see Figure 3-7). One of the gears, the *worm* driver, has its teeth set a very shallow angle, much like a screw. The follower can be a helical gear, or in some cases, a simple spur gear. Worm gears are capable of very high gear ratios and, like helical gears, tend to be quiet in operation. One particularly useful

Figure 3-5
Spur gear

Figure 3-6 Helical gear from an automotive transmission

Figure 3-7 Worm gear

feature of worm gears is that while the worm driver can easily move the follower output gear, it is almost impossible to move the worm by driving the follower gear. This means that a worm drive has a strong tendency to stay in place even when no power is applied.

Bevel gears Bevel gears are conical in shape. Their teeth may be straight and perpendicular to the axis of rotation, or they may be curved (see Figure 3-8). Bevel gears typically mate at right angles to one another, although other angles are possible. Bevel gears are capable of carrying fairly high loads, although not as high as helical gears.

Figure 3-8 Bevel gear

Planetary gears Planetary gears are distinguished by the fact that one or more gears (termed *planets*) move around and along the surface of a central gear (called the *sun*). The planet gears are often linked to the central axis via an *arm* or *planet carrier*. Often the sun gear is "turned inside out" and implemented as an *internal* gear, which makes for a very high torque, yet a compact gear train.

Planetary gears can tolerate significant loading, but are usually less efficient and produce more noise. These gears are commonly found in power tools or anywhere high torque is needed in a small amount of space (see Figure 3-9).

Rack gears Rack gears have their teeth set on a flat, as opposed to a curved, surface. When driven by a pinion, they are used to translate rotary motion to linear motion (see Figure 3-10).

Figure 3-9 Planetary gear

Figure 3-10 Rack and pinion

Gearheads and Gearboxes

In order to gear down a DC motor, you'll need to construct a gearbox from scratch, purchase a gearbox to use with your motor, or use a gearhead motor, a motor with a gearbox already attached (see Figure 3-11). As a rule, most amateurs will choose the latter option, although gearbox construction and standalone gearboxes are also an option in some cases.

Gearbox Construction—Sealed Versus Open Frame Whether standalone or permanently attached to a motor, the construction of a gearbox can give you a number of clues as to its suitability and performance when used to drive your robot. One of the most obvious distinctions in construction you will come across is *open-frame* versus *closed* or *sealed* gearboxes. Figure 3-12 shows an example of each type.

Figure 3-11 Gearhead motors and a standalone gearbox

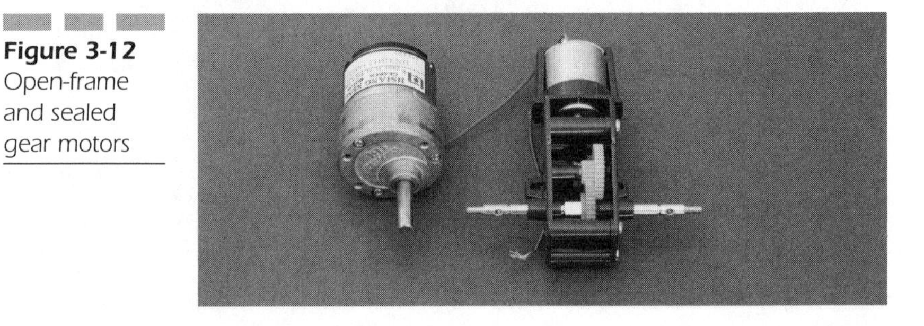

Figure 3-12 Open-frame and sealed gear motors

By far, the sealed gearbox is the most commonly used and offers a number of advantages. Sealed gearboxes tend to be less noisy, since the gears are fully enclosed. More importantly, all gears perform more efficiently—and thus last longer—with lubrication. In the case of a sealed gearbox, the gear mechanism is sealed in with a generous quantity of lubricant. As a rule, you'll never need to apply lubrication within smaller sealed gearboxes—in fact, a sealed gearbox can often not be easily opened at all. Larger units, such as those intended for industrial use, may have external grease ports.

Hobby servos are good examples of sealed gearhead motors. If you have a chance to modify a hobby servo for continuous rotation, you'll be sure to notice the messy lubricant inside—which you'll need to preserve on the gearing mechanism so as not to shorten the life of the servo.

Open-frame gearboxes are somewhat less common on the surplus market, but may still be found occasionally. They turn up frequently attached to motors scavenged from toys. Also, gearbox kits are available from educational supply outlets and some robotics specialty stores that offer more than one reduction option. The open-frame gear motor pictured in Figure 3-12 was constructed from such a kit. Although interesting, these plastic gearboxes are often not sturdy enough for anything other than moderate loads and are somewhat expensive compared to other options. In general, most open-frame boxes will be lighter duty, although there are exceptions.

If an open-frame gearbox is built to take a lubricant—and some aren't—you'll want to be careful in the application of the grease or other lubricant. You'll also want to be sure that any lubricant stays in the gearbox itself, rather than soiling the robot, carpet, pets, or children. The latter can be surprisingly annoying to less understanding spouses.

One advantage to open-frame gearboxes that should be mentioned is that unlike closed gearboxes open-frame boxes can be inspected. If you're purchasing your gearhead motor at a retail establishment, this gives you the option of determining what material the gears are made of, how well the mechanism is constructed, and, in the case of used or surplus units, whether the gears have suffered excessive wear or damage. Closed gearboxes, on the other hand, will require more guesswork to determine their suitability for your application.

Gearbox Construction—Metal Versus Plastic Gearboxes you'll run across can be constructed of plastic, metal, or a combination of the two. Metals commonly used in constructing gears include steel, aluminum, and brass, with steel being the most expensive. Steel gears tend to have more load-carrying capacity than brass gears, but at up to twice the cost.

Plastic or nylon gears are significantly cheaper than brass—by about a third in the case of Delrin™ gears—and when run with appropriate loads they can be much quieter than their metal counterparts. Plastic and brass gears can often be used together in the same gear train. Plastic gears cannot generally take the same kinds of loads as steel or even brass gears.

Unless your gearbox is open frame, or you have specifications available for your sealed gearbox, you will likely have to guess at the internal details of the materials used for gear construction.

A gearbox with an exterior made of plastic may or may not use plastic gears. If you can run the gearbox, you should be able to tell immediately by the amount of noise produced if the gearbox uses plastic gears since they will run relatively quietly.

Despite the fact that plastic gears have the potential to run more quietly than their metal counterparts, many plastic gearheads you are likely to see for purchase are made for use in toys, and as such they have been designed and built in such a way as to minimize cost. This means that a plastic gearbox is likely to run less efficiently, produce more noise, and have a shorter service life than a metal counterpart of equivalent quality.

Shaft loading, in particular *radial loading*, can also be a problem with these cheaper gearboxes. Radial loading refers to the amount of force exerted on the gearbox output shaft perpendicular to the shaft. If you choose to mount your robot's wheels directly to the gearbox output shaft, then radial load will become something you'll need to take into account, since the weight of your robot will be borne in large part on the gearbox shaft. Excessive force can cause the gearbox to run poorly and wear out early. In general, the amount of axial load that can be borne by any gearbox is a function of the bearings used to support the output shaft. In the case of most plastic gearboxes, shaft bearings are often cheap, with little load-bearing capacity.

On the other hand, plastic gear motors can be quite inexpensive. The motors depicted in Figure 3-13 were manufactured for use in children's electric ride-on toys and are commonly found on the surplus market. As of this writing, these motors, with a stall torque of well over 60 inch-pounds, were selling for less than $7. That's roughly $1 per 10 inch-pounds of torque, a hard price to beat.

Gearboxes made of metal are likely to be of stouter internal construction, although nothing guarantees that this will be the case. In general, we prefer these for robots, especially larger ones. You will likely find the

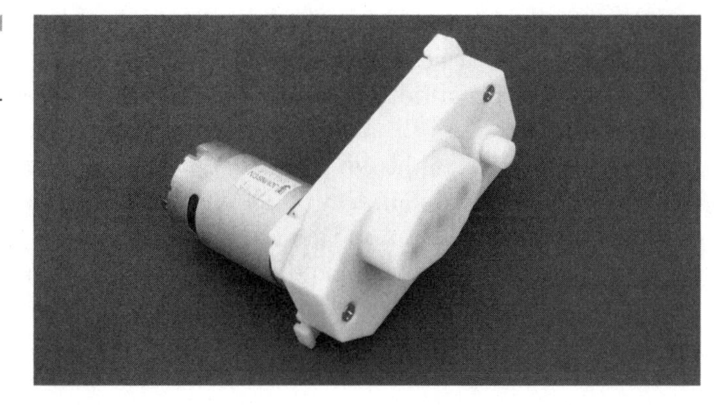

Figure 3-13
Plastic gearbox

extra outlay required is justified by the longer service life and improved performance of the motor. Moreover, most gearhead motors on the surplus market tend to be made of metal, although quality will vary widely.

Backlash

Backlash—or, more picturesquely, *slop*—is the amount of space between an engaging tooth and the tooth space of the mating gear. Put another way, backlash is the difference between the tooth width of the engaging gear versus the tooth space width of the engaged gear. Backlash will allow play in the shaft in both directions of rotation. Backlash is often specified in angular units, such as degrees or, occasionally, radians. It's not unusual to come across gearboxes with substantial play, up to several degrees, on the shaft.

We should be clear here: In general, backlash is written about as if it were always a bad thing. However, some backlash is built in to most gear systems. Space will be required for the lubricant to flow between meshing gear teeth. Moreover, machining quality may vary from shop to shop. As the machining error factor rises, so must the amount of backlash built into the machined gears to prevent lockups. More precisely, manufactured gears can be manufactured with less backlash. Some backlash may need to be built in order to accommodate gears that are expected to expand and contract at different rates, as may be the case when the gears are made of different materials.

When backlash rises to objectionable levels, then we term it *slop*. Gearheads or gearboxes with high amounts of slop will tend to be noisier and less durable. The latter is especially true when the direction of motion is suddenly reversed and gear teeth fly across the backlash interval to engage—or rather crash into—the opposite tooth. Since a sudden reversal is a common (although inefficient) means of braking used by many amateurs, slop is a factor worth considering when purchasing a gearhead or gearbox, although it need not be a decisive factor.

Checking for backlash is easy to do. With a pair of vise grips, grip the output shaft of the gearbox and gently rock the vise grips back and forth, checking for play—there should be little to none (see Figure 3-14). If you have good specs on your gearbox, you may actually be able to get a precise figure. In either case, given two motors substantially alike in most other respects, we prefer the motor with less backlash.

Obtaining Standalone Gearboxes

As a rule, most amateurs will not want to go this route. Standalone gearboxes when new can run into the hundreds of dollars. It *is* possible to remove gearboxes from used AC motors, which often cost far less than their used DC counterparts (see Figure 3-15). Often, these AC motors seem inexpensive enough to justify the cost of purchasing them for the gearbox alone.

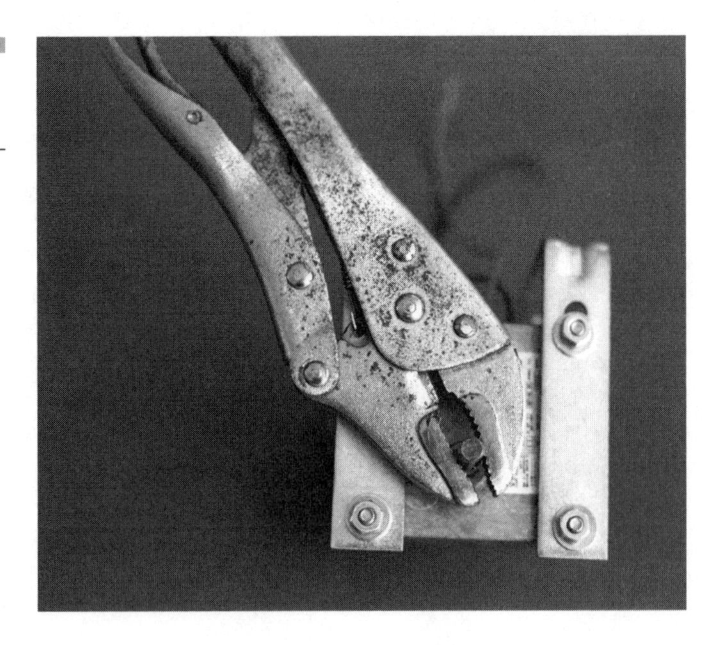

Figure 3-14 A gearhead and vise grips

Figure 3-15 Scavenged gearbox

This approach to gearing has a few potential problems, however:

- **Input shaft size** You will need to match the input shaft size to your motor. This may be impossible to do without removing the gearbox from the AC motor to inspect it. The shaft will also probably need a certain type of mating gear attached in order to work with the gearbox. If you can't get this gear off the shaft of the old motor, you may be in for a great deal of difficulty.
- **Motor attachment** You'll incur extra hassle mounting the motor and gearbox in such a way as to get the motor output shaft correctly aligned with the gearbox input. You may be the beneficiary of the happy coincidence of a motor sized and tapped to mate perfectly with the gearbox, but this is unlikely. Flexible couplings may be helpful in this regard.
- **Gearbox input speed** All gearboxes have a maximum acceptable input speed, beyond which they shouldn't be run. Motor output should not exceed this speed. Unfortunately, you will most likely not have this rating available to you.

As always, consider the value of your time along with any possible savings realized by scavenging.

Wheel Diameter and Gearing

The final force and speed delivered by your motor is dependent not only on the output of your gearbox, but also on the diameter of the attached wheels. As the wheel diameter lengthens, speed increases and force diminishes. As wheel diameter shrinks, speed decreases and force increases. Note that when we use the term *speed*, we aren't referring to shaft radial velocity (RPM or radial/second), but actual robot velocity. The methods used to compute wheel force and velocity work equally well whether your wheel will be connected to a gearbox output or coupled with an ungeared motor.

To figure the effect of the wheel diameter on gearing, divide the torque by the wheel radius (diameter/2). Be sure that you are using the same distance units for both the wheel radius and torque (for example, inches and inch-pounds). This will give you the amount of force delivered by the wheel.

To compute your robot's speed, you'll need to first figure your wheel *rollout*, or circumference. The circumference of any circle, in this case, your wheel, is simply

$$\text{Circumference} = \pi d = 2\pi r$$

where d is diameter, and r is radius. Multiply this figure by the shaft speed in RPM and divide by 60 to get the robot velocity in the specified units per second.

For example, assuming your wheel radius is 3.5 inches (diameter = 7 inches) driven by a gear train outputting 10 inch-pounds of torque at 35 RPM:

$$\text{Force} = 10/3.5 = 2.9 \text{ pounds}$$

$$\text{Speed} = 2 \times 3.5 \times 3.141 \times 35/60 = 12.8 \text{ inches/second}$$

If we shrink our wheels down to a radius of 1.5 inches (diameter = 3 inches), force rises and speed falls, as follows:

$$\text{Force} = 10/1.5 = 6.67 \text{ pounds}$$

$$\text{Speed} = 2 \times 1.5 \times 3.141 \times 35/60 = 5.5 \text{ inches/second}$$

Gearing Down Without Gears Given that the force exerted by a wheel increases as the wheel diameter shrinks, one simple technique for increasing the amount of force delivered by a motor without resorting to gears is to shrink the wheel size to almost nothing. This is a technique commonly used in small robots powered by pager motors. In general, it's difficult to find gearhead motors in these smaller sizes. BEAM roboticists have come up with the novel solution of getting useable torque from

Figure 3-16
BEAM wheel

these tiny motors by simply capping the motor shafts with a rubber sheath or shrink-tubing "tires," thus shrinking the "wheels" to the smallest diameter practical (see Figure 3-16).

This method has some obvious disadvantages. The small but rapidly spinning shafts and the small area of contact between the wheel and the floor mean substantial power losses to friction. Note that this arrangement also places both axial and radial loads on the motor shaft, and many motor bearings are not meant to take more than very light axial loads (see "The Motor Shaft" section for more information on axial and radial loading). Nevertheless, you may want to experiment with this technique on your own.

Gearing and Efficiency

Our discussion of gearing so far has for the most part often assumed "perfect" gears. That is, we assume no losses to mechanical friction in our gear train. However, in the real world, any gear train will incur losses between gears. This loss in efficiency occurs at each point of contact between gears as well as at the gear bearings. Thus, the loss effect is additive throughout the entire gear train. With the exception of worm drive gearboxes, most motors use increasing numbers of gears to provide increasing gear ratios, so you should not be surprised to discover that overall motor efficiency drops substantially as you increase the gearbox reduction ratio. Table 3-1 shows typical figures for an inexpensive gearhead motor. Note that the efficiencies listed pertain to the gearbox itself, not the motor. Given that most small DC motors will have maximum efficiencies in the 40 to 75 percent range (and that only when run at about 10 to 20 percent of maximum torque), losses can be quite substantial.

You can see how dramatic efficiency losses become as gear reduction rises. A motor that runs at 50 percent efficiency connected to a gearhead

Table 3-1 Gearbox efficiency versus gear ratio

6:1	30:1	75:1	100:1	180:1
81%	73%	66%	66%	59%
300:1	**500:1**	**800:1**	**1000:1**	**3000:1**
59%	59%	53%	53%	48%

with a reduction ratio of 100 will have a final efficiency of 33 percent. The implication is that you will do better (from an efficiency standpoint) by avoiding higher gear reductions. When you have a choice, you can look for slower motors connected to lower reduction gearboxes. Lower input speeds have the added advantage of prolonging gearbox life and decreasing noise. Unfortunately, you are unlikely to have the luxury of mixing motors and gearboxes unless you are ordering new motors in quantity.

Efficiency can also worsen with some gear types, particularly planetary gears. Planetary gears are able to handle high loads by virtue of the fact that the load is spread over multiple gears at each stage. The increase in the number of gears, however, means a corresponding loss of efficiency as well as increased noise. Planetary gears are frequently used in cordless power tools, the motors of which are sometimes scavenged by amateurs for use as high-torque robot motors. Of course, a well-made planetary gear train would certainly outperform a poorly constructed spur gear train—clearly, a meaningful comparison of gear types must consider the quality of the mechanisms being compared.

The Motor Shaft

Motor output is usually via a cylindrical shaft. This will not always be the case, however, when motors are purchased surplus or used and removed from existing equipment. In such cases, you may need to cope with a variety of odd shaft shapes and sizes. Worse yet, some motors may lack output shafts altogether or end in output gears or other widgets that defy all reasonable attempts at removal.

Motor output shafts will also only support limited loads. In the case of smaller robots, shaft load may never be a problem. For larger robots, however, failure to take into account shaft loading may drastically shorten motor life and result in noisy and inefficient operation.

Types of Shaft Loading

Two types of shaft loading exist: *radial* shaft loading, the variety that concerns us most, occurs perpendicular to the shaft. *Axial* loading occurs parallel to the shaft. Figure 3-17 illustrates both types of loading.

Motor spec sheets may specify maximum shaft loads in ounces, pounds, grams, kilograms, or newtons (about 100 grams). Maximum radial loading specs are usually accompanied by the distance along the shaft from the motor housing. Think of the motor shaft as a lever, with the motor bearing acting as both fulcrum and load, and you can see why this distance figure is critical. As the point of application of radial load moves away from the bearing, the load on the bearing itself increases proportionately, as in Figure 3-18.

In a wheeled robot, radial load can become a problem particularly when wheels are attached directly to motor or gearbox output shafts. Such designs, while mechanically simpler, place a large portion of the

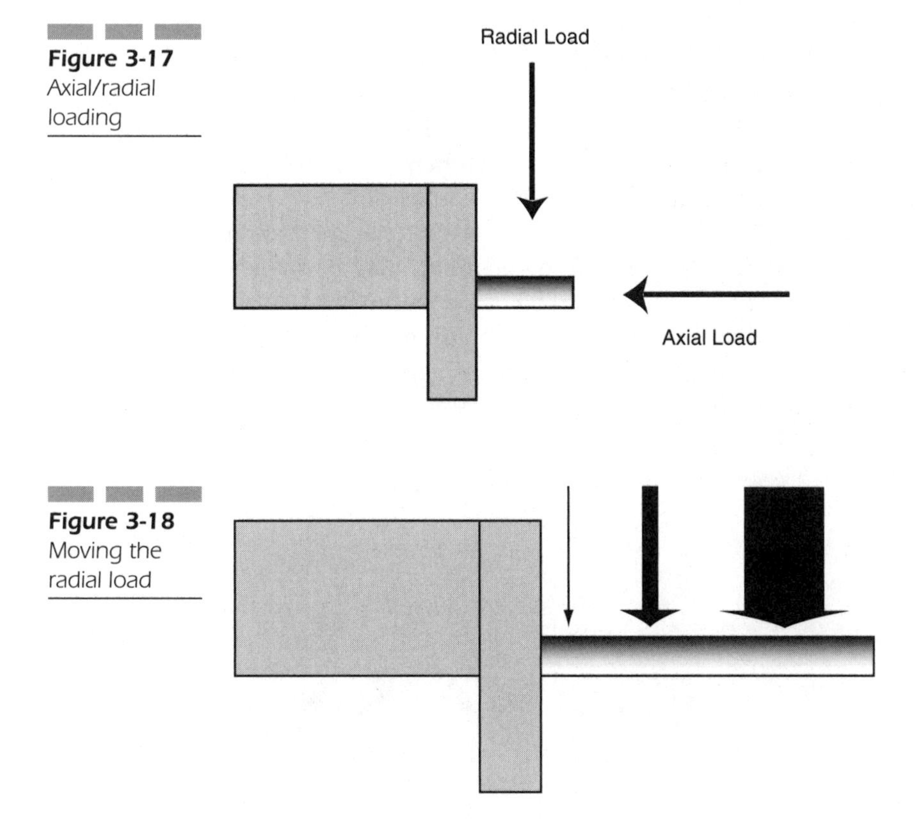

Figure 3-17 Axial/radial loading

Figure 3-18 Moving the radial load

robot's weight on the motor shaft. To mitigate the effects of shaft loading, robots with wheels attached directly to the shaft should place the wheels as close to the bearing as possible. Where possible, avoid extending the motor shaft or mounting the wheel at the end of a long shaft. Other sources of radial loading include overly tight belt or chain drives, where a pulley or sprocket connected to the shaft is under excessive tension.

Although it comes at a somewhat higher cost in terms of mechanical complexity, a design that mounts wheels on a separate pillow block and drives them via a flexible coupling to the motor shaft can vastly reduce radial loading. Such an approach should be considered for heavier robots, especially when driven by less expensive gearhead motors. One simple method for implementing flexible coupling will be detailed in the next section.

The amount of radial load a motor shaft can take is generally a function of the type of bearing used to support the shaft. In the case of a gearbox, the bearing protects the gears themselves from undue forces that might cause noise or shorten service life. The most common bearing types you are likely to encounter are *sintered metal* bearings and *ball* bearings (see Figure 3-19).

Sintered bearings start out as a powdered metal, usually bronze or a bronze alloy, which is compressed to a near solid, formed into a sleeve, and heated in a furnace. This process of heating, or *sintering*, causes the metal particles to bond to one another. The result is a porous metal sleeve, which is impregnated with a lubricant. As the shaft spins in the bearing, a film of lubricant forms between the shaft and the bearing surface. As a rule, this film forms more reliably as the shaft spins faster, which makes sintered metal bearings somewhat less effective than ball bearings when used with slowly spinning shafts. Sintered metal bearings are, however, substantially less expensive than ball bearings and often find their way into lighter duty gearhead motors.

Figure 3-19
Ball bearings

Ball bearings consist of an array of balls arranged in a circular sleeve through which the motor shaft fits. The balls are arranged so that the shaft turns on the array of balls, which themselves roll freely with the motion of the shaft. Although substantially more expensive than sintered bronze bearings, ball bearings can take far more load than the latter, and they perform better at the relatively low velocities of robotic gearhead drives. Ball bearings are common on larger motors with average mechanical outputs greater than 5 watts, but they are something of an expensive bonus on smaller motors. A ball bearing can often take loads five times or greater than its sintered metal cousin. If possible, you should generally choose ball bearings over sintered metal, especially when you expect higher radial shaft loads.

Axial loading primarily needs to be kept in mind when attaching or removing things from a motor shaft. Use common sense: Avoid hammering things on to the shaft or pulling things off with a pair of vise grips and all the strength you can muster. Excessive axial force can be especially damaging to shafts that use preloaded ball bearings, knocking the bearings out of alignment and causing slow but steady degradation of both the bearings and the gear train.

Radial loading can be a pernicious problem, because it typically takes a while for damage to show up. Often a gearbox with excessive radial load will run fine for months, or even years. Then the day will come when you notice that gearbox noise has risen to positively annoying levels, and black goo seems to be leaking out around the shaft (and perhaps getting on the carpet, pets, or children). At this point the motor will not have much longer to live and will need to be replaced soon.

Shaft Coupling

If you're not planning or able to mount your wheels directly to the motor shaft, or if you otherwise need a way to mate something to the motor shaft, you'll need some form of *shaft coupling*. For example, you might want to extend the motor shaft with a bit of threaded rod, to which you could mount a wheel or pulley. You may find it necessary to adapt a larger shaft to a smaller size or vice versa. In either case, you have more than a couple of options.

Rigid Coupling Shafts of the same size can be coupled together with rigid couplings. You may either purchase these new or create your own fairly easily. Manufactured couplings can be bought directly from large suppliers such as McMaster-Carr or Small Parts, Inc., although they are somewhat expensive. A few varieties are shown in Figure 3-20.

These couplers are generally just short tubes of metal, drilled and tapped with two holes for setscrews, which are often included with the couplers. They can be manufactured from a number of different materials, including iron, steel, stainless steel, and aluminum. Some types are available with differing inside diameters on either end for use in mating shafts of dissimilar sizes. The setscrews keep the shaft from rotating in the coupling and work best if the shaft has been flatted. If the shafts you are attempting to join aren't already flatted, you can use a file to flatten them yourself.

Because these manufactured couplers can't usually be found at your local hardware outlet and will most likely have to be ordered, you may want to make your own. This can be done by using commonly available *coupling nuts*, which can be bought from most hardware stores (the bigger depots will have the largest selection). Coupling nuts are just long nuts normally used to join lengths of threaded rod. Just drill and tap two holes for setscrews and you're ready to go (see Figure 3-21).

Tips on Tapping *Tapping* involves threading a hole to accept a screw. A hole of a diameter dependent on the tap size is drilled in the work piece. The tap, which looks something like a drill bit, is then worked slowly into the hole using a *tap wrench*, which threads the hole. Taps and

Figure 3-20 Rigid couplings

Figure 3-21 Coupling nut with setscrews

Figure 3-22
Tap assortment

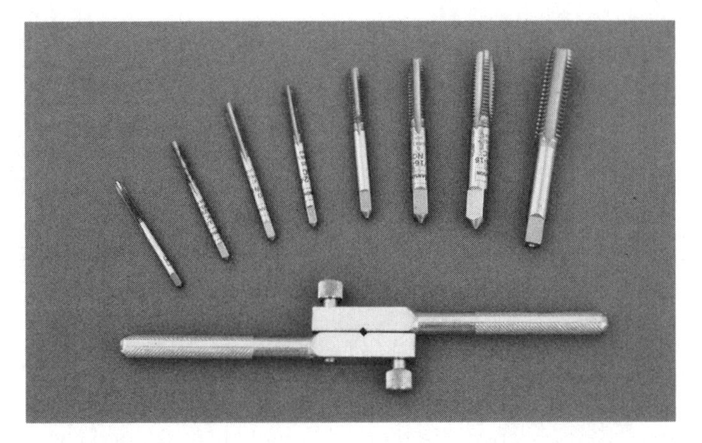

tap wrenches can usually be found at good hardware stores and are a required tool for the roboticist (see Figure 3-22).

It is quite easy to break off the tap if the hole drilled is not appropriate for the tap size. If this happens, your work piece can be rendered useless and you'll have to start again. To help prevent breakage, be sure to drill a hole with a diameter appropriate to the tap size. The recommended hole diameter should come with the tap or tap set.

When tapping, take care to avoid excessive force. It's best to work the tap in, back it out, and work it back in again gradually. Repeat until the hole has been tapped. You should also use a lubricant as you tap. WD-40 or any light oil will do in a pinch, but tapping fluids are also available that will do a much better job and help prevent tap breakage.

Most coupling nuts are made of galvanized steel, which pretty much requires that you use a tap for threading. However, when using softer materials such as aluminum, you may be able to get away with self-tapping machine screws. Just drill a hole slightly smaller than the screw diameter and drive in the screw. As you do so, the screw will cut its own threads into the softer metal walls of the hole. Unfortunately, threads cut into aluminum are significantly less durable than threads cut into steel. Over time, the thread quality will begin to degrade with frequent assembly and disassembly.

Flexible Coupling Flexible shaft coupling can be used to mate shafts together when some minor degree of shaft misalignment is expected. As is the case with standard shaft couplers, these can be bought ready-made (see Figure 3-23). These are handy to have in your toolbox, since it can be difficult for amateurs to get two shafts to mate precisely. When a

Figure 3-23 Commercial flexible couplers—light- and heavy-duty

motor output is rigidly coupled to another shaft and the alignment of the two shafts is poor, the motor life can be shortened considerably.

Do-It-Yourself Flexible Coupling One great use of flexible shaft coupling is to drive a wheel not directly mounted to the robot body, as might be the case when a directly mounted wheel would place excessive radial load on the motor output shaft. In this case, a wheel and shaft can be mounted on a *pillow block*, a housing containing a bearing, and the wheel shaft can be driven through a flexible coupling. Although a rigid coupling can also be used, the flexible coupling permits a less exact alignment between the wheel shaft and the motor shaft, thus simplifying construction substantially. One such complete arrangement is shown in Figure 3-24.

Figure 3-24 Flexible motor-wheel coupling

In the figure, the flexible coupling is just a short length of braid-reinforced vinyl tubing secured with hose clamps. Such tubing is readily available from most hardware stores. Select a thick-walled length of tubing with an exact inner diameter (abbreviated *ID*) of your shaft diameter. Thicker-walled tubing will have less play along the axis of rotation. You can also avoid excessive play by using shorter coupling lengths. Note that a small amount of play may actually be somewhat beneficial, providing damping (shock absorbing) of hard motor stops. These hard stops can be a major source of wear on gearboxes, especially where large amounts of backlash are present.

To attach the tubing to the motor shaft, slip it on and secure it with a hose clamp if needed. You may also heat the tubing with a hair dryer or heat gun (set on low), so that the tubing expands a bit, making attachment easier and providing a firmer grip once cooled. Use the same approach to attach the other end to the wheel shaft. Because the tubing can be made to expand with heat, you can use it to connect shafts of slightly differing sizes. Be aware that small ID tubing fit over a wider shaft using the heat expansion method can be somewhat hard (though not impossible) to remove later. When attaching and removing anything from a motor shaft, try to avoid axial loading (pulling and pushing) on the motor bearing by clamping the motor shaft (*not* the motor itself) securely.

Although vinyl tubing is an easy and inexpensive way to mate shafts, the amount of torque that this arrangement can take is limited. The exact value depends on the thickness of the tubing wall, the amount of offset or angular displacement between shafts, and the distance between the shafts. In general, we've found this method workable for up to about 8 inch-pounds of torque, with significant wear occurring at the coupler beyond that. Note that beverage supply houses often carry thicker-walled tubing meant for carrying gas used in beverage dispensing systems. Tubing meant to carry compressed air and braid-reinforced tubing can often be found in hardware outlets with inside diameters as small as a quarter of an inch (6.4 millimeters). These varieties also have thicker, stronger walls than garden-variety tubing.

Frankly, we feel the uses of vinyl tubing in robotics are generally underappreciated. The thin-walled varieties can be slipped over a small motor shaft and used to enlarge the effective shaft diameter. When a piece of tubing is split along its length, it can be slipped over the edge of a robotic platform and clamped or glued to provide a rough and ready bumper. The possibilities are, well, not exactly endless, but worth exploring nevertheless.

Coupling High Loads At high loads, shaft couplers can become an annoying cause of repeated failure, especially when setscrews are relied upon to keep the attached shafts from spinning in the couplers. Frequent disassembly of the coupling mechanism may exacerbate the problem.

Shaft marring can occur over time at the point where the set screw contacts the flat of the shaft. The resulting damage can eventually wear a groove in the shaft that causes excessive play to occur even when the setscrew is tightened to its maximum (see Figure 3-25). You can mitigate some of the effects of shaft marring by using setscrews made of softer metal, such as brass. If self-tapping screws have been used in construction, make sure that the pointed ends have been filed flat.

You may also consider spreading the load out among more than one setscrew. A particularly effective arrangement is to add a second setscrew 120 degrees along the axis of the coupler from the first, as in Figure 3-26.

Avoid disassembling the coupling frequently, especially couplings made of softer metals, such as brass or aluminum, as the setscrew threads may begin to degrade.

For very heavy leads, you may want to consider the use of an expansion pin, which is inserted into the top of the coupler, passes through a hole drilled lengthwise into the shaft, and out the other side of the coupling, as in Figure 3-27.

Figure 3-25 Setscrews and the shaft cross-section

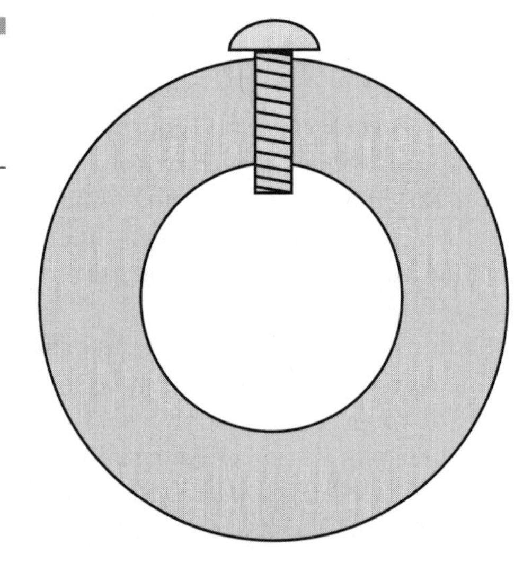

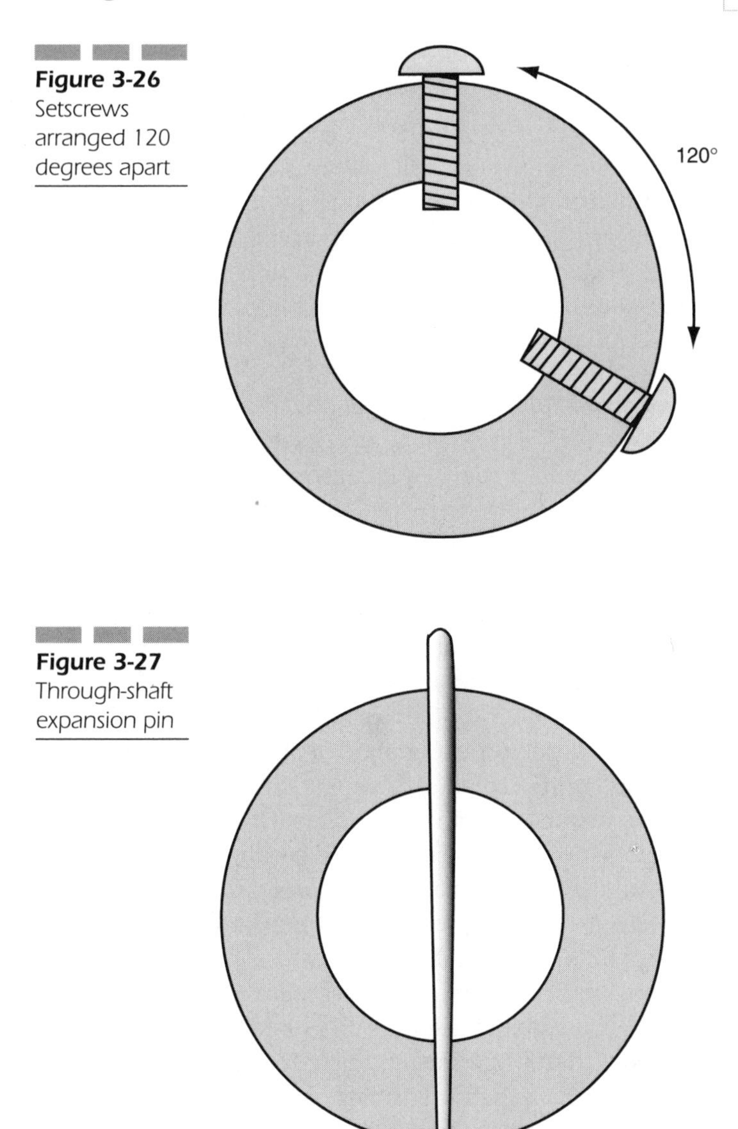

Figure 3-26 Setscrews arranged 120 degrees apart

Figure 3-27 Through-shaft expansion pin

Take extreme care when drilling the motor shaft. Be sure that the hole is drilled dead center in the shaft—a drill press will be virtually required to drill the hole properly. Since the drilling will apply a great deal of force to the shaft, support the shaft with a sturdy clamp to avoid placing stress on the motor bearing.

Shafts: Standard and Strange

When purchasing motors surplus, you may need to deal with odd shaft sizes or nonstandard output types. Even when the shaft is a standard size, the standard might not apply locally, such as when a buyer in the United States purchases a motor with a shaft size specified in millimeters, or when an Australian finds him- or herself with a gearbox having a quarter-inch output. In this case, the shaft may need to be modified to accommodate the locally available hardware.

Mating Different Sized Shafts Mating shafts of dissimilar sizes can be done in myriad ways. It is possible to purchase shaft couplers with differing bores at either end made for just such a purpose. Some types even come in two pieces, allowing you to mix and match bore sizes as necessary. These are sometimes available on the surplus market, but you may have some difficulty finding the sizes you want. McMaster-Carr and other big parts outlets stock them, but they can be somewhat expensive. Fortunately, you can roll your own shaft adapters without a lot of difficulty, although when the sizes of the shafts don't really differ by *too* much, you can get away with using simpler strategies.

As mentioned earlier, you can use a short length of vinyl tubing to mate two shafts of different diameters. This works well when the shaft diameters are within about 1/8 of an inch (about 3.2 millimeters). Normally, the shaft to which you are mating should be supported by its own bearing, since the tubing really can't support much weight and would cause the entire shaft assembly to sag. The tubing ID should be pretty close to the smaller shaft diameter, if not a little wider. You can use a hose clamp, available from most hardware stores to secure one end of the hose to the smaller shaft. Heat the tubing on the other end to get it to slide on to the larger shaft. Note that when the larger shaft gets to be close to the maximum diameter of the expanded tubing, expect the tubing, once cooled, to be *very* difficult to remove from the larger shaft, short of cutting it away with a utility knife.

Another approach is to use a standard shaft coupler with a bore size equal to the largest of the two shafts. You then enlarge the smallest of the two shafts until it fits in the other end by slipping a sleeve or sleeves over the smaller end until it fits snuggly into the coupler. The sleeves act as cylindrical *shims*.

The sleeves may be brass (or other metal) tubing stock, available from most hobby stores. You can add successively larger sizes until the shaft has been enlarged to the coupler bore size. You can also try using vinyl tubing as a sleeve in a rigid coupling, which yields a semiflexible coupling.

We've had limited success using masking tape cut to the length of the shaft and carefully rolled around its circumference until the shaft diameter is larger than the target diameter. We then begin trimming lengths until the shaft fits snuggly into the coupling. This technique works reasonably well when the differences in shaft sizes are small (less than 1/16 of an inch or about 1.5 millimeters), but when the masking tape layer becomes too thick it becomes susceptible to damage under load, eventually causing the shaft to slip in the coupling.

The coupling nut coupler from the previous section can be modified by adding additional setscrews along the opposite sides to allow for the alignment of a smaller shaft inserted into one end (see Figure 3-28). The size difference really shouldn't be much more than 1/8 of an inch. If you have a manufactured coupler, it's not too hard to drill and tap a new setscrew along the section in which you intend to place the smaller shaft. The new setscrew should be placed opposite the existing screw. Getting the smaller shaft true with respect to the larger shaft is simply a matter of adjusting the setscrews supporting the smaller shaft. This is done by trial and error.

Finally, it is possible to manufacture your own dual-bore coupler (see Figure 3-29). This will, however, require the ability to drill very straight holes and thus a drill press. Although this may be possible to manage with a hand drill, we haven't ever attempted it. To start with, you will need a short length of aluminum rod stock, which can be purchased from

Figure 3-28 Coupling nut shaft coupler with six setscrews

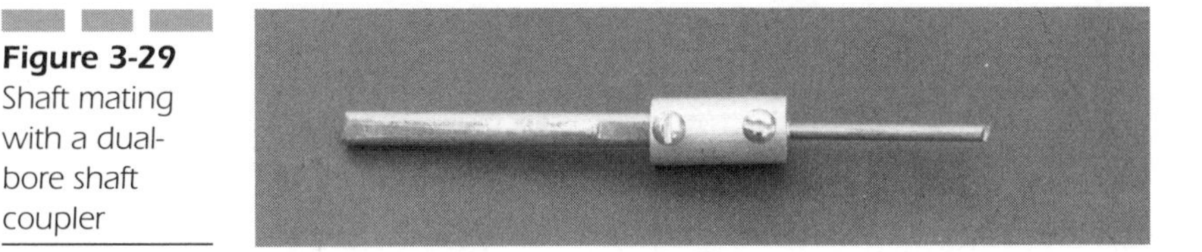

Figure 3-29 Shaft mating with a dual-bore shaft coupler

large hardware depots as well as many hobby shops. The stock diameter should exceed the size of the larger of the two shafts you intend to couple by 1/8 of an inch (around 3 millimeters) or so to leave a wall thick enough for the setscrews to sit firmly. For a more durable coupler, you can use steel rod, but aluminum is easier to work with.

Cut a small length of the metal stock—around 3/4 of an inch (19 millimeters). Those of you with time and muscles to spare will want to use a hacksaw to do this; we, however, prefer to use a *reciprocating saw* with a metal-cutting blade, a marvelous instrument for cutting almost anything. For neatness' sake, try to make your cut as straight as possible, although the face of the stock really need not be perfectly planar. You can use a file to remove any residual burs.

Mark a hole in the exact center of one end of the cylinder. Next, clamp your work securely and carefully drill out the marked hole using your drill press, as in Figure 3-30(A). The hole should extend the length of the cylinder—which should be a tube when you're done. The hole bore should be equal to the diameter of the smallest of the two shafts to mate, although you may want to start with a fairly small hole initially and drill it out with a larger bit to its final size. You may want to use cutting oil as you drill, but if you're using aluminum stock, cutting oil is not required.

Next, drill out about half the length of the hole to the bore matching the diameter of the larger shaft, as in Figure 3-30(B). Provided the hole was drilled straight, the two shafts should line up perfectly when placed

Figure 3-30 Drilling the dual-bore shaft coupler

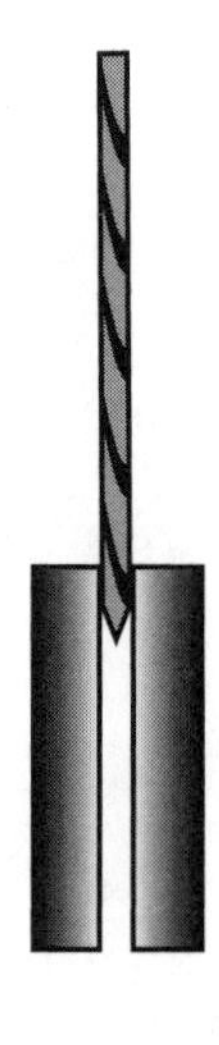

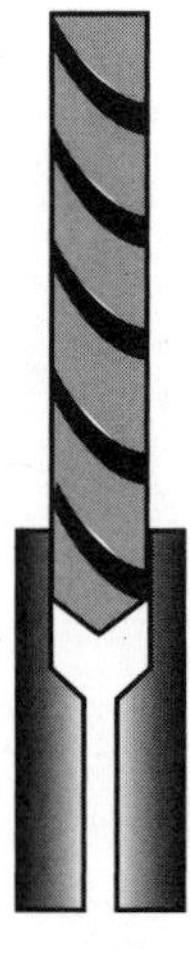

in the coupling. All that remains is to drill and tap two setscrews along the side of the coupling. Because we're working with relatively soft aluminum stock, you may find that you can bypass this last step and instead drill starter holes and use self-tapping machine screws to thread the holes.

Dealing with Odd Output Types When shopping the surplus market, it is not uncommon to find what appears at first to be a real motor bargain—high torque, ball bearings, built-in encoder circuitry, attractive faux wood housing, nifty leather carrying case, and a built-in CD player all for under $10—until a close reading of the description reveals that the motor requires a 67-volt power supply and provides output in the form of a 1-millimeter kidney-bean-shaped shaft. Before taking on such a motor, carefully consider the cost of both the time and tools you'll need to render this beast usable in your design.

Of course, we would recommend punting on *any* motor that wants a 67-volt power supply. However, with some thought, it is possible to adapt strange output types with just a little extra cost and hassle. Unfortunately, so *many* varieties of output types exist that any attempt to cover them all would require more space than we have to offer here and would be hopelessly out of date in short order in any case. Instead, we'll offer a few tips and then two case studies on a couple of common surplus bargains.

Tip 1: Go Retail, Where Possible Retail surplus outlets have a distinct advantage over their mail-order and Internet counterparts in that it is possible to check out a prospective purchase in person. Even when a mail-order firm provides detailed descriptions of a given part, it is always possible that they may accidentally omit a small detail that makes the motor shaft harder (or easier) to couple to a load than would be gleaned otherwise. This is *not* to say that you should avoid going mail-order—we purchase surplus by mail on a regular basis—but we merely want to point out that retail has a distinct advantage if you're lucky enough to have a good establishment within a reasonable distance, especially when dealing with nonstandard parts.

Tip 2: An Amazing Variety of Shaft Shapes Can Be Made to Fit in a Cylindrical Coupler, Given Enough Setscrews Square, double-D, or hexagonal shafts are not uncommon on some motors—these will often work fine with regular round-shaft couplers. For additional hold, you may drill and tap extra setscrews. When the shaft is both a nonstandard shape and size, consider constructing a dual-bore coupler as detailed earlier.

Tip 3: Make Sure That Items Attached to the Shaft Can Be Removed Although a small spur gear attached to a motor shaft can sometimes be dealt with according to Tip 2, very large items, such as larger gears, hubs, or eccentric weights, will pose a problem if they can't be removed. Often an attached item might be connected via a setscrew or spring clip, or be press-fitted in such a way as to be removable with reasonable force. Other items may be virtually impossible to remove.

In some cases, items attached to a shaft can be crushed in a vise, and the broken pieces removed from the shaft. Some success has been reported using this method to remove the eccentric weights attached to pager motor shafts. Clearly, however, you'll need to use considerable caution to avoid crushing the shaft along with the attached object. The judicious use of a hobby mototool and a cutoff wheel will remove almost anything that is attached to the shaft, if you can reach it with the tool.

Tip 4: Take Care to Avoid Damaging the Motor Bearing when Removing Items from Shafts Clamp the motor shaft securely whenever you intend to apply force to the shaft. As mentioned earlier, motors are typically not constructed to handle large axial loads.

The motor shown in Figure 3-31 is built to power a child's toy. Such motors are widely available on the surplus market as of the writing of this book and provide a large amount of torque for very little money (typically less than $7). The motor has no shaft at all; instead, output is provided through two rotating hexagonal sockets. Although at first such a motor might seem to be a lost cause, the half-inch hex socket lends itself to a number of fairly straightforward means of attaching shafts.

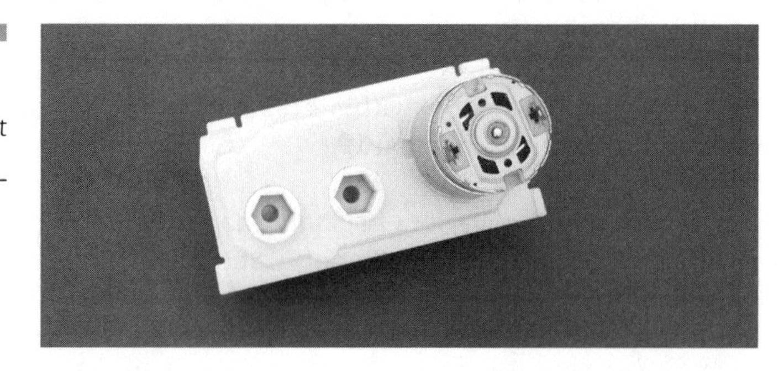

Figure 3-31 Surplus motor with hex socket output

In our case, we happened to have on hand a pair of coupling nuts that fit exactly into the hexagonal output socket. The nuts were cleaned with a degreaser, roughed up a bit with a file, and permanently secured in the socket using a “cold weld” epoxy product. The degreasing step is especially important; many steel parts are packaged with a thin, protective layer of oil, which will defeat even the most effective bonding agents if not first removed with a good solvent. These can be found at good hardware stores under brand names such as Gunk.

Be sure to verify that the coupling nut is seated straight in the socket before the epoxy sets. Should the epoxy harden with the nut at even a slight angle, the entire motor will be rendered useless.

Before attaching each coupling nut, we drilled and tapped it for a pair of 6-32 setscrews. This enables the attachment of a variety of output shafts. The motor shown in Figure 3-32 is fitted with a coupling nut adapter that holds a short length of threaded rod.

In the case of our hexagonal output socket, we could also have used other means of mating with the socket, including flat-top hex bolts, hexagonal aluminum or steel stock, hex standoffs, and so on. Metal rod stock can be purchased in a number of noncylindrical shapes. If our output had been a square-shaped socket, we could have used a bit of square metal rod as our shaft adapter.

Another motor frequently found on the surplus market is shown in Figure 3-33. Manufactured for use as a power window motor for cars, these motors provide high output torque via an integral worm gear. Models providing 50 inch-pounds of torque at 45 to 70 RPM can often be had

Figure 3-32 Surplus motor with coupling nut shaft adapter with threaded rod attached

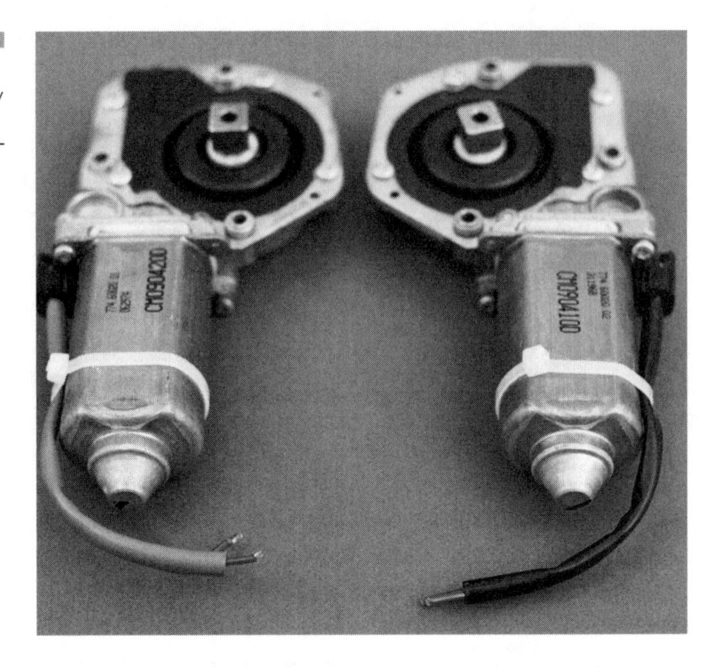

Figure 3-33 Power window motors

for less than $25 on the surplus market. The torque and speed ranges on these motors, together with supply voltages of 12VDC make them seem perfect for robotics use at first glance. These motors, however, have outputs specifically designed to mate with automobile window raising/lowering mechanisms. In the case of our example motors, the shaft output is a 9/16 inch steel square bolt with a thickness of about a 1/4 inch.

One solution that immediately suggests itself is to fashion a dual-bore coupler as described in "Mating Different Sized Shafts," where one side of the coupler is large enough to accommodate the square motor output, and the other side is drilled to fit a smaller, more convenient shaft diameter. The coupler will need to be made from a fairly large piece of stock, since the large bore must be wide enough to match the diagonal of the output square. Our 9/16-inch square will require a bore of about .8 of an inch. Still another approach might be to fashion some kind of shaft adapter from square metal tubing, assuming the appropriate size is readily available.

However, upon examining our example motors—purchased sight-unseen through a mail-order house—we found ourselves with an unexpected bit of luck. In the center of the square output shaft on each motor was a 1/4-inch hole, shallow but still useable. By drilling through one side of the square shaft all the way through to the center hole and tapping

Figure 3-34 Power window motor with attached shaft

6-32 thread, we were able to secure a 1/4-inch threaded shaft with relative ease, as in Figure 3-34. Because the square shaft is constructed of extremely hard steel, we had to use a tungsten-carbide drill bit. We recommend keeping an assortment of these bits around if you own a drill press. Despite their excellent drilling capabilities, however, tungsten-carbide bits are quite brittle and the smaller diameters can break when used with a regular power drill.

As always, if you need to work directly on the motor shaft, take care to clamp it securely to guard against undue load on the motor bearings.

Attaching Things to Shafts Although the ins and outs of attaching wheels will be covered in a subsequent chapter, we would like to introduce some basic fastening techniques here. If you've made it this far, you've probably gathered that many items fasten to shafts via setscrews, which assume flatted shafts. If you need to flatten a shaft, a simple file can be used. Try to keep the flat straight, rather than sloping as you file.

Shaft collars can be used to provide arbitrary stops along the length of a shaft. In Figure 3-35, a shaft collar acts as a retaining ring to keep a free-running wheel on its axle. A shaft collar is often constructed as a simple ring with a setscrew. Shaft collars are widely available from large parts outlets, such as McMaster-Carr, surplus mechanical parts vendors, and from hobby stores. The latter are usually a good source for smaller bore collars. You can make your own shaft collars using some of the same techniques used to make couplings, but they are relatively inexpensive to purchase.

Figure 3-35 Shaft collars keep this steering wheel on its axle and provide stops at key points along the steering shaft.

Figure 3-36 A wheel attached to a shaft via a coupler

A common strategy for attaching wheels to shafts involves press-fitting a shaft collar (or coupler) into the wheel hub and drilling the hub so that the collar setscrew passes through the hub, as in Figure 3-36.

When threaded, a motor shaft can be fitted with a variety of items secured by nuts and washers. This can be useful for pulleys, sprockets, wheels, or encoders. Threads are cut into shafts with a tool called a *die*. A die is threaded onto a shaft with a die wrench (see Figure 3-37). As the

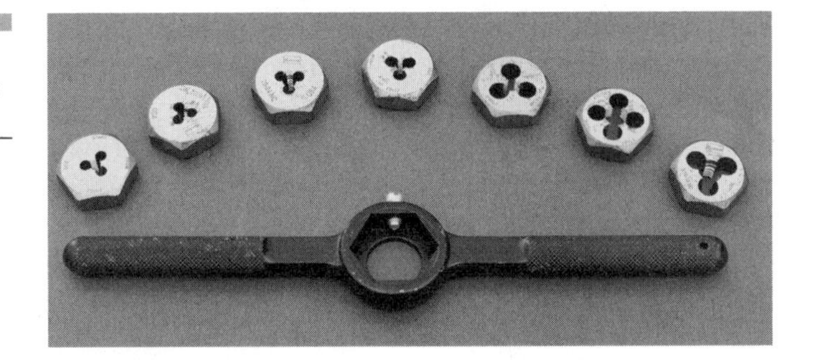

Figure 3-37 Dies and a die wrench

die is threaded on, it cuts threads into the shaft. As when tapping holes, we recommend using tapping fluid and working slowly. Dies and wrenches are available from most hardware stores.

You'll need a good method of clamping the motor shaft, which can be difficult to the point of being impossible if the shaft size is excessively small. In such cases, threading the shaft may not be an option. Threading is also not an option when a shaft has been flatted or if the shaft is not a standard size.

If you find yourself in need of a threaded shaft but have no means or desire to thread your motor shaft directly, you may consider the option of coupling a short length of threaded rod to the motor shaft with a shaft coupler (see Figure 3-38). When going this route, however, remember

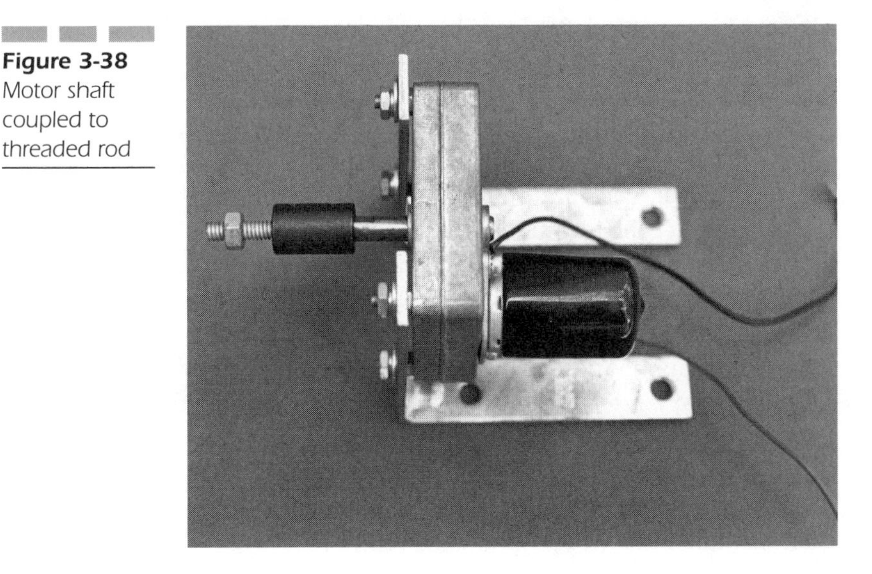

Figure 3-38 Motor shaft coupled to threaded rod

that the coupler will add extra length to the shaft. If the threaded end is to bear a great load, any extra length will increase the radial force on the motor or gearbox bearing.

Other Issues—Service Life, Noise, and Braking

We have chosen to group these items together in the same section for the simple reason that they are related in many respects. A noisy motor will tend to be shorter lived than a quiet motor. Proper braking can increase the lifetime of a motor; improper braking can shorten it.

Maximizing Motor Life

Whether your motor is a gearhead motor or not, you will need to take care to avoid excessive radial loading. The lighter a load you put on the motor bearing, the better. If at all possible, choose a ball-bearing motor over one that uses a sleeve bearing.

The additional mechanical complexity introduced by a gearbox brings many factors into play that can shorten a motor's life. If you have a choice, choose a slower motor coupled with a lower-reduction gearbox, assuming such an arrangement can meet your torque needs. The slower a gearbox is run, the longer it is likely to last. Unfortunately, mixing and matching motors and gearboxes is usually only an option for commercial buyers or those with unusually deep pockets.

Any gearbox will have a tendency to last longer when delivering less torque. Since you will almost certainly have minimum torque requirements, however, you may want to consider oversizing your gearbox so that you can get the torque you need while still running the gearbox well below its maximum rating. As we have pointed out frequently, oversizing motors is a good practice in general.

You may also choose to go with a gear type other than spur gears for high-torque applications. Worm gears can be a good choice in this regard. Planetary gears, because they distribute developed torque among multiple gears, are also frequently chosen for high-torque applications, which is why they are utilized in power screwdrivers and similar tools. Both of these gear types, however, tend to be less efficient than spur gear mechanisms.

The lubricants used in gearboxes can be adversely affected by temperature extremes. This is primarily a concern for robots designed to operate outdoors. Avoid operations during periods of extreme cold or heat, just as you'd avoid operations during thunderstorms or snow.

Gearboxes used in mobile robots are usually subject to substantial vibrations as the robot travels over various bumps, holes, and other terrain pitfalls. Outdoor robots in particular can be subject to truly punishing vibrations, even on sidewalks and driveways. You may want to consider taking measures to lessen motor vibration. Often this can be as simple as a couple of rubber washers (available from almost any hardware store) between the motor mount and robot body, as shown in Figure 3-39.

If you want to get *fancy*, you can purchase materials specifically meant for damping vibration from specialty outlets such as Small Parts, Inc., and place the materials between the entire body of the motor mount and the robot chassis. If you want to get *chintzy*, try cutting up a bit of old mousepad if you have a firm rubber one on hand.

Hard stops and sudden reversals are a major source of gear wear. The problem becomes worse with gear trains that exhibit a great deal of backlash. Avoid the common practice of braking a moving robot by reversing the supply polarity and throwing the motor into reverse. In addition to wasting precious battery power, you are sure to shorten your gearbox life. If available on your motor controller, consider using a *dynamic* or *regenerative* braking scheme.

A coupling that exhibits some torsional flexibility (the capability to twist slightly along the axis of rotation) can be used to damp sudden stops (see Figure 3-40). The vinyl tubing coupling discussed earlier is ideal for this purpose. A flexible coupling, however, generally doesn't support much weight, and so may not be usable for some applications.

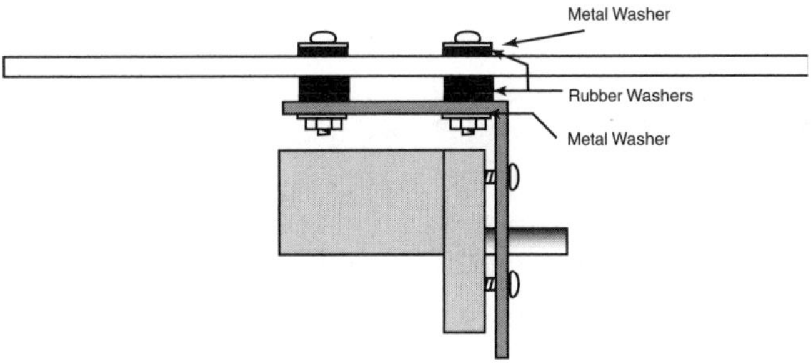

Figure 3-39 Shock mounting with rubber washers

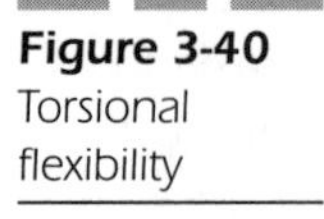

Figure 3-40
Torsional flexibility

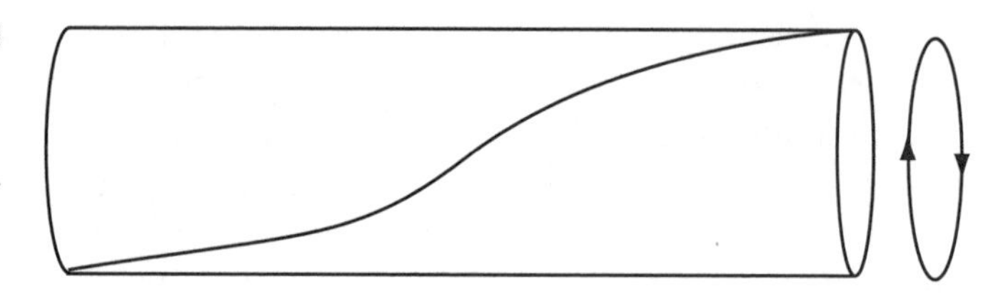

Finally, any motor will last the longest when run at its maximum efficiency point. Unfortunately, this is usually at about 10 to 15 percent of maximum load—yet another reason to oversize your motor relative to your projected needs. In the same vein, avoid any prolonged periods of high current draw from your motors; stall-detection or current-monitoring controllers can be a great help in this regard.

The practice mentioned earlier of throwing a motor in reverse to affect braking is detrimental to motors as well as gearboxes. To understand why this is the case, remember that as a motor runs, back *electromotive force* (EMF) voltage (E) is generated with the opposite polarity. As the speed of the motor increases, so does E. The actual voltage in the coils is equal to the applied voltage (V) minus E. Using Ohm's Law, we can then compute the total current flowing through the motor coils as

$$I = (V - E)/R$$

By reducing the effective coil voltage, you can see that E has the effect of limiting the current through the motor coils. Now consider what happens when the supply voltage is suddenly reversed. Until the armature actually reverses, our voltage term goes negative and our current equation effectively becomes

$$I = (-V - E)/R$$

Were our motor spinning at top speed when reversed, the effective voltage in the coils would be momentarily doubled, which would have the effect of producing a current that is double the motor stall current. Frequent current surges such as these will degrade motor performance over time as well as waste battery power.

Braking

It will frequently be the case that simply cutting power to your robot motors will not bring it to a halt fast enough. As we touched on, many beginners choose to reverse the supply polarity briefly to the motor to cause braking action, which has the detrimental effect of causing a powerful current surge through the armature coils.

Regenerative and Dynamic Braking Instead, it is possible to use the motor back EMF to effectively and safely brake the motor. Consider again the effective coil voltage of a motor running at full speed. If our supply voltage is 12 volts, then an idealized motor running at top speed will have a back EMF voltage of −12 volts, leaving an effective coil voltage of 0.

If we remove the supply voltage, there will no longer be anything to oppose the back EMF. If we can provide something to sink the reverse current produced, the motor will brake. The lower the resistance of the load, the more rapid the braking will be, although with lower resistances come higher coil currents.

Some motor controllers are capable of returning the current produced to the battery, which acts as the current sink. This is termed *regenerative braking*. A simpler method of braking shorts a resistor across the motor terminals once power has been cut. In this case, the EMF current is dissipated as heat instead of being sent back to the battery, making this method less efficient than regenerative braking. Some texts term this second type of braking *dynamic braking*, although you will also often see it referred to as regenerative braking as well.

A regenerative or dynamic braking capability is built into many commercial motor controller boards and can be found as a feature on some motor controller *integrated circuits* (ICs). We strongly suggest using (or building) controllers that support at least dynamic braking where practical.

Mechanical Braking Although they are less power efficient, it is possible to find gearhead motors on the surplus market with built-in electromechanical brakes (see Figure 3-41). These are quite common on surplus wheelchair motors. Usually, these motors require that power be applied to the break to release it; thus, you will not be able to run the motor without energizing the brake as well. As soon as power is cut to

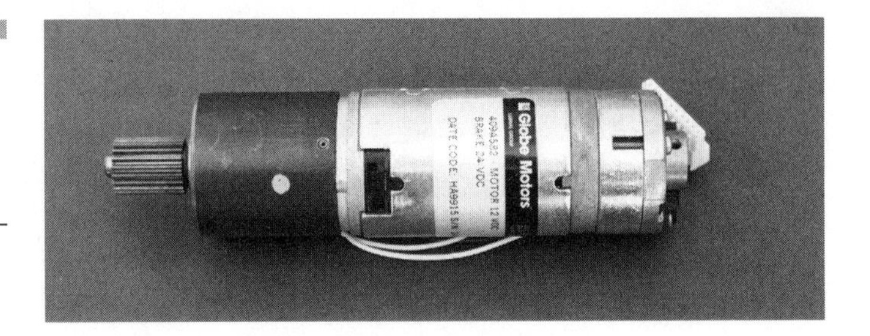

Figure 3-41 A motor with electro-mechanical brakes

the brake, the brake engages and the motor stops. Of course, it helps if you remember to also cut power to the motor.

Unfortunately, it is often the case that the motor brake requires a different supply voltage than the motor itself. The motor shown in Figure 3-41 runs off of 12 volts, but the brakes require 24 volts to disengage. Even when the voltages required are identical, we prefer regenerative or dynamic braking for both efficiency and simplicity. If you find yourself in possession of one of these motors, you may want to remove the brake altogether.

Minimizing Noise

Motor noise is produced by a number of common sources, both mechanical and otherwise. With their high mechanical part count, gearhead motors are often egregious offenders in this area. However, even when a motor is by nature particularly noisy, you may use simple techniques to reduce problem noise to more acceptable levels.

Pulse Width Modulation (PWM) Whine Motors controlled via *Pulse Width Modulation* (PWM) can be prone to high-pitched whines when PWM frequencies are in the audible range. PWM is a method for controlling the effective motor coil voltage by pulsing the motor supply voltage at a fixed frequency. The power level to the motor is varied by changing the *duty cycle*, or width, of the pulse train. PWM will be discussed in detail in Chapter 7, "Motor Control 101: The Basics."

When a PWM frequency is chosen that is within the audible range, an audible tone will often be emitted by the motor. The volume of the tone will depend upon the construction of the motor. One way to eliminate PWM whine is to raise the PWM frequency beyond the audible range, to about 20 kHz or higher.

It may also be possible to modify the offending frequency by a relatively small amount and see a reduction in volume. This is because motors often tend to be noisier at sympathetic frequencies dependent upon motor construction. When the frequency is changed even by a small amount, the amount of noise can be lessened. Of course, you might hit on a frequency that causes *increased* volume, so you'll need to experiment a bit to find the point closest to the desired frequency that produces the least noise. With some luck, whine can be reduced to a level below other sources of motor noise, effectively eliminating it.

Gearbox Noise If your gearbox is producing a clattering, chattering, ratchet-like noise, it's likely broken. The reduction of problem noise here is not an issue. Instead, the gears or bearings have probably become damaged and the entire unit should be replaced.

Normal mechanical gearbox noise can be reduced in a couple of ways:

- Avoid noisy gear types. Planetary gears, which are often scavenged from power tools by amateur roboticists for their capability to provide high torque in relatively little space, tend to be significantly noisier than a spur gearbox of similar construction quality. The higher part count for a given reduction also means these gears, while better at handling high loads, run less efficiently. Worm gears tend to be quiet indeed, although the relative silence of a worm drive often comes with a high price in efficiency.
- Gearboxes machined to higher tolerances will be quieter than less well constructed units. If noise is really an issue, you may want to consider spending extra money on a higher-quality gearhead motor.
- Just as lower input speed improves gearbox efficiency and service life, it also reduces mechanical noise. This implies that when possible you should choose a slower motor with less reduction for quieter operation.
- Plastic gears, while often less durable under load, are less noisy than comparable metal gears. It is often the case, however, that high-quality plastic gear assemblies may be hard to find. Once again, however, if noise is an issue, you may want to hold out for a decent plastic or nylon gearbox.
- High radial loads will increase gearhead noise. When quiet operation is a priority, wheels should not be attached directly to the gearbox output shaft. Instead, support the wheels on a separate bearing and transfer the gearbox output to the wheels through a coupling, a sprocket and chain, or a belt and pulley.

Damping Motor Noise When motor brackets are mounted directly to the robot chassis, vibration can be transmitted from the motor to the chassis, which acts like a loudspeaker cone, amplifying the motor noise. Isolate the chassis from the motor by using rubber washers between the bracket and chassis, as described in the section "Maximizing Motor Life." More damping material will provide better results, so you may want to use a sheet of rubber to completely separate the motor mount from the chassis. If you happen to have a collection of unused mouse pads just lying around and taking up space—not an uncommon situation for the technically inclined—you may find that those made of firmer material serve well as damping stock.

Finally, remember that any gearhead motor becomes noisier under load. A motor that runs as quiet as a whisper on the lab bench can become surprisingly noisy once installed and driving your robot. Unfortunately, you'll not likely be able to test a motor under load before you purchase it.

Selecting and Purchasing a Motor

Motor selection for most amateur robot builders is a compromise between requirements and availability. Chapter 2 guided you through the process of determining your torque and speed requirements. Here we will try to help you take those same requirements and use them to make an informed purchasing decision.

Understanding DC Motor Data

Motor spec sheets, as useful as they can be, are rather hard to come by for surplus, used, or out-of-stock motors. Since surplus or used motors are the kinds of motors most of us will be purchasing, we often have to make do with what information the vendor provides us. However, spec sheets can occasionally still be had for newer motor models. The seller may provide the material online or, if you have the make and model of the motor, you might have some luck with the motor manufacturer's web site. In the case of electronic parts, most manufacturers now provide complete data sheets online; sadly, this is not usually the case for motor manufacturers, especially with older models. Still, we suggest spending a few minutes of time online to try to find out what you can. It may even be possible to find a spec sheet for a motor similar enough to your prospective purchase that you can make do with it.

Understanding Vendor Data Of course, having a motor spec sheet is nice, but not critical. Many will simply oversize based on the information provided by the motor vendor and enjoy great success with their project. Data provided by surplus motor vendors is necessarily sketchy. Motor specifications are often unavailable, and even when they are available, catalog space considerations often prohibit reprinting them in much detail, although good online catalogs may provide links to the spec sheets, and retail vendors may have copies on hand for perusal. In any case, look for the following information:

Motor supply voltage Almost any seller should be able to tell you what kind of voltage a motor requires. Common voltages are 6, 12, and 24 volts. Consider avoiding motors that use other values. We prefer 12-volt motors in cases where both 5-volt logic and motors are to run from the same supply. Twelve volts provides adequate headroom for older and cheaper linear converters, which require a minimum of 7.4 VDC for 5-volt conversions, whereas 6 volts cuts things rather close even with modern low-headroom converters, especially as the battery wears down.

Twenty-four volts is usable, although you may need to use two 12-volt batteries in a series and tap 12 volts off the center (as illustrated in Figure 3-42) to provide power to 5-volt logic and other

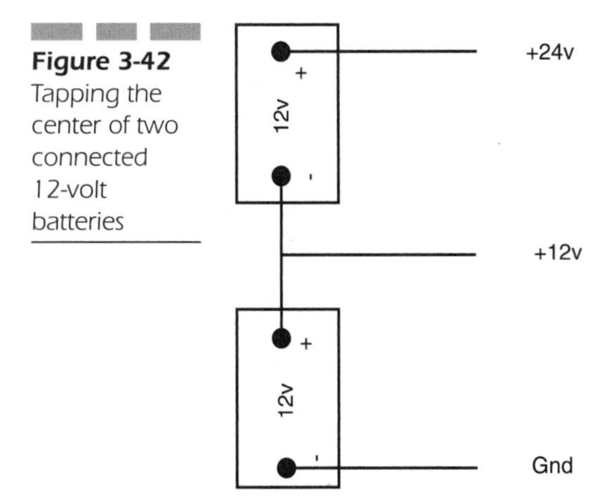

Figure 3-42 Tapping the center of two connected 12-volt batteries

peripherals. If these other nonmotor loads are substantial, one battery may deplete faster than the other, which can eventually shorten the life of both.

Of course, newer low-headroom and switching power converters can render these objections entirely moot, or you may simply choose to run your motor from a separate supply, a common strategy to avoid logic problems due to motor-induced power supply glitches.

Be aware that mechanical output (torque multiplied by speed) is purely a function of electrical input (voltage multiplied by current). Thus, there is nothing to be gained or lost by using a lower or higher voltage motor; a 6-volt motor will require double the current of a 12-volt motor to provide the same power output, and a 24-volt motor will require half the current. In any case, total battery mass will remain the same.

Finally, it is possible to run a motor at less than its rated voltage, although there will be a corresponding reduction in output speed and torque. You will generally be safe running a motor at 50 percent rated voltage. Motors are tolerant of excessive voltages in varying degrees; in general, try not to exceed 10 to 12 percent.

Motor speed It's unusual to *not* see this provided. Although some motor data sheets may show this as *radians per second*, usually speed is provided in terms of RPM. You can compute the speed of your robot in millimeters or inches per second by the following formula:

$$s = (v/60)\pi D$$

where v is RPM, and D is the wheel diameter in inches or millimeters. For indoor robots, we like a top speed of about 1 foot (330 millimeters) per second or slower. In some cases, a slower gear motor may be less efficient than its speedier counterpart.

As we will see later, gearing motor output down has the effect of dropping the total motor efficiency by a factor directly proportional to the gear ratio used. As slower motors will tend to use higher gear ratios, the slower motors will tend to be less efficient. Of course, this assumes similar input speeds to the gearbox, which will not always be the case. Furthermore, gearbox efficiency is not simply a function of gear ratio, but of the gearing method used as well as the machining quality of the gearbox and parts. Among gear motors of the same make and similar size, however, the slower motor will almost certainly be the least efficient.

Motor torque As stated in Chapter 2, motor output torque is a function of current multiplied by the torque constant, k_{torque}. Although k_{torque} is normally only provided on the motor spec sheet itself, many vendors will provide at least *stall torque.* This is the torque a motor can provide just at the point the shaft can no longer rotate, and normally it is the point of maximum current draw for the motor. Stall torque is also sometimes referred to as *starting torque* or sometimes just *maximum torque*. Your running torque requirement as computed from our guidelines in Chapter 2 should be well below this value, ideally no more than 20 percent of the motor stall torque. This will maximize both motor service life and efficiency. Remember that if your robot uses dual motors, the total torque the motors can provide is effectively doubled.

Some catalogs unfortunately fail to provide quantitative torque information at all, instead describing motors as "very strong" or the ubiquitous "perfect for robotics." Always view the latter with a healthy dose of skepticism, and in any case, never assume that motors without stated torque ratings will work for your robot. This doesn't mean that you shouldn't ever purchase such motors; just don't pay much for them—they may well end up in your spare parts bin.

Motor current Some vendors commonly omit current draw from motor descriptions, which is a shame because *no-load* current (current drawn by a motor when the shaft spins freely) and *stall* current (current drawn by the motor when stalled) are relatively easy to determine. Motor current requirements, along with the desired continuous runtime, will determine how much heft you'll need from your batteries. The actual amount of current you'll use in operation depends on a variety of factors. Maximum torque, and thus current, will be required when accelerating the robot from a standstill. Under some circumstances, sudden motor reversal can cause brief surges that nearly double the stall current. Negotiating uneven terrain and becoming snagged on obstacles can all temporarily raise current requirements. In general, plan on being able to supply 20 to 50 percent of the stall torque over the desired period of continuous operation. When taken together with torque, the motor current draw can give you an idea of how efficiently a motor converts electrical power into mechanical power.

Physical dimensions Although an obvious point, it still bears mentioning: Be sure any motor you are considering purchasing will fit comfortably on your robot's chassis. Of course, some builders will actually design a chassis around a prized pair of motors, in which case the dimensions of the motor itself will be less of an issue. Pay attention to the length and description of the motor shaft (assuming the motor actually has a shaft, which is by no means guaranteed).

Reading a DC Gear Motor Data Sheet If you are fortunate enough to come across a motor for which a data sheet is available, you'll have access to a wealth of information, some of which you may find relevant, some of which you may not. Motor specs tend to be for ungeared motors, with data for gearbox options often included in a table following the main motor data.

Some highlights from a typical motor spec sheet might include the following:

Nominal supply voltage This is the motor supply voltage. In general, you'll not want to exceed this value by more than 10 to 15 percent.

No-load speed This is the maximum speed of the motor specified in RPM. Speeds of 5,000 to 15,000 RPM are not unusual for ungeared motors.

Stall torque This is the maximum torque that the motor can provide. At the stall torque, the motor speed is 0 and the current draw is at maximum. This can be specified in newton-meters, inch-ounces, gram-centimeters, or any other common torque units.

Friction torque This column represents the amount of torque lost to friction in various parts of the motor mechanism. These include bearings, commutator brushes, and so on. Clearly, lower is better.

Power output This is the motor's maximum power output in watts. As mentioned in Chapter 2, the maximum motor power is found by multiplying the maximum motor speed and stall torque and dividing by 4. This bit of data simply saves you a little arithmetic.

No-load current This indicates the amount of current the motor uses when unloaded. No-load current will be the minimum current draw of the motor when the shaft is spinning freely and the motor is running at full speed. At this level of current draw, no torque, other than frictional torque, is being produced. As the load on the shaft increases, so will the current draw.

Back EMF constant Usually specified in millivolts/RPM, or volts/1,000 RPM, this constant is used to compute the amount of back EMF generated by a motor at a given speed. As back EMF rises, speed increases (and vice versa) and current draw decreases as the effective voltage in the motor coils drops. For an unloaded motor running at top speed, back EMF will be at equilibrium with the supply voltage. At this point, the armature voltage will be effectively 0 volts, and a perfect motor will draw no current. Of course, no motor ever really runs unloaded due to frictional torque.

Velocity constant This number, usually expressed in RPM/volts, is used to compute the speed of the unloaded motor for a given voltage. For example, assume you have a motor with a velocity constant of 1,200 RPM/volt. If you power the motor from a 9-volt supply, the unloaded shaft should rotate at 10,800 RPM (9×1200).

Torque constant This is used to compute the amount of torque the motor produces at a given level of current consumption. This number is expressed as torque/amp or torque/milliamp, where torque can be in gram-centimeters, inch-ounces, millinewton-meters, or in whatever units are used elsewhere on the spec sheet. Imagine a motor with a torque constant is 1.50 ounce-inches/amp. Thus, at a current draw of 2 amps, our motor would produce 3.00 ounce-inches torque (2×1.50); remember, these numbers are for ungeared motors. Of course, you can also use this same constant to find the current required to produce a given torque by dividing the torque produced by the torque constant. So to produce 1 ounce-inch torque, our imaginary motor will require .667 amp (1.00/1.50).

Armature resistance Armature coil resistance is measured in ohms. The current pulled through the coil is a function of back EMF, supply voltage, and coil resistance. Using Ohm's Law, $I = (V - E)/R$, where I is current, V is supply voltage, E is back EMF voltage, and R is armature resistance.

Armature inductance This is usually specified in millihenries (mH) for small motors. Inductance comes into play particularly when we use PWM to control a motor. As the power supply is pulsed off, the armature coil inductance causes a large voltage to appear in the controller circuit. This voltage goes by a number of names, including *flyback voltage* and *inductive kick*. The formula for determining the voltage produced is

$$V = L(dI/dT)$$

where V is voltage, L is the inductance in henries, dI is the change in current, and dT is the change in time. When current is switched off rapidly, dI/dT tends to be very large, making L a critical factor in determining the amount of flyback voltage—which even for modest motors can range into the hundreds of volts. Fortunately, flyback can be dealt with by using clamping diodes. We'll discuss this in more detail in Chapter 7.

Maximum efficiency This value is the maximum efficiency you can expect to obtain from this motor. For small motors, this can range from 50 to 75 percent. Very large DC motors can reach up to 90 percent. Maximum efficiency is usually obtained at around 10 to 15 percent of the motor stall torque. Expect gearing to reduce this number even further.

Maximum radial shaft load As discussed, this is the maximum force that should be applied to the motor shaft in the plane perpendicular to the shaft. It may be specified in grams, newtons, or ounces for small motors.

Maximum axial shaft load This is the maximum load that should be applied to the shaft in the plane parallel to the shaft. As is the case with radial load, it is usually specified in grams, newtons, or ounces.

Often a datasheet will be accompanied by a graph such as the one shown in Figure 3-43. These graphs illustrate the total power output, efficiency, current draw, and speed versus torque. Note that although the

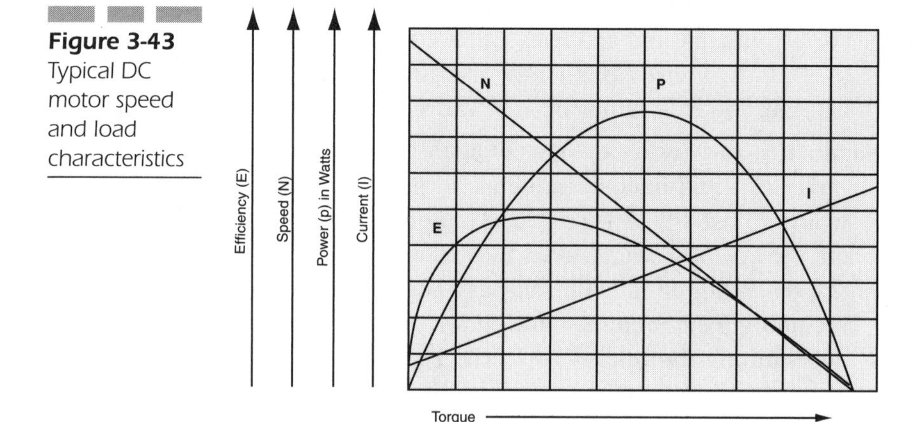

Figure 3-43 Typical DC motor speed and load characteristics

details of the graph may vary from motor to motor, the shape of the curves tends to be similar across motors of similar construction.

Do-It-Yourself Current Draw and Torque Testing

Often, written information provided by a vendor or manufacturer is adequate for determining motor suitability. Sometimes, however, you'll find yourself with an unknown motor on your hands, perhaps purchased at a ham fest or flea market, or just dug up from some forgotten corner of your parts bin. In such cases, estimating critical operating parameters such as torque and current draw are relatively simple to do.

Testing Motor Current Draw Testing no-load and stall current are so easy that it's surprising that many vendors don't provide this data:

Stall current A stalled motor can be modeled as a simple resistor having a resistance equal to the coil resistance. It should be easy then to use Ohm's law ($I = V/R$) to find the stall current draw using only an ohmmeter. Simply measure the motor coil resistance by hooking an ohmmeter to the motor terminals, and divide the result into the motor supply voltage. For example, a motor having a measured coil resistance of 4 ohms and a supply voltage of 12 volts will have a stall current of 3 amps (12/4).

This method will require a meter that can accurately measure low resistances. Less expensive multimeters may be off by an ohm or two —a significant amount of error, when measuring the coil resistance of larger motors, which will tend to be fairly low.

No-load current This can be measured by connecting a resistance in a series with the motor and measuring the voltage drop across the resistor, which will be proportional to the current drawn by the motor, again according to Ohm's law, $I = V/R$. Figure 3-44 illustrates a typical setup.

The best resistors to choose for this purpose are low-value, high-wattage types. These are widely available on the surplus market. Lower resistances tend to dissipate less heat, but the voltage drop will be smaller, which will require a multimeter capable of measuring low voltages. We like the 1.4 ohm, 75-watt resistor shown in Figure 3-45, affectionately nicknamed "the Behemoth."

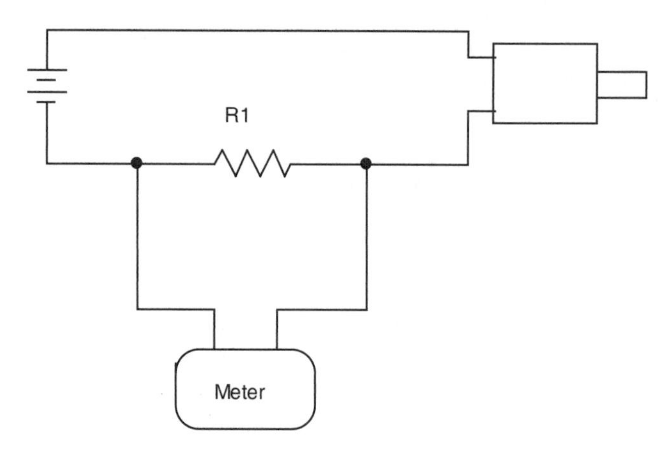

Figure 3-44 Measuring no-load current

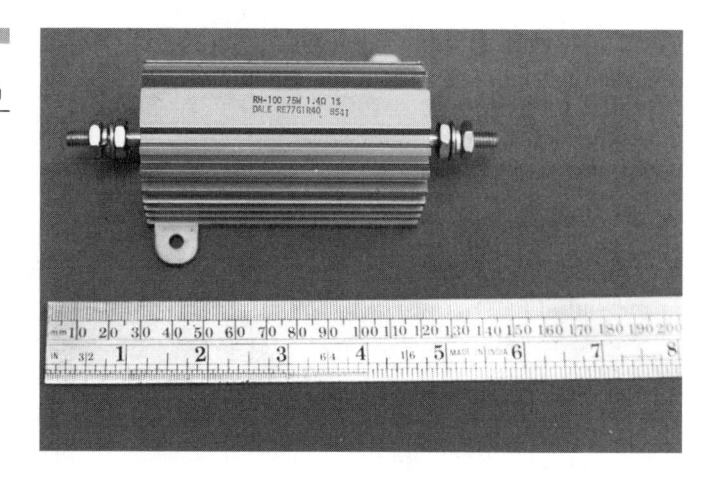

Figure 3-45 The Behemoth

By way of example, consider a smaller gear motor. When connected to a 12-volt supply, the voltage drop measured across our 1.4 ohm resistor is 0.084 volts. Using Ohm's law, we calculate the no-load motor current draw as 0.060 amps (0.084/1.4).

You can also use this same method to measure motor stall current by clamping the motor (if necessary), stopping the shaft (if possible), and checking the voltage drop across the resistor. This can be useful when your meter is incapable of accurately measuring low resistances, but be sure the resistor can handle the current, and try to keep the stall time as short as possible.

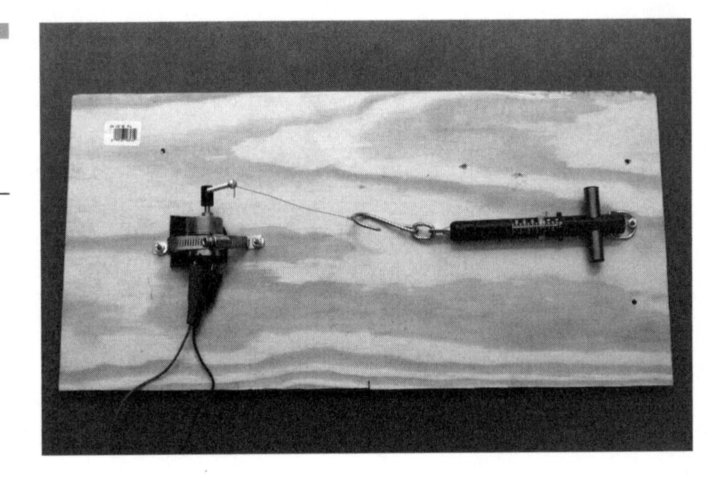

Figure 3-46 Simple torque test setup suitable for small motors

Testing Motor Torque You can get an idea of motor torque by using the simple setup shown in Figure 3-46. You'll need something that can clamp firmly to the motor shaft to use as an arm. You can use a shaft collar and long screw (as shown), or a pair of small vise grips will also do.

With the motor firmly clamped to a board—one or more worm drive hose clamps can be useful in this regard—tie a short-length, high-test fishing line or other suitably strong cord around the makeshift arm at a convenient distance, and connect the free end to a spring scale of a suitable range. These scales are sometimes available from sporting goods stores or local hardware stores, but mail-order scientific and educational supply outlets will usually have a wider variety of ranges available.

The distance component of the measured torque will be the point along the improvised arm at which you attach the fishing line. Multiply this value by the scale reading to get the measured torque. For example, if the line is tied 2 inches from the shaft center, and the motor stalls at a reading of 10 pounds, your motor is capable of providing 20 inch-pounds of torque.

A couple of words of caution: *Be sure everything is well secured*. Should the motor mount break, or the vise grips or scale come loose during the test, the flying debris could be a real hazard. Consider using a longer arm on the shaft; this will produce substantially less force on the spring scale, making the whole arrangement somewhat less dangerous should things fly apart unexpectedly. *Wear eye protection*.

As the motor stalls, it will draw a large amount of current. You should keep the stall time as short as possible. If you are concerned about the excess current draw or fear that the torque produced by the motor will be excessive, you can halve the voltage applied to the motor, which will halve both the maximum current and torque, and simply multiply your results by 2 to get the actual maximum torque.

Finally, this is ultimately a fairly crude method of determining torque. However, we've found it a reasonably effective way to get a ballpark estimate of motor power output. As always, your mileage may vary.

Other Factors to Consider

You should factor a few other things in to your purchasing decision. Price obviously will be uppermost in the minds of many purchasers, but don't forget to look at overall motor quality and condition, whether the motor is designed for continuous usage, and any extra features offered.

Motor Quality and Condition The quality of construction will be difficult to judge when purchasing via mail order. The manufacturer name can give you a hint. Motors from well-known manufacturers such as Globe, Escap, MicroMo, and others tend to be of uniformly good quality. Pictures may give you an idea of how a motor is constructed—plastic versus metal, for instance—and the catalog description may tell you whether a motor is new or removed from working equipment. Of course, price is always a good clue; as is the case with everything else, you'll pay more for quality.

The ability to inspect a motor before buying it is a big advantage. Check for excessive backlash as outlined earlier. Look for rust on the motor casing. Check the bearings and gearbox seals for signs of lubricant oozing out. Bring a battery with you and a couple of patch cables terminated with alligator clips, and give the motor a spin. A quieter gear motor is generally a better-made gear motor. Check again for signs of lubricant ooze after the motor has been running for a bit.

Continuous Versus Intermittent Duty Not all motors are meant to be run continuously. Then again, not all robots will run continuously, so you may not find this to be a problem; it all really depends upon your design and intent. Without a data sheet, it may be hard to determine whether a motor is designed for continuous use, but if you know what kind of equipment a given motor is designed to power, you may be able to make a reasonable guess. For instance, it is common to find automotive electrical motors on the surplus market. Some are made to power windows, sunroofs, and power seats. These motors (in addition to using quite a bit of current) are not designed to be used for hours on end without pause. Windshield wiper motors, on the other hand, are built to run continuously.

In our experience, most robots seldom run for extended periods without a break, so we don't usually fret over whether a motor is designed with continuous duty in mind. As always, however, your mileage may vary.

Extras Certain things can be nice to have in a motor, depending upon your design requirements:

- **Integrated pulse encoder or analog tachometer** The former in particular is quite handy. Usually (but not always), an integrated pulse encoder will provide output in the form of 5-volt logic pulses, making its output ready-made to interface with a microcontroller or other circuitry. A tachometer will provide a voltage that varies with motor speed, which can make interfacing to a microcontroller a little more complicated if the controller does not have built-in analog-to-digital conversion capabilities. Either feature is worth looking for. We prefer built-in encoders, because they are easier to interface with just about any microcontroller and can provide a great deal of accuracy.
- **Mechanical brake** Although in general we find ourselves *removing* these from motors, situations occur in which a mechanical brake can be handy. A robot that may be negotiating steep inclines or sumo combatants will find advantages in the capability to lock itself in place. Motor brakes normally engage when *not* powered, so there will be some additional (but usually modest) current draw when the motor runs compared to a motor without a brake. Be sure that you are capable of providing the correct voltage to disengage the brake; the brake voltage may differ from the motor voltage.
- **Armature shaft** In addition to a gear motor's output shaft, the driver motor may also have a shaft directly connected to the armature protruding from the rear of the motor that spins at the motor speed, as opposed to the geared-down speed. If long enough, this second shaft can be used in the construction of a high

resolution shaft encoder. The encoder disk need only have a couple of divisions; because the shaft rotates at such a high velocity (10,000 RPM would not be unusual), the resulting encoder has a fine resolution. Motors that have mechanical brakes often reveal long armature shafts once the brake is removed.

CHAPTER 4

Using RC Servo Motors

The most convenient motor package you can get for your small robot is the hobby *radio-controlled* (RC) servo motor. It has a DC motor, an efficient and quiet gearbox, and a motor speed and direction controller that requires a single *input/output* (I/O) line to operate. In this chapter, we demonstrate how to use this nice little package, how to make the servo into a continuous rotational drive motor, and how to control it. We also show you some code while we're at it.

Choosing the Right Servo to Use

Many manufacturers of RC hobby servos can be found out there. All of them manufacture a variety of shapes, sizes, and strengths of RC hobby servos. Which should you choose? In most cases, the servo you choose will depend on what you can get for the best price that fits into your budget. If you are looking for a particular power range, this will also limit the playing field. Here are the criteria that we look for in an RC servo, in no particular order:

- Can I hack it for continuous rotation?
- As far as power, what is its torque rating and how much can I *overdrive* its rated voltage?
- Looking at speed, which converts to *revolutions per minute* (RPM), how fast is it?
- What is its strength? Does it break or strip gears easily?
- How do I control a hobby servo?

Hacking a Servo for Continuous Rotation

If you are *hacking*[1] your servo for continuous motion, you will want to use a servo with plastic gears rather than metal ones, unless you relish hacking brass gears on a more expensive servo, that is. Another consideration, one that only applies to micro servos and really high-powered servos, is *can* the servo be hacked for continuous rotation? Some can't because their

[1]Hacking refers to modifying an RC servo for continuous rotation.

spline gear does not have teeth around its entire perimeter. Hundreds of servos are available out there and we didn't go buy one of each to give you the list of which servos can and can't be hacked; you will need to ask the manufacturer or reseller for that information before you buy. We *can* tell you that all of the midrange RC servos *can* be hacked; it's only the high-powered and micro servos that we've found to have partially populated spline gears. Well, that's all fine. Now let's hack some servos!

Tools You Will Need Some essential tools will be needed here and some really nice tools will make your life easier. Pick your level of comfort.

Essential Tools The following are tools necessary for servo hacking operations (see Figure 4-1):

- A razor saw. We like the Exacto® razor saw, with the most number of teeth we can find.
- A hobby knife with a sharp blade.
- A #1 Posidrive screwdriver, which is like a Phillips but with a sharper point.
- A *really* small Phillips screwdriver, such as an Xcelite® P12S.
- A drill with a selection of drill bits.

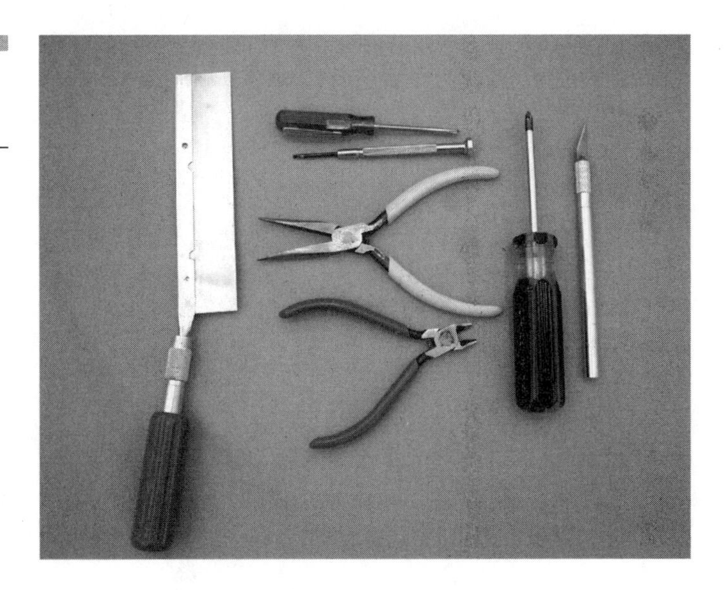

Figure 4-1 Essential hacking tools

Nice-to-Have Tools The following are tools not essential to the exercise but can make the process easier:

- Drill press
- Vice

Hacking the Futaba® F3003 Standard Servo The following section outlines the steps for hacking the F3003 servo and leaving it mostly intact with minimal fuss and mess. To make it a "normal" servo again, just buy a replacement gear set for a new spline gear.

1. We start with a servo and the tiny Phillips screwdriver (see Figure 4-2). Hold the control horn while removing the screw. You can't get the top off without removing the control horn (see Figure 4-3).

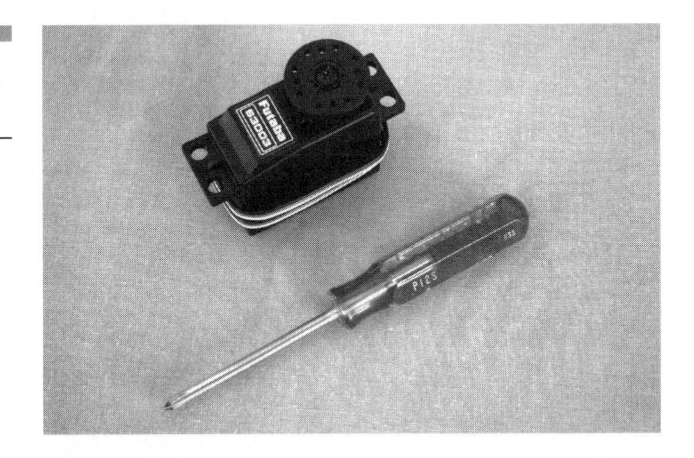

Figure 4-2 The servo and our first tool

Figure 4-3 The control horn removed

2. Loosen, but do not remove the case screws (see Figure 4-4). Use the #1 posidrive screwdriver for this. You could use the tiny Phillips screwdriver, but the posidrive is more comfortable (see Figure 4-5).
3. Remove the top of the servo by holding the case and pressing down on the spline (which the control horn was bolted to). Watch out for other gears sticking out of the top instead of being properly docile as you see in Figure 4-6.
4. Remove the spline gear (the black one) and the white gear that sits on top of it (see Figure 4-7). Carefully note where the gears go. That metal rectangle on the right side of the servo is the feedback

Figure 4-4 The bottom of the servo showing the case screws

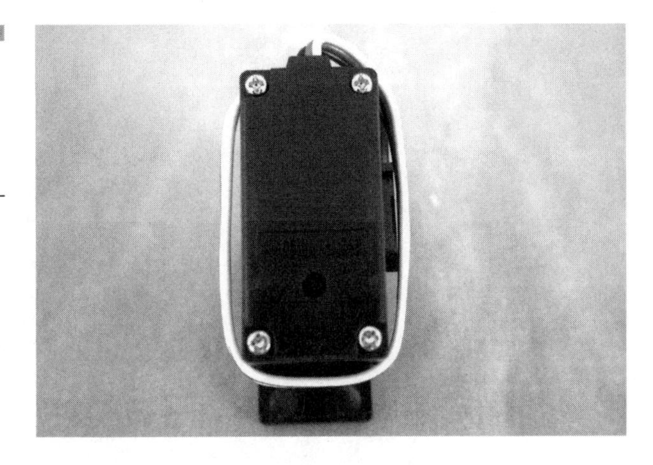

Figure 4-5 The case screws loosened

Figure 4-6 The gear arrangement

Figure 4-7 The spline gear removed

pot tab. This is what you will be adjusting later when setting the center point of the servo.

5. Place the spline gear on a hard, flat surface and carefully place your razor saw on the gear so it's flat to the disk. You're going to cut this end-stop tooth off (see Figure 4-8). Cut only up to the hub of the gear (see Figure 4-9).
6. Use the hobby knife to trim your cut from the hub and clean it up. Do not be tempted to cut the end-stop tooth off with diagonal cutters. Many of these gears will shatter if you try to snip this off. You never know which ones will take the snip gracefully and which ones will end up scattered about your workshop. Figure 4-10 shows how your gear should look if you've done this properly.

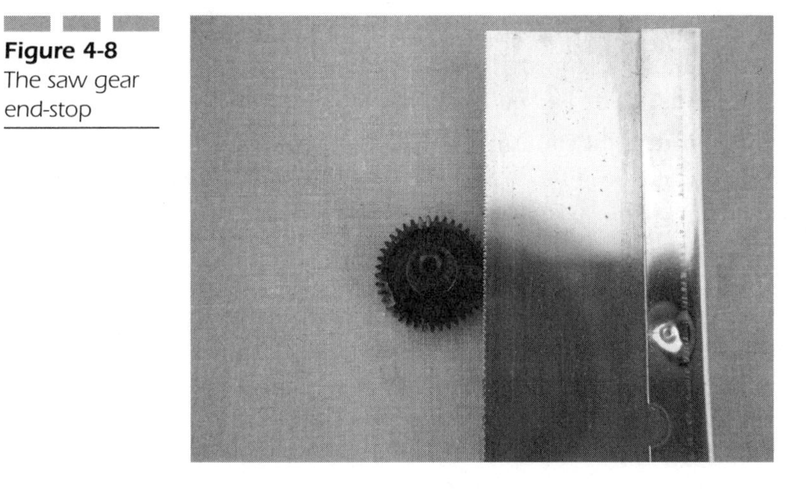

Figure 4-8 The saw gear end-stop

Figure 4-9 Cut this far and no farther.

Figure 4-10 How it should look now

7. Look at the bottom of the gear and you'll see a "+" shape molded into the hub of the gear. To allow the spline gear to spin on the pot, this + must be drilled out (see Figure 4-11).
8. Pick your drill bit carefully to drill this hole. In our case, the perfect fit was a 3/16-inch bit. We use the smooth end of the bit to fit into the hole to drill, so we know it's the correct size (see Figure 4-12). You can hold this gear down in a variety of ways while you are drilling, but we prefer a vice. Gently snug the gear upside down into the vice. If you snug it too hard, it will deform. A drill press comes in pretty handy right now, but you can do this with a plain old hand drill if you want. When you are done, your gear should look like Figure 4-13.

Figure 4-11
The retainer in the spline gear

Figure 4-12
The gear, vice, and drill bit

Figure 4-13
A finished gear

9. Before reassembling the servo, you need to set the center position on the feedback pot. To do this with Parallax® Stamp II (for instance), enter, compile, and download this code:

```
'{$STAMP BS2}
'This generates a 1.5ms pulse approximately
'every 20ms.

SERVOPIN con 15

Loopit:
    pulsout SERVOPIN,750
    pause 17
goto Loopit
```

Connect the servo to I/O port 15 on your Stamp II and adjust the pot nub until the motor stops turning. Now put the spline gear back on and the intermediate gear on after that (see Figure 4-14).

Figure 4-14
The reassembled gear train

When it's set, you can put a small blob of glue on the pot hub to hold it in place.

10. Carefully replace the top of the servo and tighten the screws back in until they are just snug, no tighter. You are done!

Hacking the Hitec® HS303 Standard Servo Hacking the HS303 servo will render it a permanently hacked servo. The spline gear is molded to the feedback pot, so we need to destroy the feedback pot to allow it to freely rotate. However, this is a blessing in disguise because we'll install a trimmer pot to the outside of the case that will enable us to tweak the servo when it's completely assembled. To complete this hack, we need a soldering iron, a solder, needle-nosed pliers, and diagonal cutters, but we don't need a drill. We also need a little more care as we're taking this little guy completely apart and messing with the electronics.

Steps 1 through 5 are the same with the HS303 as they are with the F3003 servo previously, so we'll leave those out and move right to the unique procedures that we need to do:

1. Remove all the gears. To remove the center bottom gear, we will need to pull the post it rides on as well. Carefully arrange the gears in such a manner that you will know where they have to go back in.
2. Remove the bottom plate and screws. You now have a servo that looks like Figure 4-15.
3. Remove the nut that holds the pot in. This is the nut you see on the right side of the servo in Figure 4-15.
4. With a really small Phillips screwdriver, carefully push down on the motor, which is tucked under the hood on the left side of the

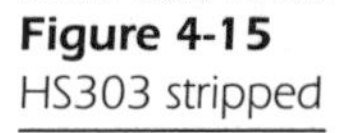
Figure 4-15
HS303 stripped

servo, as shown in Figure 4-15. Keep pushing down on it until the motor, PC board, and pot pop out of the bottom (see Figure 4-16).

5. Snip off the leads to the pot as close to the pot as possible. *Carefully note which color wire goes to the center pad of the pot.* You will be soldering that center wire to the center conductor on your trimmer pot.
6. Pry off the three retainer tabs that hold the back of the pot on. Then snip these tabs off. In Figure 4-17, you can see an indentation that prevents the pot from turning continuously; turn the pot back and forth to find it. Use a hammer and something blunt to *carefully* smooth this dent out until the pot shaft turns freely. Try not to deform the pot body in the process of smoothing it.
7. Reinstall the pot into the servo body and tighten the nut to hold it in place. You need to align a tab that sticks up into the top of the servo body to do this (see Figure 4-18).

Figure 4-16 The guts of the servo

Figure 4-17 The pot with its back removed

Figure 4-18 The pot reinstalled

8. Using your diagonal cutters and hobby knife, enlarge the hole that the servo control wires exit. The three wires you snipped from the pot will also exit here. Reinstall the motor and PC board, routing the pot wires outside the servo body above the servo control wires. See Figure 4-19 to see how this should be done. Put the bottom plate on and adjust your hole until it's a nice fit.
9. Reinstall the gears so that they all are in their correct places, as Figure 4-14 shows. The first gear to install is the last one removed, the bottom center gear with the post you removed.
10. Solder the center wire from the servo pot to your 5K trimmer pot, and solder the other two wires to the end contacts of the trimmer

Figure 4-19 The exiting wires

Figure 4-20 The finished product

pot. Reattach the top of the servo and tighten up the four screws. Now glue the trimmer pot to the part of the servo case you want it attached to. Figure 4-20 shows where we chose to put our trimmer pot. Using that really small blade screwdriver, center the trimmer.

11. Now you need to tweak the pot to center it so that the servo is stopped at a 1.5 millisecond pulse. Let's use a BasicX program (for instance) for the Netmedia® BX-24 to do this. Create a project with the following code and download it to your BX-24:

```
'BX-24 calservo.bas
'This generates a 1.5ms pulse approximately
'every 20ms.

Const SERVOPIN as byte = 5
Dim tstart as single

Sub main()
    do
        tstart = timer()
        call pulseout(SERVOPIN,1382,1)
        do while((timer() - tstart)< 0.020)
        done
    done
End Sub
```

Also connect your servo to pin 5 of the BX-24 chip.

Hacking the Hitec HS225MG Mighty Mini Servo Sometimes you want a smaller servo footprint. The Hitec HS225MG Mighty Mini servo is really nice because it's almost half the size of the previous two servos and it is actually more powerful. It's also the easiest one of all three to hack. The HS225, however, is also more than twice as expensive as the prior two servos. You don't get anything for free.

Two versions of this servo are available: the HS225BB and the HS225MG. The MG has metal gears. Remember when we said don't hack MG servos? Well, we lied. This is one case where the MG is actually *easier* to hack than the nylon-geared unit. The HS225BB has a weird oval on the top of the spline gear instead of a nice neat tab like all the other servos we've looked at. If you carefully modify the oval, you have a hub that is very flimsy. The HS225BB can be hacked, but its strength might not be very good. The HS225MG has instead a simple pin you can yank out with a pair of pliers—the easiest servo yet to hack!

To hack the HS225MG, follow these steps:

1. Remove the control horn and the top cover using the tiny Phillips screwdriver (see Figure 4-21).
2. Remove the top center and spline gear (see Figure 4-22). They will have to come out at the same time.

Figure 4-21 The top removed

Figure 4-22 The gears removed

3. Remove the end-stop post from the spline gear with pliers (see Figure 4-23 and 4-24). Be careful not to scratch the spline gear very much; scratches might snag on other gears. Save the post; we can return this servo to stock functionality later on.
4. A plastic clip is attached to the feedback pot knob that slides into the spline gear (see Figure 4-25). Remove this clip and save it (see Figure 4-26).

Figure 4-23 The removing post

Figure 4-24 The post removed

Figure 4-25 The potentiometer clip

Figure 4-26 The clip removed

5. Load and run this code for the Savage Innovations OOPIC® controller (for instance) to set the center point where the servo stops turning. Note that the servos are controlled on I/O lines 11 and 12, *not* pins 11 and 12 on the OOPIC connector.

```
//Test Servo with OOPIC using Java syntax

oServo right = New oServo; // Make a servo object.
oServo left = New oServo;  // Make a servo object.

Sub void main(void)
{
  // Setup I/O bits
  right.Ioline = 12;        //Set the servo to use I/O Line 31.
  right.Center = 22;        //Set the servo's center to 22.
  right.value = 32;         //Center position
  right.Operate = cvTrue;   //Last thing to do, turn the Servo on.
  //--Set up the second servo-reversed because it's on the other
side.
  left.Ioline = 11;         //Set the servo to use I/O Line 30.
  left.Center = 22;         //Set the servo's center to 22.
  left.Value = 32;          //Center position
  left.InvertOut = cvTrue;  //Tell the servo to move in reverse.
  left.Operate = cvTrue;    //Last thing to do, turn the Servo on.
}
```

6. Now replace the gears. A lip on the spline gear requires both gears to be reinstalled at the same time (see Figure 4-27).
7. Put the top cover back on. Note that ball bearings support the servo hub on the pot shaft and on the top cover. This makes this a stronger, faster servo (see Figure 4-28).

Figure 4-27 Gear spacing and the lip

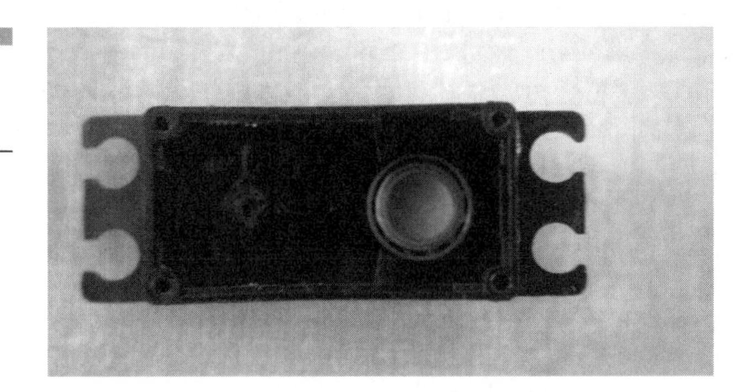

Figure 4-28 The top cover bearing

Determining and Increasing the Hobby Servo's Power

In Chapter 2, "Motor Types: An Overview," we thoroughly discussed how to calculate the power of a hobby servo, so we're not going to cover that topic again. The standard servos we hacked previously are powerful enough to handle up to a 2-pound robot with ease and power to spare. If you want to have more power, get a servo with a higher torque rating. If you want to go faster, get a servo with a faster transit time, but you will pay more for a stronger, faster servo.

If you want a comfortable combat margin in your choice of hobby servos, you can move about 1 pound of robot (0.45 kilograms) per 40 ounce-inches (3.1 kilogram-centimeters) of servo torque. This is a rough rule of thumb that assumes a more or less flat surface and a sumo-style pushing contest with two servos of this size per robot using wheels that are about 3 inches (7.6 centimeters) in diameter. This rule of thumb is for servos that run at their nominal voltage of 4.8V to 6.0V. If we run them at 7.2V or higher, their power will be even greater.

Another way exists for getting a stronger and faster servo: overdrive it with a higher voltage than it's rated for. Won't that ruin the servo? Usually, no; sometimes, yes. You pay your money and you take your chances. We've never toasted a Futaba, Hitec or Expert® servo at up to 7.2V. On the other hand, we've found that the GWS® S03 servo will stop functioning when the voltage goes much over 6V. On yet another hand, we've driven the Futaba S148 and Hitec HS300 servos at up to 12V for 5 minutes at a time and they just got a little warm. We offer this information for those who wish to experiment. We don't recommend that you overdrive your servos; that would void their warranty. On the other hand, hacking them for continuous rotation certainly voids their warranty, so do as your spirit moves you. Raising the voltage will increase the servo's torque and it will increase the servo's speed.

We use both 7.2V and 9.6V battery packs with our hacked servos on a regular basis. These packs are common and easy to find on the surplus markets. With the lone exception noted earlier, we've not had any of our servos fail, even at these voltages. Prudence dictates that you test your servos cautiously before designing in an overdriven servo motor. You might hit the jackpot and get servos that won't work above their rated 6.0V.

Determining the Speed of Your Hobby Servo

Assuming that your robot is powered adequately, how fast will it go? Does the wheel size matter to speed? We calculated a servo's rotational velocity in Chapter 2, but it's worth showing this again because this is a useful bit of information when building a robot sumo, a maze runner, or perhaps even a robot racer. Servos are rated for torque and for the transit time of a 60° arc. A common speed is 0.22 seconds. If it takes 0.22 seconds to transit 60°, then it will take 1.32 seconds to transit a full 360°, which translates to 45.5 RPM:

$$\text{RPM} = \frac{1}{1.32} \times 60 = 45.5$$

To determine how far your robot will go at 45.5 RPM, you need to know the *rollout* of your wheel. This is a fancy term that simply means the distance a single rotation of the wheel will take you.

It's easy to calculate your rollout. Remember that the circumference of a circle is πd, or pi multiplied by the diameter of the circle. So if we use a 3-inch (7.62-centimeter) wheel, we will travel 9.4 inches (24 centimeters) per revolution (which takes 1.32 seconds), or 7.1 inches per second (18.1 centimeters), which is 35.7 feet (10.9 meters) per minute, which is about 0.41 miles (0.65 kilometers) per hour. Not very fast it seems, but this is a comfortable speed for a small indoor robot.

The Strength of a Hobby Servo

Most hobby servos have plastic gears. The most expensive ones have brass metal gears, but even the metal-geared servos have one gear that is plastic. When a servo is jammed, it's going to fail somewhere. The manufacturers put that plastic gear in there to choose where the point of failure will be. The micro servos will be the weakest, their gears are the smallest, and the quarter-scale servos (called that because they are designed to move the control surfaces of quarter-scale aircraft) are the strongest. Everything else falls somewhere in between.

Another place that determines the strength of a servo is the hub. The low-end servos will have no bushing beyond the plastic in the servo case at the hub. These would be the Hitec HS303 and Futaba S3003 servos, for instance. Take a step up, like the Futaba S148 or Hitec HS402, for instance, and they will have oilite bushings. More expensive servos, those with BB designations in their names, will have ball bearings supporting their hubs. These steps up the quality ladder each add strength and longevity to the hubs. To improve the chances of our servos surviving their duty as robot motors, we don't recommend connecting wheels directly to the hubs if the robot is over 2 pounds. Use a shaft coupler on a better supported axle instead. We'll discuss these issues in greater detail in Chapter 6, "Mounting Motors," and Chapter 10, "Wheels and Tank Tracks."

Controlling the Hobby Servo

When testing your carefully done hacks, it would be nice to have a tester that is easy to use. You can quickly hack out some code on your favorite microcontroller, which we show examples of for each servo hack described, or you can build a nice simple hardware circuit that will move your servo to the most common servo endpoint extremes, 1 millisecond to 2 milliseconds every 20 milliseconds, as shown in Figure 4-29.

Hardware Servo Control

Figure 4-30 illustrates a 555 stable multivibrator circuit that will move your servo to the most common servo endpoint extremes. We've tried to use parts that you could find even at your local Radio Shack®.

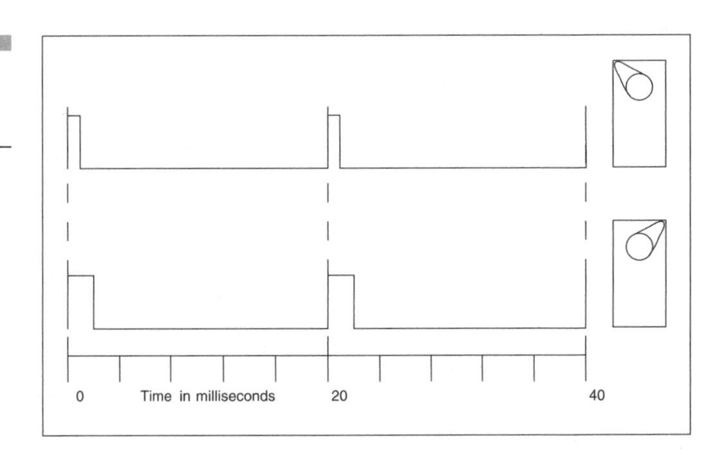

Figure 4-29 Servo control signals

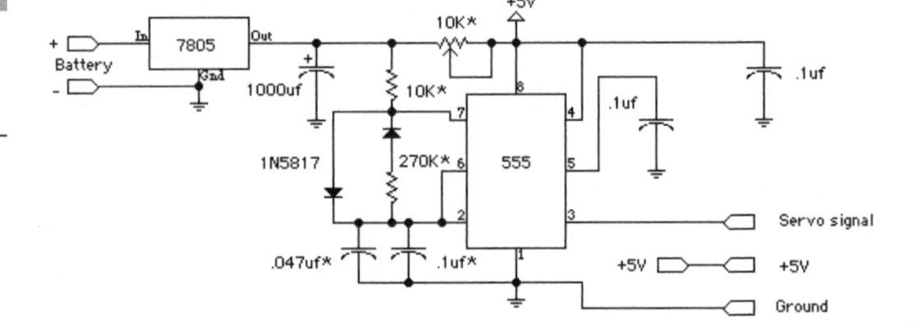

Figure 4-30 A servo pulse generator

When building this circuit, use 5 percent tolerance resistors and 10 percent or better tolerance capacitors where noted with an * for the best accuracy. Since a 0.144 uF capacitor is not going to be easy to find, we put a .1 uF and a 0.047 uF capacitor to work in a parallel manner in order to build a 0.147 uF capacitor, which will work almost perfectly. When completed, a turn to one extreme of the potentiometer will yield a 1-millisecond pulse and the other extreme a 2-millisecond pulse.

Unless you like having useful circuits built onto permanent boards, we recommend that you build this circuit onto a solderless breadboard (see Figure 4-31). A solderless breadboard uses 22-gauge solid wire to make connections between components you can simply (and nonpermanently) press into the board for prototyping or creating temporary circuits.

Enter, compile, and run the following BasicX code (for instance) to calibrate your 555-based servo testbed hardware. Open the COM port for the debug output display. When the value printed is 150 (1.5 milliseconds), then mark this position as *Center* for the potentiometer.

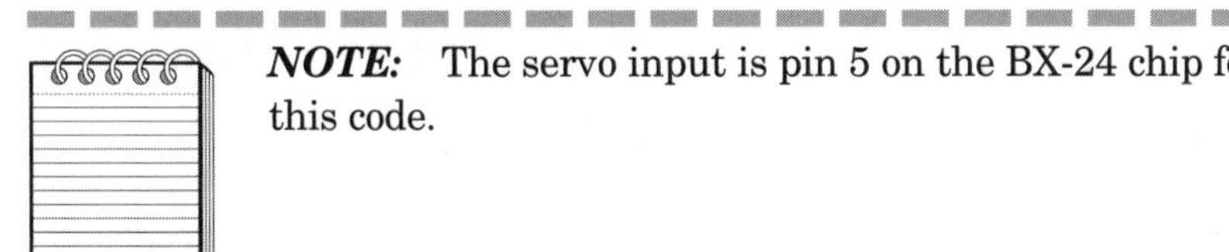

NOTE: The servo input is pin 5 on the BX-24 chip for this code.

```
'BX-24 555 servo circuit calibrate

const SERVOPIN as byte = 5
dim tstart as single
dim temp as integer

Sub Main()
    debug.print "Testing pulse."
    do
        call pulsein(SERVOPIN,1,tstart)
        temp = fixI(tstart*100000.0)
        debug.print "pulse= ";CStr(temp)
        call delay(0.5)
    loop
End Sub
```

In Chapter 7, "Motor Control 101, The Basics," and Chapter 8, "Motor Control 201—Closing the Loop with Feedback," we will discuss microcontroller-based servo controllers and interfacing in much more detail.

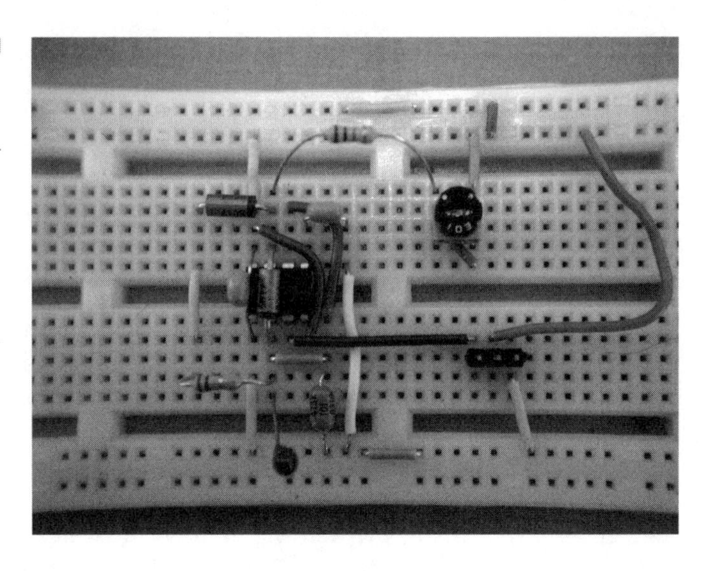

Figure 4-31 A solderless breadboard

An Alternative Futaba S3003 Servo Hack for Fully Variable Speed Control

The Seattle Robotics Society's September 2000 issue of *The Encoder* describes an electronic modification to the Futaba S3003 servo that gives fully proportional speed control to the hacked servo. We have confirmed that this does indeed work quite well to create a reasonably well regulated variable speed gear head motor from a hobby servo.[2]

The hobby servo has a narrow input control range and it is difficult to impossible to control its speed accurately, although it has adequate speed and torque. The main reason for this is that the servo electronics do not have the resolution for speed control because it's not needed. This drawback can be corrected for our hobby robotics use.

To improve the resolution of the servo speed variation, we need to do two things. The first is to increase the scale factor (gain) of the speed control feedback so that the maximum motor speed range corresponds to the full 1 millisecond (1 to 2 millisecond) signal input range. Since this will increase the total servo loop gain by the same amount, the servo may

[2]Lee Buse. "Variable Speed Control Modification to the Futaba S3003 RC Servo." *The Encoder*, August 2000. Reprinted here by permission of the Seattle Robotics Society and Lee Buse.

become unstable. We can solve that problem by decreasing the pulse stretcher gain by a similar amount. The result will be a fully variable speed-controlled hobby servo with good torque and stable operation. Figure 4-32 shows the deduced servo electronics and what should be modified to get what we want. This circuit is deduced because we have been unable to obtain any documentation for the BA6688 and BAL6686 devices that make up the heart of the servo electronics. Long live the reverse engineers such as Lee Buse!

You can either replace the servo pot with a trim pot placed external to the servo like our HS303 hack or leave the original pot intact inside the servo; take your pick. We like minimizing the hassle and leave the stock feedback pot inside.

Hacking Components The components needed for the hack are as follows:

- One 1/4-watt or 1/8-watt 220K to 240K surface mount or 330K shunt resistor
- One short length of wire insulation to shield the resistor lead (if using shunt resistor)
- One 0.1 uF, 50V ceramic capacitor with 10 percent or better tolerance

Necessary Hacking Tools The tools needed for the hack are as follows:

- Very fine tipped soldering iron
- Solder sucker (We recommend a spring-loaded one.)
- 60/40 electronics solder
- Tiny Phillips screwdriver
- Needle-nosed pliers
- Diagonal cutters
- Some kind of servo driver, like the circuit we give you in this chapter

The Hacking Procedure To hack the Futaba S3003, follow these steps:

1. Disassemble the servo as described in the Futaba S3003 procedure. Remove the servo feedback pot and servo electronics board by carefully pressing down on the motor gear under the

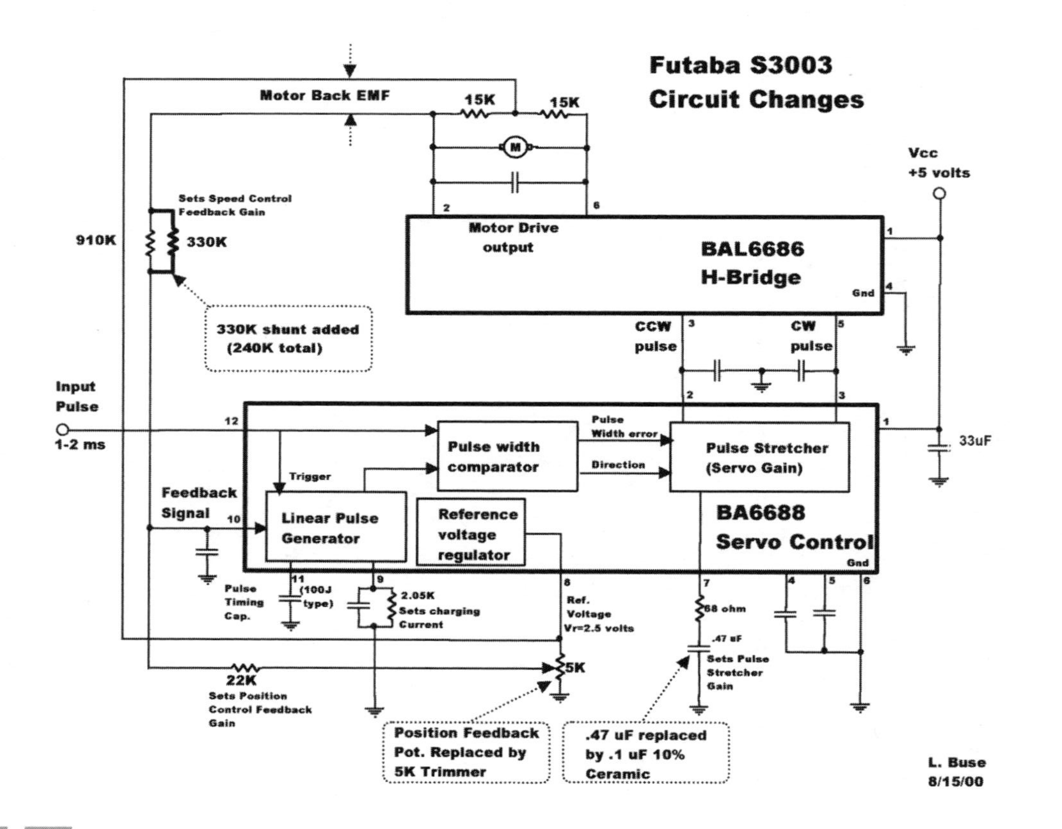

Figure 4-32 S3003 electronics

hood (as in the HS303 procedure) and pressing down on the pot shaft (see Figure 4-33).

2. Either replace the 910K surface mount resistor in the speed control feedback path with a 220K to 240K surface mount resistor or add a 330K (orange, yellow, or gold) shunt resistor (in parallel with the 910K surface mount resistor). Be sure to insulate the long lead on the shunt resistor (small-diameter, heat-shrinkable tubing and insulation stripped from 22 gauge wire work well here) and push the resistor down close to the board. Figure 4-34 shows the proper resistor placement and solder locations.

Figure 4-33 S3003 electronics

Figure 4-34 The resistor location

3. Replace the 0.047 uF capacitor in the pulse stretcher circuit with a 0.1 uF 5 percent or 10 percent dipped tantalum capacitor. *Do not use a 0.1 uF bypass capacitor.* A bypass capacitor has a tolerance of about −20 to +80 percent. You just won't know what kind of performance you will get with this type of capacitor. The tantalum capacitor has one lead longer than the other; the longest lead is marked on the cap with a stripe and a +. The + side of the cap is nearest the motor. Figure 4-33 shows the proper capacitor placement.
4. Connect the servo electronics to your servo driver. Adjust the servo driver to a 1.5-millisecond pulse width and then tweak the S3003 servo pot until the motor stops turning. Reassemble the servo, taking care to make sure you don't squish anything into the wrong location. Make sure you did the continuous rotation hacks!

Now when you vary the servo control pulse you will get a smoothly changing motor speed variation. Similar hacks may be possible with other servos if you can figure out the drive electronics, especially if the board is similar to this servo.

Other RC Hobby Servo Controller Options

We've seen code to control RC hobby servos with a Basic Stamp II®, BasicX BX-24®, and an OOPIC microcontroller. Usually, it's a waste of resources to have our main microcontroller directly control an RC hobby servo. Here are some products and projects that offload that duty to what we call *slave* processors. Let's add some more variety!

Parallax Stamp II Controlling a Scott Edwards MiniSSC® Servo Controller

This board can control up to eight servos at once and can daisy-chain another board off of it to control up to 16 servos from a single serial port connection:

```
'SCAN.BS2 Scans a servo slowly back and forth
'Servo is on Port0.  Configure the MiniSSC for
'9600 baud

servo     con    0              'Servo 0
sport     con    0              'Port to use
sync      con    255            'Required sync byte
pos       var    byte           'Servo value sent
n9600     con    84             '9600 baud 8N1

again:
for pos = 0 to 254 step 1     'Clockwise scan
  serout sport,n9600,[sync,servo,pos]
next

for pos = 254 to 0 step 1     'Counter-clockwise scan
  serout sport,n9600,[sync,servo,pos]
next
goto again

end
```

Parallax Stamp IISX® Controlling a Ferrettronics FT639® Controller Chip

This is a single-chip solution that can control up to five servos at the same time using a single serial port connection.

```
ldrive     var     byte             'left drive parts
lupper     var     ldrive.nib1      'We need to break these up
llower     var     ldrive.nib0      'with this controller.
servo      var     byte             'general servo value holder
SC         con     0                'Servo port number
I2400      con     17405            '2400 baud inverted serial

for ldrive = 0 to 255 step 1          'just scan the servo
  gosub act
next

stop
act:                                   'This runs the servo wheels
       servo.nib1 = %0000             'left wheel
       servo.nib0 = llower
       serout SC,I2400,[servo]        'only do this if different
       servo.nib1 = %1000
       servo.nib0 = lupper
       serout SC,I2400,[servo]        'only do this if different
return
end
```

BASCOM/AVR® Controlling a TTT SSC Chip with an Atmel® 90S2313 Microcontroller

This is a nice compiler free for use with the 90S2313 chip. The TTT SSC chip is Dennis' own simple servo controller that controls four hobby servos from a single serial line:

```
Dim Lmotor As Byte
Dim Rmotor As Byte

Const Rrfast = 47                      'fast forward on the right
Const Lffast = 16 + 64                 'fast forward on the left

'Configure to talk to the servos
Open "comb.3:2400,8,n,1" For Output As #1 'On I/O port B.3

'Stop all motors
Printbin #1 , 0                        '0 turns off servos 0 and 1

Wait 5

'set initial motor speeds
Lmotor = Lffast
Rmotor = Rffast

'Output data to TTT servo chip
Printbin #1 , Lmotor
Waitms 10
Printbin #1 , Rmotor

Printbin #1 , 64                       '64 turns on servos 0 and 1
End
```

One More Way to Use the RC Hobby Servo

If we remove the electronics of the RC hobby servo, we still have a nice, small, quiet DC gearhead motor that is easy to mount. We can use another driver chip, such as the TI® 754410 or National® L298 H-bridge driver chip, and *Pulse Width Modulation* (PWM) to control this package for accurate speed control. We can also hack the servo driver board to use PWM on the BAL6686 H-bridge chip in the S3003 servo directly. PWM and DC motor control electronics will be detailed in Chapters 7 and 8.

CHAPTER 5

Using Stepper Motors

Stepper motors are the most precise motors that we can use with our robots. However, they are the most complex motors to use and are not very powerful for their size. Once again we see a trade-off in action. Chapter 2, "Motor Types: An Overview," described the two kinds of stepper motors: *unipolar* and *bipolar*. In this chapter, we'll discuss the specifics of how to interface with these motors. Here too you will find out how to sort out the wiring for each kind of stepper. It's really not all that difficult once you know the secret . . .

What We're Not Going to Discuss

Let's say right now that you can get greater step rates from your stepper motors if you overdrive the windings with much higher than the specified voltage while at the same time limiting the current and using a form of *Pulse Width Modulation* (PWM). This is called a *chopper-stepper* driver and we're not going to talk about them. This is a much more complex stepper driver and is commonly used on *Computer Numerically Controlled* (CNC) mills and lathes.

Another topic we aren't going to cover is microstepping stepper motors to get much more precision movement from them. This is highly complex with very specific stepper driver chips, and we feel it is over the top when discussing mobile robotics' use of stepper motors.

Choosing the Right Stepper

Remember, a stepper motor is very precise, but it is *not* very powerful for its size. We don't recommend that you use a stepper motor if your robot needs to push other robots around (as in sumo competitions) or if it has to climb hills. If, however your robot needs to be really accurate in its maneuvers and will operate on a flat, level surface, you can use a stepper drive system successfully.

In Chapter 2, we figured out how much power we needed to move our robot, and how to match that power to the correct stepper motor. We're not going to go through those calculations again here. In general, a big-

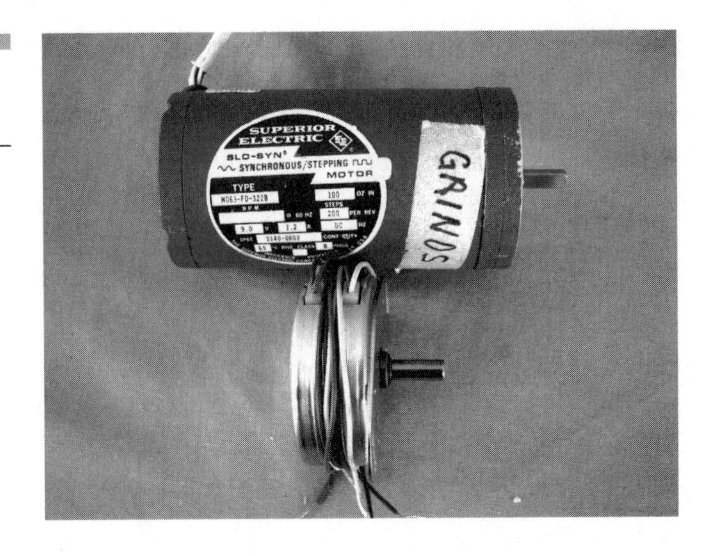

Figure 5-1 Bigger is stronger.

ger stepper motor is stronger than a small stepper motor, and a bipolar stepper motor is stronger than a unipolar stepper motor (of the same size). Beyond that, it's all in the stepper construction. In Figure 5-1, the really big stepper is a 9V, 200-step (1.8° per step) unipolar model with a dynamic torque of 100 ounce-inches (0.71 newton-meters). The much smaller stepper is a 24V, 48-step (7.5°) unipolar stepper with a dynamic torque of about 22 ounce-inches (0.17 newton-meters). The larger stepper specifies a winding current of 1.2 amps, and the smaller one specifies a winding current of 333 ma.

You can't divide anything by *anything* to get those torque values from any of the other attributes of these steppers. Believe us, we've tried!

You need the spec sheets to find the power of a stepper motor; there just isn't any other way. Okay, there is one other way. Put a pulley on the shaft of the stepper that is 1 inch (2.54 centimeters) in diameter, hang a weight from that pulley, and run the stepper. Then increase the weight until the stepper starts to skip. Now you have the dynamic torque of the stepper, so go back to Chapter 2 and figure out the power. That doesn't sound like fun, and you need the stepper to do the experiment, so you've already bought it! Let's discuss some rules of thumb for what you can expect from the stepper motors you buy and the ones that you glean from scrapping disk drives and printers.

Dynamic Torque and Other Specs You Want to See

On those rare occasions when you actually get to see the specifications of a stepper motor, you are usually overwhelmed with data, usually data that makes no sense. This data consists of the specifications that you need to know to properly determine how to use your stepper and if it will be strong enough (see Table 5-1).

It seems redundant to specify voltage, current, and winding resistance when you can calculate any one of those specifications by using the other two. These ratings are all given because sometimes (with bipolar stepper motors usually) a user will overdrive the voltage to a stepper winding to get its current to ramp up more quickly. This is called a *chopper* driver. The chopper driver will use a much higher voltage to get current to quickly reach its maximum in a stepper winding. We do this to get a higher torque at high step rates. A normal speed-torque curve for any stepper will show that as the step frequency increases (as the motor's revolutions per minute [RPM] increases), the winding current falls off, and since torque is directly proportional to torque, so too does torque fall off and our stepper gets weaker.

Figure 5-2 shows the relationship graphically between speed and torque. This curve is often called the L-R curve because it displays the inductance and resistance relationship to the current of the stepper motor at various step rates. The lower torque curve represents the maximum torque load that the motor will start and stop without losing steps (pull in). The upper curve represents the maximum torque that the

Table 5-1 Important stepper motor specifications

Specification	Definition
Voltage	The normal voltage the stepper was designed for.
Winding resistance	The DC resistance of the winding *per phase*.
Current	The maximum current the winding will safely accept.
Number of steps	7.5° = 48 steps, 1.8° = 200 steps. You can see the pattern.
Dynamic torque	The maximum torque generated while the stepper is moving.

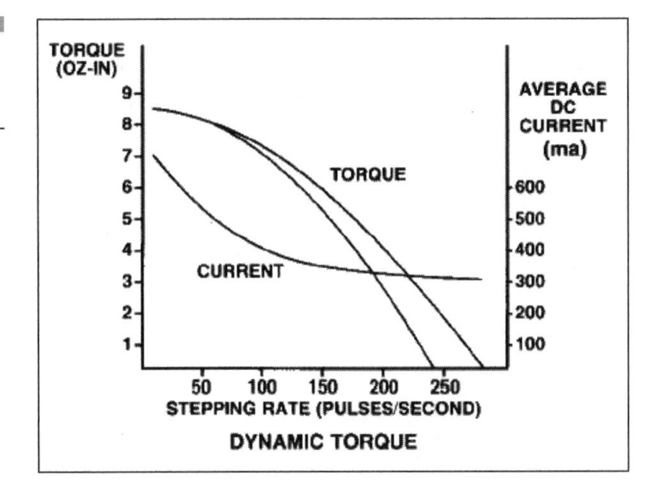

Figure 5-2 Stepper L-R curve[1]

motor can develop at a given pulse rate, or the maximum rate to which a given load can be accelerated (pull out).

Often all these specifications are not available, and it can be difficult to determine if the stepper is strong enough for your needs. A rule of thumb useful for steppers is as follows: Dynamic torque is approximately equal to the holding torque multiplied by 0.65. "Where did you get that number?" you ask. You've twisted my arm; I got that SWAG[2] by reading the specifications of dozens and dozens of stepper motors.

Most stepper motors are rated in gram-centimeters, but for us Americans, ounce-inches is more familiar. To convert gram-centimeters to ounce-inches, multiply your gram-centimeters by 0.01388. To convert gram-centimeters to newton-meters multiply your gram-centimeters by 0.0000981.

But Will It Move the Robot?

The rules are a little different for robots that don't have to climb hills or have power to spare. Just as long as our little motors can pull the weight of the robot around, it will be fine. Since we had no clue on how to figure

[1]Courtesy of Allegro Microsystems Inc.

[2]The etymological origins of this word are not appropriate for those younger readers or those with delicate sensibilities. Please do a Google® search for its definition. I guarantee it's out there.

in all the variables to achieve this lofty goal, we decided to hunt down empirical evidence. We did some experiments using steppers whose specifications were known (www.google.com is a great tool!). Table 5-2 shows the stepper motors we tested and how much they would bear before becoming unpredictable. The weight carried includes the weight of the motors. The first two steppers are unipolar that we got from old printers. The last motor is bipolar and we ordered that one from www.jameco.com.

These numbers seem to make sense compared to each other. The two unipolar steppers have a 24V maximum winding voltage. Since we ran them at 10V, we derated their power by 50 percent. The bipolar stepper could go much faster than a unipolar stepper when its speed is ramped up slowly. The 200-step bipolar we could successfully clock at up to 625 *pulses per second* (PPS); the two unipolar steppers would not run reliably past about 150 PPS.

Based on these empirical studies, we recommend that you need at least 8 ounce-inches' (576 gram-centimeters) of dynamic torque per pound of robot. These are weak as kittens though; any obstacle in the path will stop a robot so minimally powered, as these numbers suggest.

Our test bed (shown in Figure 5-3) consists of light Lexan™ base and wheel mounts with a 1-inch (2.54 centimeter) castor wheel. The electronics are a 555 timer clocking two UCN5804B unipolar stepper/driver chips or two MC3479 bipolar stepper/driver chips all wired onto two solderless breadboards. These driver chips will be discussed in greater detail in Chapter 7, "Motor Control 101, The Basics." The wheels are 2.6-inch, (6.5 centimeter) low-drag plastic wheels with rubber rims. Power is supplied by a 9.6V 8-cell *Nickel-Cadmium* (NiCads) battery pack. This test bed would not climb a 5-degree incline without stalling, but the Jameco bipolar stepper happily chugged along on medium-pile shag carpet. Go figure.

Table 5-2

A selection of stepper motor strengths

Stepper motor	Dynamic torque (pull-out)	Weight carried (@10V)
PM42L-048 (48 step)	288 g-cm (4 oz-in)	500 g (1.2 lbs)
PM55L-048 (48 step)	866 g-cm (12 oz-in)	1600 g (3.6 lbs)
Jameco® 163408 (200 step)	504 g-cm (7 oz-in)	1000 g (2.24 lbs)

Figure 5-3 The (ugly) test bed

Determining Stepper Motor Types and Wiring

The most common stepper motors used today are known as *hybrid* stepper motors. These steppers are known as hybrids because they use the best of two other types of stepper motors: the *variable reluctance* stepper, which has no permanent magnets, and the *permanent magnet* stepper, which has very few (as few as four) step positions. We're not going to talk about these other steppers, just the hybrid stepper motor.

The two most commonly available stepper motors to the hobbyist are *unipolar* (also known as *bifilar* or four-phase) and *bipolar* (also known as *unifilar* or two-phase). One other, less common hybrid stepper motor is the *universal* stepper. It's called universal because it can be configured to be unipolar, bipolar with windings in parallel, or bipolar with windings in a series. More on this uncommon stepper will be discussed later.

The Unipolar Stepper Motor

This stepper motor is called unipolar because current flows in only one direction in each winding. It is also called bifilar because it contains two coils whose polar orientation is opposite each other while wrapped on the same core with a center tap. The unipolar stepper is also called four-phase is because it has four windings to energize. Unipolar stepper motors have either five or six wires to connect. They are most commonly found with 3.6° per step (100 steps per full rotation) or 7.5° per step (48 steps per rotation).

If your stepper has five wires, one is a common wire (connected to your V+) and the four other wires are for each phase of the stepper. If your stepper has six wires, it is a *can-stack* unipolar and has two sets of windings, each with a center tap wire. The six-wire unipolar is the most common stepper motor to find in printers that you may be scrapping out for parts. But how do you sort out the wires? We'll tell you; read on!

Sorting a Five-Wire Unipolar Stepper's Connections To find the correct wiring configuration for the five-wire unipolar stepper, you will have to do some sleuthing, but it's pretty simple. Figure 5-4 shows the basic wire configuration of the five-wire stepper motor.

To find the proper order of windings in order to get this stepper to rotate, you will need a battery and a piece of tape (and a five-wire stepper, of course). Use a marking pen to note which winding is which when you find them. Then follow these steps:

1. Using a *digital volt-ohmmeter* (DVM) find the common wire. This will be the wire from which all other windings measure the same value. Figure 5-5 shows the resistance from the common to a

Figure 5-4
A five-wire winding layout

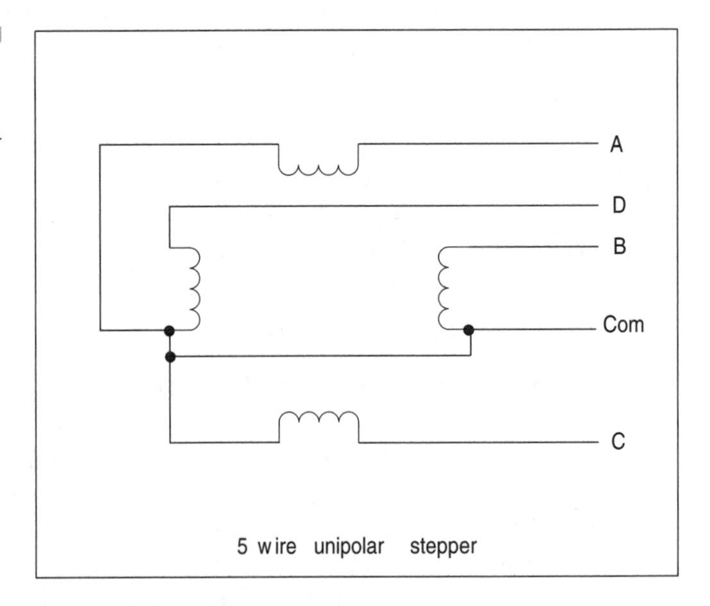

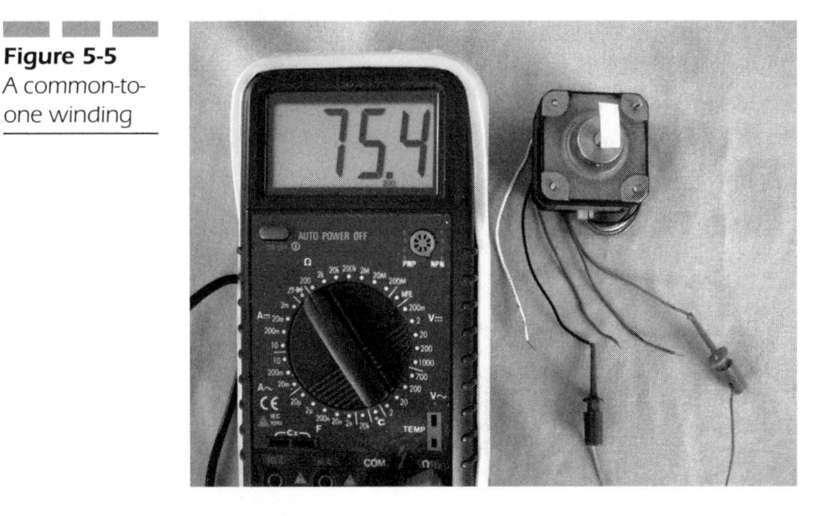

Figure 5-5
A common-to-one winding

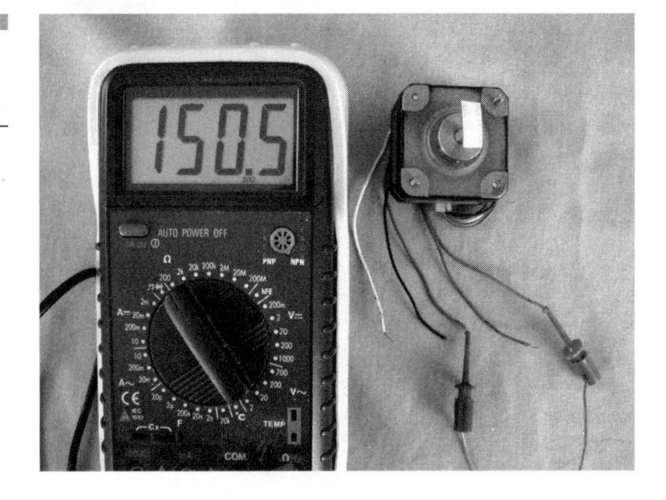

Figure 5-6
Between two windings

single phase. Figure 5-6 shows the resistance if you are measuring between two phases and not a common and a single phase. Connect this wire to V+ on your battery or power supply. A supply of 5V to 6V will be enough for these tests.

2. Put a piece of tape on the shaft of the stepper so that it sticks out perpendicular to the shaft like a flag. We will use this *as* a flag so we can see when the stepper moves.

3. Arbitrarily pick one wire and call this *phase 1*. Ground this wire. The shaft may move a slight amount. The stepper is now locked at phase 1. See Figure 5-7.
4. Pick another wire and ground it while carefully watching the piece of tape on the shaft. If the shaft turns a tiny bit to the right, this wire is *phase 2*. See Figure 5-8.

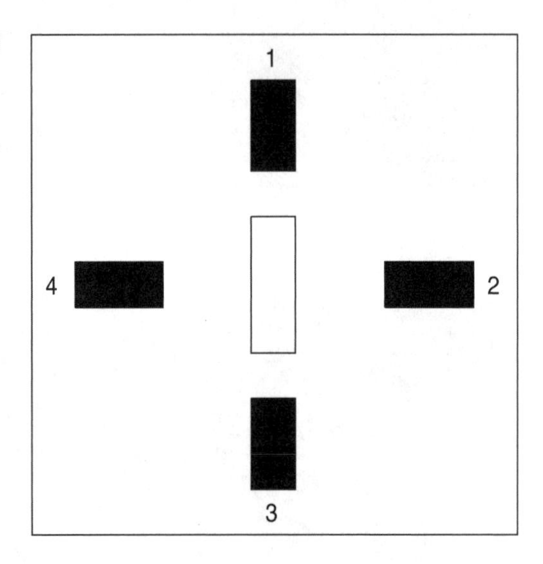

Figure 5-7 The first phase location

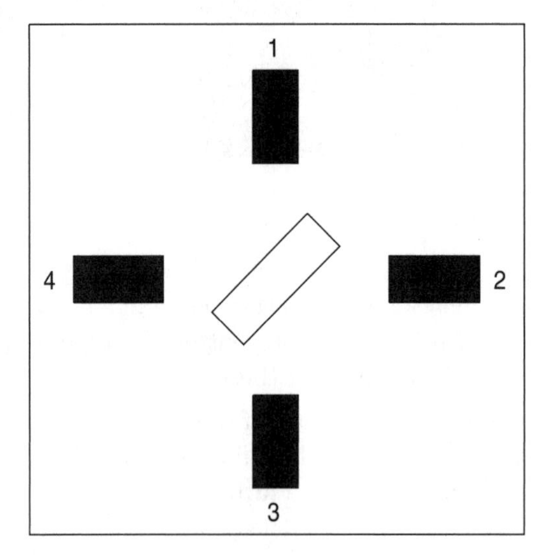

Figure 5-8 The second phase found

5. Pick another wire and ground it while carefully watching the flag on the shaft. If the shaft turns a tiny bit to the left, this wire is *phase 4*. See Figure 5-9.
6. Pick another wire and ground it while carefully watching the flag on the shaft. If the shaft doesn't turn, this is *phase 3*. See Figure 5-10.

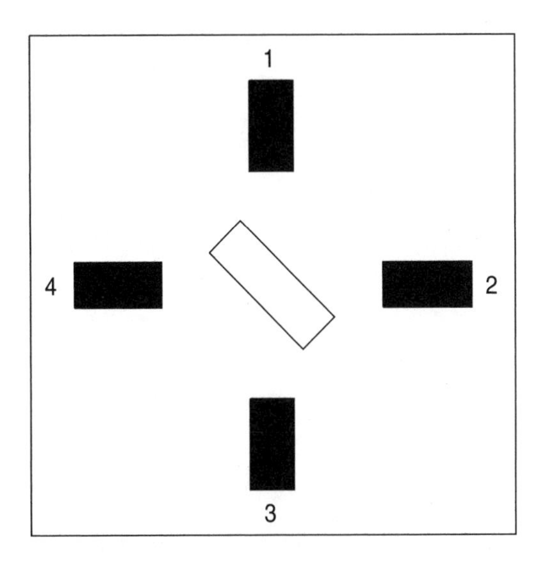

Figure 5-9 The fourth phase found

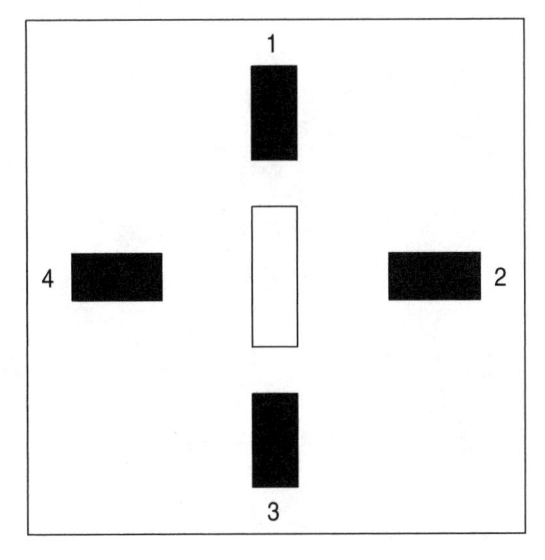

Figure 5-10 The third phase found

Sorting a Six-Wire Unipolar Stepper's Connections This is the most common unipolar stepper to encounter in a printer-scrounging job. The six-wire unipolar stepper usually looks like two skinny cans stuck together, each can having three wires issuing from it (see Figure 5-11). This is a very easy stepper motor to sort wires on.

Finding the wiring order of the six-wire stepper is really simple. If you have a can-stack type (see Figure 5-12), then the wires are already

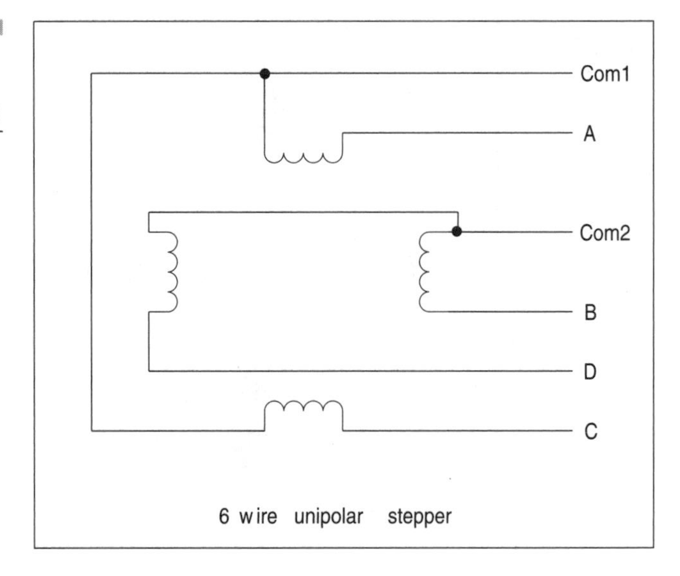

Figure 5-11 A six-wire winding layout

Figure 5-12 Can-stack sorted wires

sorted. Just use your DVM to find the common wire for each pair of windings (Figures 5-5 and 5-6 show what to look for). As long as you keep the winding pairs together, the order of them is not important; it will just change the rotation direction.

If you don't have a can-stack six-wire stepper, follow these steps to find the winding pairs:

1. Use a DVM to find the common connection between each pair of windings. Refer to Figures 5-5 and 5-6 for examples of what you would see if you were looking for the common connection between two windings.
2. Two sets of windings will be found. Separate and mark them. These windings will be paired A and C (or 1 and 3 if you want) for one set of windings, and B and D (or 2 and 4 if you want) for the other set. Which one is which within a pair is not important, just the pairings.

Unipolar Stepper Motor Step Patterns

Three step patterns are used in unipolar steppers: *two-phase*, *half-step,* and *wave*. Wave stepping involves turning on one winding at a time. The rotor then turns to align its permanent magnet with the winding of the opposite polarity. Two-phase energizes two windings at a time, and the rotor turns to align its permanent magnet at a point halfway between the two energized windings. Two-phase also uses twice the power of wave stepping because two windings are turned on; however, 41.4 percent more torque is available. Finally, half-step involves alternately turning on two windings, and then one. Depending upon the stepper design, half-step has 15 to 30 percent less torque than the two-phase stepping pattern but has twice the resolution (twice as many steps are taken).

Tables 5-3 through 5-5 give the winding-energizing patterns for these three step patterns. To reverse the stepper motor direction, reverse the pattern.

The Bipolar Stepper Motor

The bipolar stepper motor is so named because each winding is energized in both directions, so each winding can be either a south or north pole. It's called unifilar because each pole has a single winding and it's also called two-phase because it has two separate windings.

Table 5-3 Wave step patterns

	A	B	C	D
1	On			
2		On		
3			On	
4				On

Table 5-4 Two-phase step patterns

	A	B	C	D
1	On	On		
2		On	On	
3			On	On
4	On			On

Table 5-5 Half-step patterns

	A	B	C	D
1	On			
2	On	On		
3		On		
4		On	On	
5			On	
6			On	On
7				On
8	On			On

Bipolar stepper motors have four wires, two per winding. Bipolar steppers are stronger than unipolar steppers of the same size and weight because they have twice the field strength in their poles (a single winding that isn't center tapped). Each winding in a bipolar stepper requires a reversible power source, commonly an H-bridge driver. Because a bipolar is stronger than a unipolar, it is almost always used in designs where space is a premium. This is why the head stepping mechanism in floppy drives is always a bipolar stepper.

Sorting a Bipolar Stepper's Windings This is simple—use your trusty DVM and find the two windings. If two wires measure a resistance, those two wires form a winding; the other two wires are the other winding. Bipolar steppers usually are 1.8° per step, which is 200 steps per rotation.

Bipolar Stepper Motor Step Patterns

Bipolar steppers have the same step patterns as unipolar steppers; they are just implemented differently because of their winding configuration. Their step patterns are shown in Tables 5-6 through 5-8.

Table 5-6

Bipolar stepper wave step patterns

	A_1	A_2	B_1	B_2
1	–	+		
2			+	–
3	+	–		
4			–	+

Table 5-7

Bipolar stepper two-phase step patterns

	A_1	A_2	B_1	B_2
1	–	+	–	+
2	–	+	+	–
3	+	–	+	–
4	+	–	–	+

Table 5-8

Bipolar stepper half-step patterns

	A_1	A_2	B_1	B_2
1	–	+	–	+
2			–	+
3	+	–	–	+
4	+	–		
5	+	–	+	–
6			+	–
7	–	+	+	–
8	–	+		

Did You See That? The Big Secret Behind the Step Patterns

If you are clever, or bored with too much time on your hands, you will have noticed that the step patterns are exactly the same for unipolar and bipolar. Take the unipolar tables, swap columns *B* and *C*, turn all the *On* entries into + entries, and then put a – in any column that shares a winding with the + entry. You'll see that it's the same format, just going the other direction. This suggests that any logic that controls a unipolar stepper motor will also drive a bipolar stepper motor. That's logic, not drivers. A bipolar will need H-bridges for its windings, and a unipolar will need only a simple transistor switch. The only change you need to make is to swap the *B* and *C* logic outputs to switch between controlling a unipolar and a bipolar stepper.

The Universal Stepper Motor

The universal stepper motor has eight wires coming out of it (see Figure 5-13). It can be configured as a unipolar or a bipolar stepper motor. This type of stepper can be confusing when sorting out the windings because you need to get the *polarity* of the windings correct as well as the

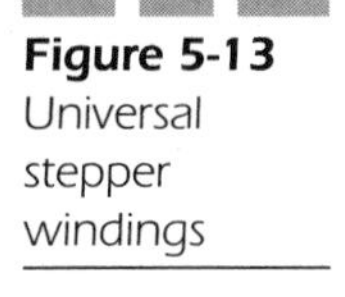

Figure 5-13 Universal stepper windings

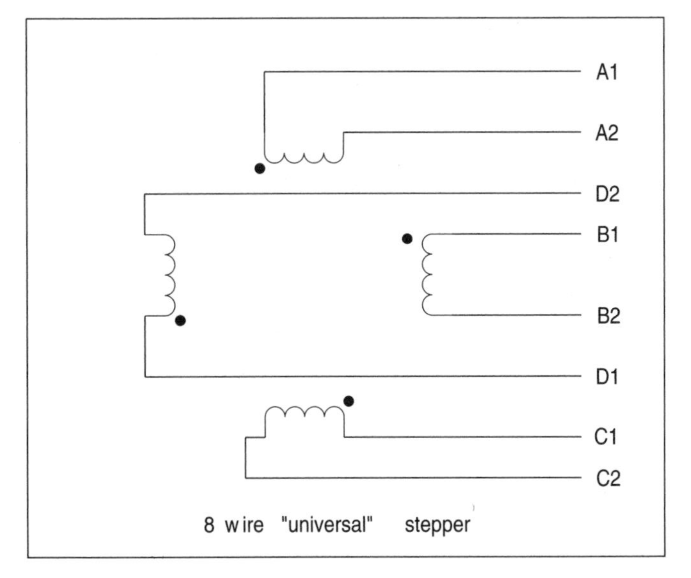

order. There is no way to find the polarity by looking at the motor, but fortunately there is a way to deduce it electronically.

Finding the Winding Polarity of a Universal Stepper Motor For this test, you will need a power supply or a battery pack and a DVM. To determine a universal's stepper motor's winding polarity, follow these steps:

1. First, find all four windings by using the DVM ohmmeter to find the proper wire pairs.
2. Choose two windings and short one lead from each together. Set your DVM to AC volts on the 20V scale and connect its leads to the end leads of the two windings you are checking.
3. Rotate the shaft and if you get a voltage reading of 2V or so you have the windings set in a series and you know their polarity (see Figure 5-14). If you get 0V, then you also know the polarity of the windings; they are opposite (see Figure 5-15). Switch to the other wire on one winding to short to the other winding to check your assumptions.
4. Mark your wires to denote the winding polarities. Which side you call the *dotted* side is not important, as long as you know how to pair them.

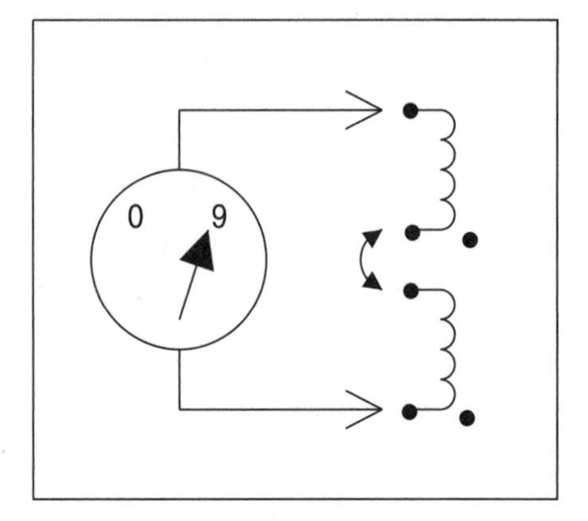

Figure 5-14 Polarities matched

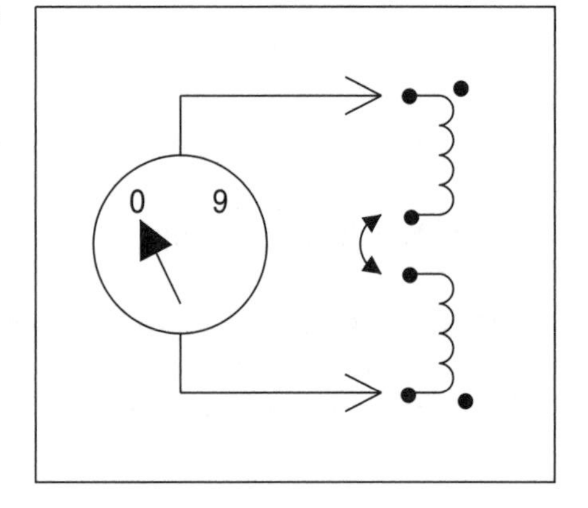

Figure 5-15 Polarities mismatched

How to Configure the Universal Stepper Motor This section shows the three ways a universal stepper motor can be configured: unipolar, bipolar series, and bipolar parallel.

The configuration shown in Figure 5-16 mimics a six-wire unipolar stepper motor.

The bipolar configuration in Figure 5-17 will deliver a higher torque at lower stepping rates. It will not step as fast because the windings in a series have a higher reluctance, so current won't peak as quickly.

The bipolar configuration in Figure 5-18 will not be as strong as a bipolar series configuration, but it has a higher stepping rate because the parallel winding configuration will have a lower reluctance and so current will peak faster.

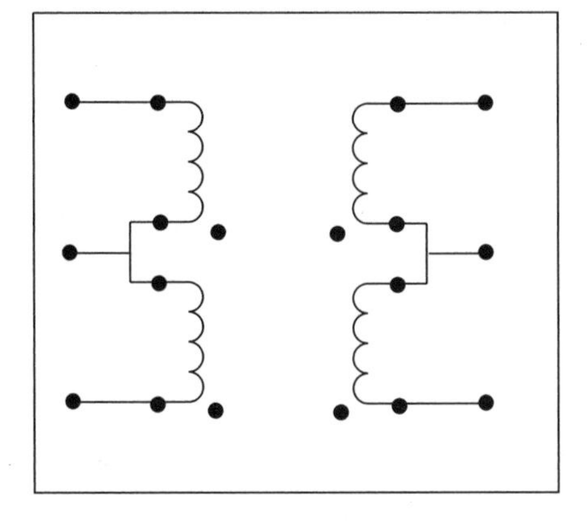

Figure 5-16 A unipolar configuration

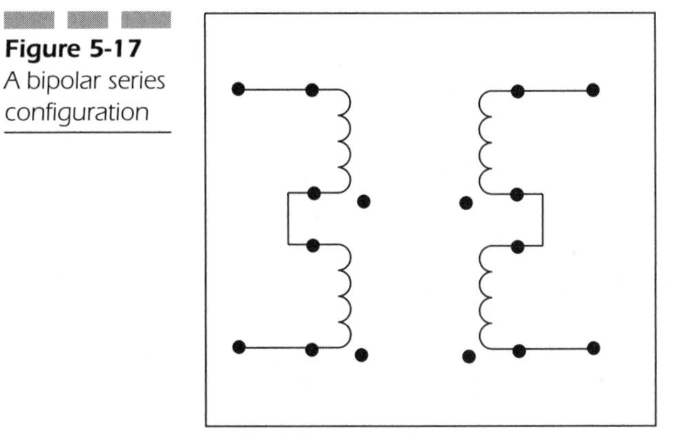

Figure 5-17 A bipolar series configuration

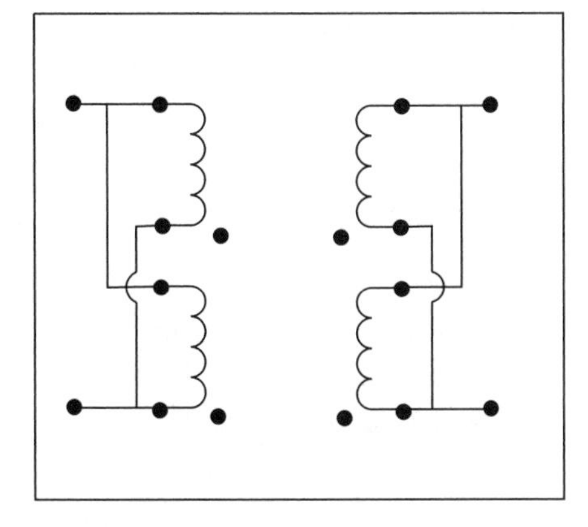

Figure 5-18 A bipolar parallel configuration

One Last Word on Bipolar and Unipolar Configurations

By now, you have probably seen the amazing similarity between the six-wire unipolar and the series-configured universal stepper. It's true that you can use a six-wire unipolar stepper as a bipolar stepper by ignoring the two common winding connections. We've tested that and it works fine.

Where to Get Stepper Motors

Stepper motors are expensive to buy new, but fortunately for us, they are common in the surplus market. Just about every surplus shop you will find will carry a selection of stepper motors for sale. Unfortunately, few surplus shops will publish specifications on their stepper motors, because they don't know them. Some do know and publish what they have for specifications.

The only mail-order shop I know that carries a large selection of inexpensive stepper motors and most of their specifications is www.jameco.com. There are other ways to get stepper motors however. C&H carries them—often with specs, but the inventory varies.

Gleaning Stepper Motors from Printers and Disk Drives

The head mechanisms for floppy disk drives are positioned by small stepper motors. Don't bother to hack open hard disk drives for steppers; they don't have them. Hard disks are so fast that they use *voice-coil* actuators to move their magnetic heads; you can get some powerful magnets from hard disk head drives, but not stepper motors. Floppy driver stepper motors are bipolar. Remember that bipolar steppers are more powerful per unit weight than unipolar, which makes them the obvious choice for a floppy stepper motor.

Figure 5-19 shows what you will get from these drives. The 5 1/4-inch floppy stepper is on the right, and the 3 1/2-inch stepper is on the left. The stepper from the 3 1/2-inch drive is less than 1 inch (2.54 centimeter) in diameter. If you find a really old full-height-style floppy and get its stepper motor, it looks like Figure 5-20. It's much larger than the modern 5 1/4-inch floppy drive's stepper motor.

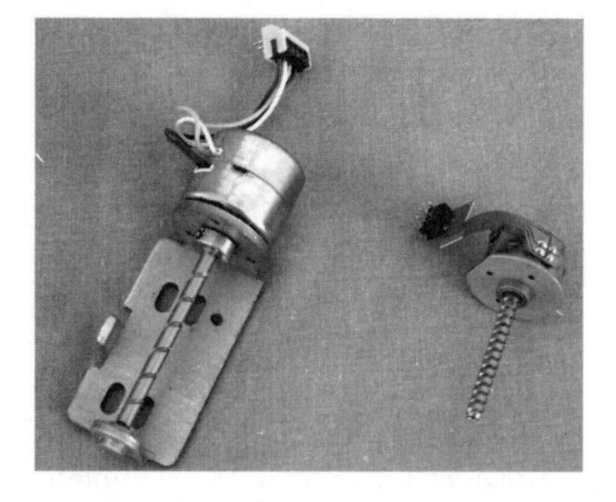

Figure 5-19 Modern floppy steppers

Figure 5-20 Older full-height stepper

A word of warning about the 5 1/4-inch half-height floppy's stepper motor. Do *not* try to remove the stepper from the bracket. It's integral with the housing and you will ruin it. Use your favorite cutting device and whack off the extra metal bracket parts you don't want. My favorite metal cutting device is a bit of well-placed C4—will that do?

Stepper motors from printers do not have the high resolution of the steppers from the floppy drives. Typically, they are 7.5° per step, 48 steps per revolution. They are, however, much larger and stronger. Usually, they are also unipolar stepper motors. These steppers look like the small motor in Figure 5-1, usually with the wire pattern shown in Figure 5-12. The vast majority of these printer stepper motors seems to be made by Minebea Hamamatsu®, and you can probably find specs for them on their web site. You can also do a Google search using the motor part number, which will be something like PM55L-048-HP01; search for the first eight digits of the part number.

Stepper Motor Control and Driver Circuits

Since the very same step patterns are used for both unipolar and bipolar stepper motors, much of the control circuitry will be identical. The main difference between them will be the driver circuitry to energize the windings. The unipolar will use a simple bank of transistors, while the bipolar will require an H-bridge for each winding. What follows are the details for creating those circuits. A few stepper motor driver chips out there can do all the control and driving duties completely on their own. Read on for such wonders of modern engineering.

Discrete Unipolar Stepper Motor Control and Driver Circuits

These circuits use TIP120 (or equivalent) transistors for high-current (5 amp) stepper motors. You can substitute lower-current transistors, such as ULN2003, into the design if you so desire. We recommend building this circuit on a solderless breadboard first before committing it to permanent use just to get used to the design. Figures 5-21 and 5-22 display both TTL (the 7400 series) and CMOS (the 4000 series).

Discrete Bipolar Stepper Motor Control and Driver Circuits

This circuitry is the basic two-chip design for controlling stepper motors in either direction using the two-phase step pattern that has been used on floppy disk controllers for years. Our one upgrade to the design is the use of the 754410 dual DC motor driver chip instead of building a discrete transistor H-bridge. This makes a simple three-chip circuit that is simple to build. You can use the L293 as a pin-for-pin compatible drop-in for the 754410, or use an L298 for much higher current stepper motors (see Figures 5-23 and 5-24).

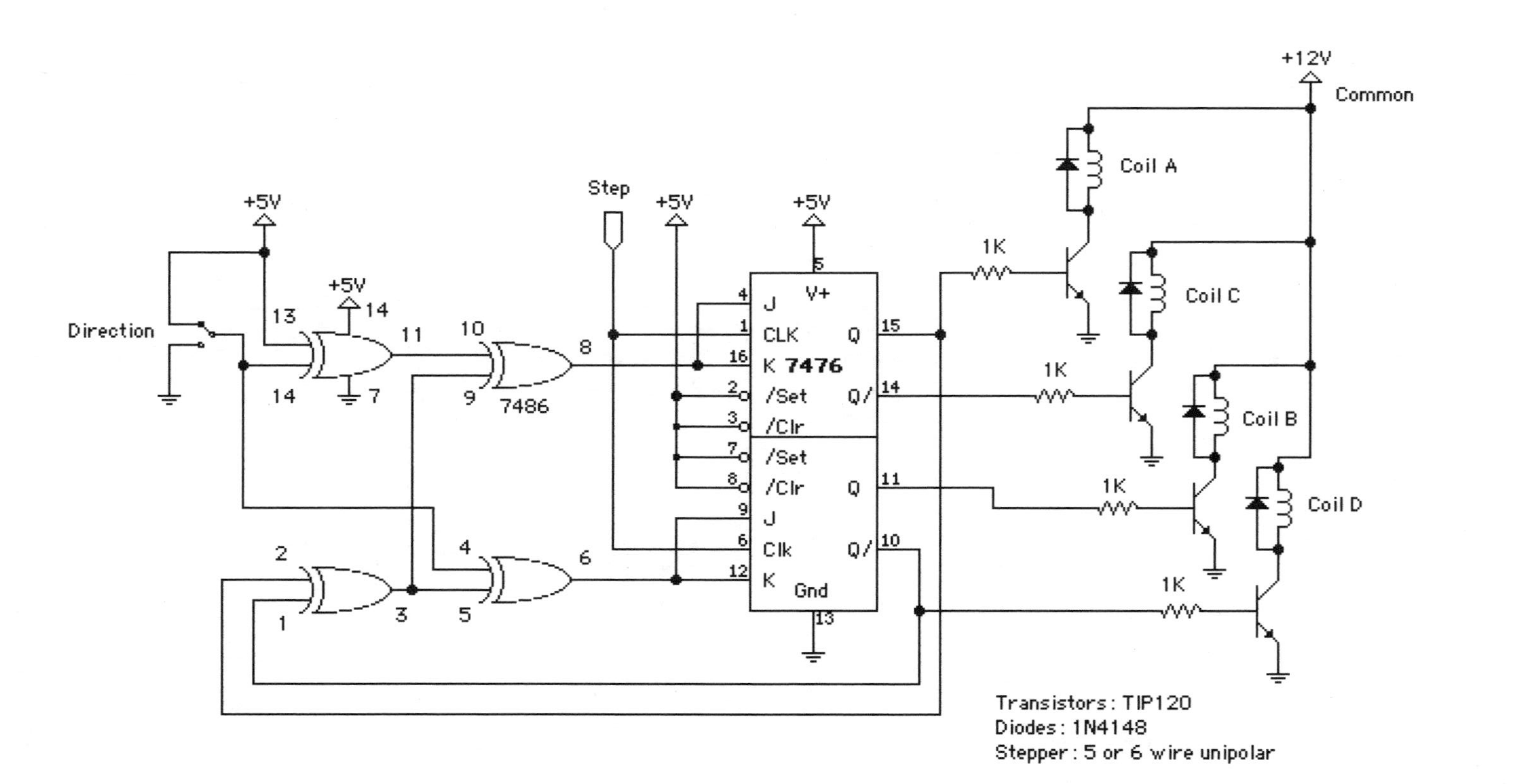

Figure 5-21 TTL unipolar controller/driver

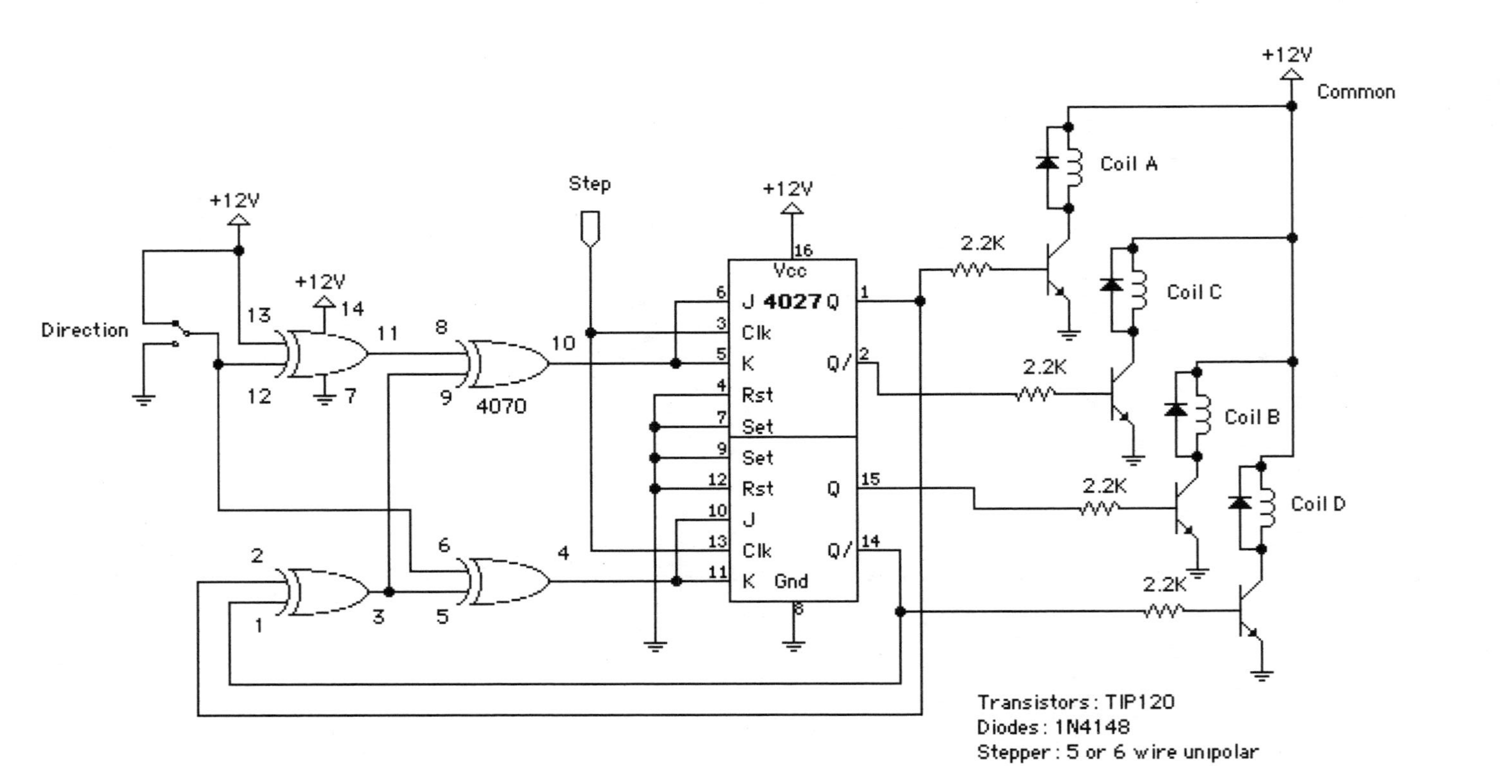

Figure 5-22 CMOS unipolar controller/driver

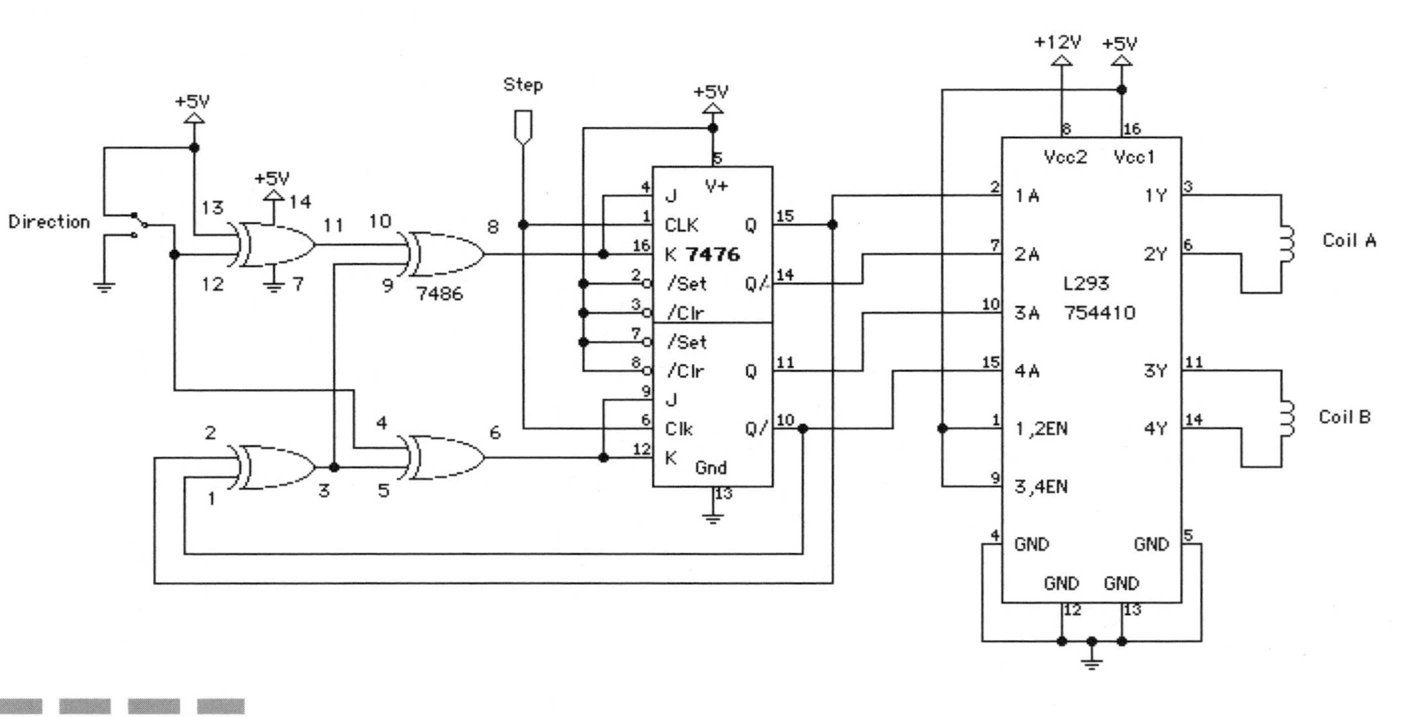

Figure 5-23 TTL bipolar controller/driver

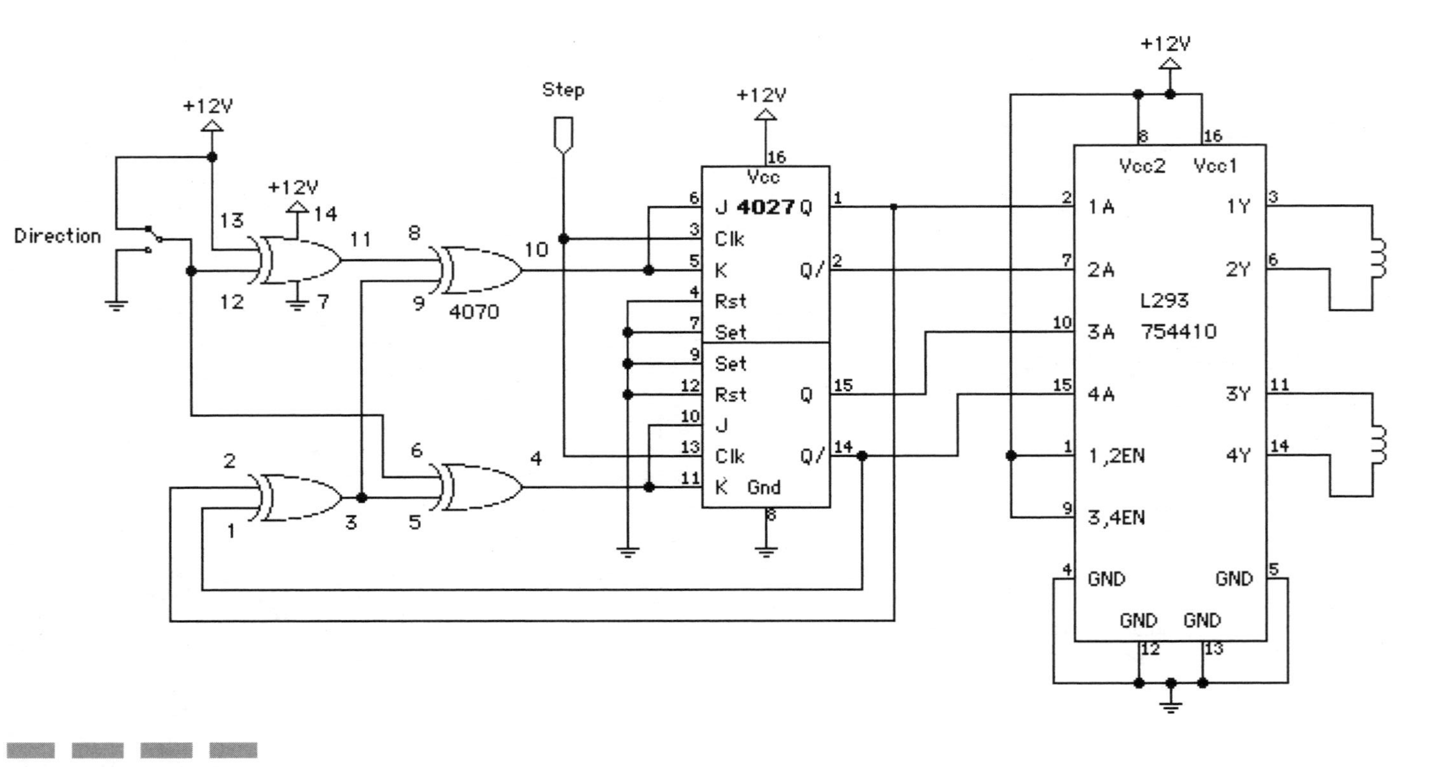

Figure 5-24 CMOS bipolar controller/driver

Test Clock for your Stepper Controller/Driver Circuits

It's useful to have some way to check out your circuits or your steppers before you commit to using them. If you build any of the previous circuits, this 555-based clock can be used to test your steppers and winding discoveries. When testing how fast you can clock a stepper, remember that it will be different when the stepper is *under load*. Most steppers can't just *go* when they are given step commands. To reach their highest speed, they need to be *ramped* up to that speed gradually. This 555 circuit can have two speed ranges. As shown in Figure 5-25, it will clock a stepper from 70 PPS to 130 PPS. To reach the 120 PPS to 550 PPS range, replace R2 with a 10K resistor.

Single-Chip Stepper Motor Driver Solutions

A few single-chip driver and controller solutions can be used for unipolar and bipolar stepper motors. We're going to discuss two of them, one for unipolar and one for bipolar stepper motors. These have limited drive current and are useful for lower-current stepper motors. If you have higher-current needs, keep reading. We'll give you solutions that are pretty cool for those too.

The Allegro® UCN5804 Unipolar Stepper Controller/Driver The UCN5804 can handle unipolar steppers requiring up to 35V and up to

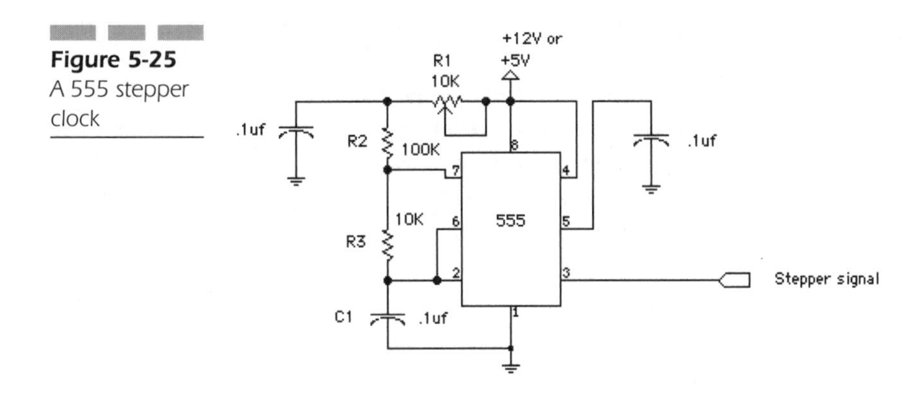

Figure 5-25
A 555 stepper clock

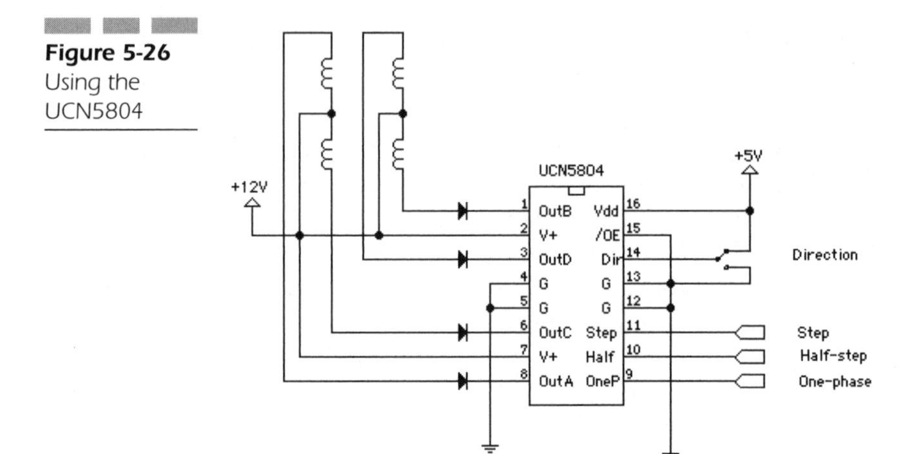

Figure 5-26 Using the UCN5804

1 amp of sustained current draw. It is reversing and enables wave, half-step, and two-phase stepping modes. It even has a power-on reset, so it's ready to run when it's turned on. The chip has onboard flyback and ground clamp diodes, so you don't need to use your own. The UCN5804 can be used directly by discrete logic (as in BEAM robotics) or by a microprocessor. It's a nice package for under $4 at many suppliers. The proper connection logic and schematic is given in Figure 5-26.

The Motorola® MC3479 Bipolar Stepper Controller/Driver The MC3479 can handle bipolar steppers whose voltage requirements are between 7.2V and 16.5V with currents of up to 350 ma. It's best suited for small bipolar stepper usage. The MC3479 is bidirectional and supports half-step and two-phase stepper patterns. It also has a resistor set current limit and a "phase A" sync output that notes when the stepper is at its first step position (once every four or eight steps, depending upon the step mode).

Figure 5-27 shows the proper setup and biasing of this controller. The current limit must be set on the MC3479, but we show you the 350 ma resistor setting to get you going right away. The specification sheet for this part details the process for determining whether a heat sink is needed. We've done this calculation for a 350 ma draw; you need a heat sink if you are really going to pull 350 ma per winding. Check your requirements by using the procedure given in the specifications documentation. It's not that hard and it may keep you from letting the magic smoke out of the chip.

Figure 5-27 Using the MC3479

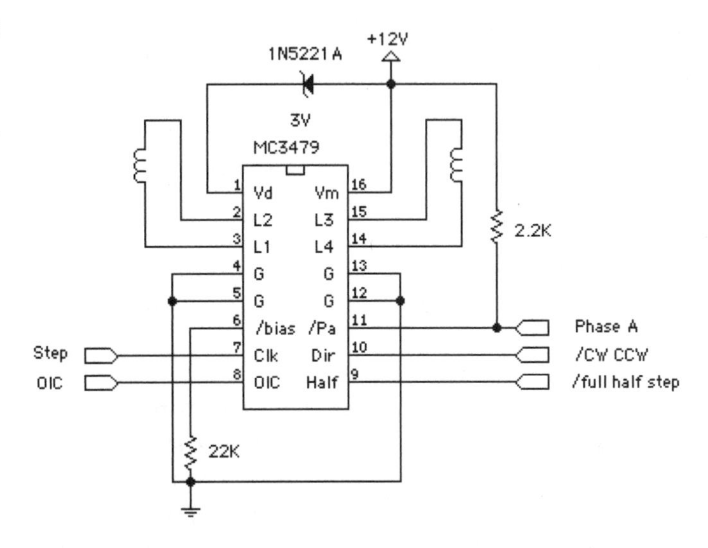

Special Microcontroller-Based Stepper Controllers

A few microcontroller-based stepper motor controllers require only the proper high-current drivers to be added to control the stepper motors programmatically. We will cover the programming and wiring of these stepper controllers in Chapter 7. We'll just list the suppliers and their capabilities here. These chips offer more options for control but are more complex to use. The MC3479 and UCN5804 controllers can also be controlled by a microcontroller. This is by no means a definitive list; we just had to stop looking and write the book.

Stepper Controllers Without Drivers These controllers interface to microcontrollers to programmatically control steppers. They do not have output drivers to power the stepper motors directly and need to be used with external driver chips.

Ferrettronics® FT609 Serial Stepper Controller This controller will control either a unipolar or a bipolar stepper motor depending upon the driver used. A computer communicates with the FT609 through a 2400-baud, 8-bit, no-parity, one-stop bit serial connection.

E-LAB® EDE1200 Unipolar Stepper Controller This controller does not use a serial link to configure it. All the control lines are on labeled pins. The EDE1200 requires a 4 MHz crystal and trim caps for operation. The E-LAB controllers fall between the FT609 and the UCN5804B in sophistication. They enable configured speed selection, step patterns, and step modes (single-step or run-mode).

E-LAB EDE1204 Bipolar Stepper Controller This controller is functionally identical to the EDE1200, but it is labeled for bipolar stepper motors.

Savage Innovation® OOPic II Microcontroller The OOPic II is a general-purpose controller that can control stepper motors, hobby servos, and PWM for DC motor control, as well as many other types of devices. Very detailed control of stepper motors is possible with the OOPic II. As with other controllers in this section, the OOPic II needs drivers to interface with a stepper motor.

Driver Chips Usable with These Stepper Controllers We've already mentioned the 754410 driver for bipolar stepper interfacing. Here are several more drivers for either unipolar or bipolar stepper motors. These are the most convenient to use when driving a stepper motor.

Allegro® ULN2003A and ULN2004A Drivers These chips have six separate drivers in them with CEMF (inductive kickback) clamp diodes on each driver. Each can sink a 500 ma constant current with a maximum output voltage of 50V. The ULN2003 is for TTL or 5V CMOS operation; the ULN2004 is for 6 to 15V CMOS operation. These are best used with unipolar stepper motors.

Allegro UDN2544B Quad Darlington Power Driver This unipolar stepper driver is a 50V, 1.8-amp quad driver with a common enabler for all four channels. The UDN2544 is a TTL or 5V CMOS interface chip only. Two inputs on this driver have inverted logic, and the FT609 has a special mode for using this chip.

Allegro UDN2540B Quad Darlington Power Driver This unipolar stepper driver is functionally identical to the UDN2544B, except that all four inputs are positive logic (instead of the two inverted inputs).

L298 Dual-Power H-bridge Driver This common DC motor driver is a 2-amp per channel bipolar stepper motor driver. It requires external CEMF clamp diodes on each output.

One Last Resource

Figure 5-28 was published in the July 1999 issue of *Nuts & Volts* in an article by Dan Mauch. In it, Dan details the wiring patterns of several common, high-power stepper motors that he has used for his projects and business. Your stepper may be listed there and, if so, Figure 5-28 could save you some time decoding your stepper motor. Dan adds that manufacturers sometimes change their wire colors, so double-check before using the stepper motor.

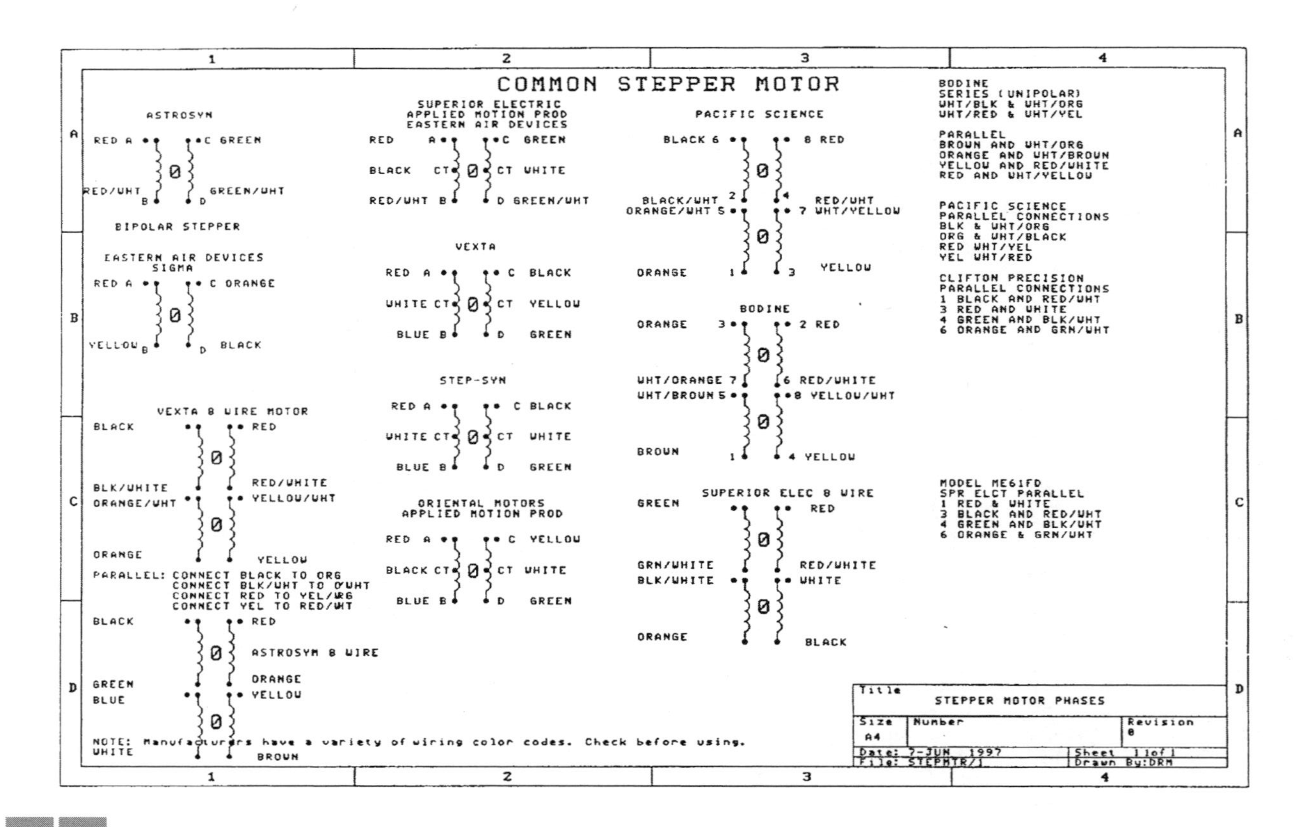

Figure 5-28 Selected stepper motor wiring lists[3]

[3]Reprinted here with permission by *Nuts & Volts* [www.nutsvolts.com] and Dan Mauch [www.seanet.com/~dmauch].

CHAPTER 6

Mounting Motors

Welcome to the least interesting chapter in the book. Let's face it: Mounting motors can be a pain, especially larger motors. At the same time, motor placement is usually one of the most critical mechanical tasks you'll need to undertake as you build any robot. In the case of wheeled robots, correct mounting will ultimately determine the quality of travel of your robot. A well-mounted motor can make the difference between a robot that moves with military precision and one that perambulates like a drunken sailor.

Fortunately, you can usually get pretty good results with a trip to the hardware store and a little patience. In general, you can make your own mounts with commonly available materials and reasonably pedestrian tools.

Balance, Symmetry, and Alignment

For most amateur robot designs where a drive wheel is connected directly to a motor, motor placement really comes down to drive wheel placement. Try to achieve the following when placing your motors:

Balance When using dual drive wheels, make sure those wheels are balanced relative to a flat surface, as shown in Figure 6-1a. In other words, make sure that one wheel is not shorter than the other, as in Figure 6-1b.

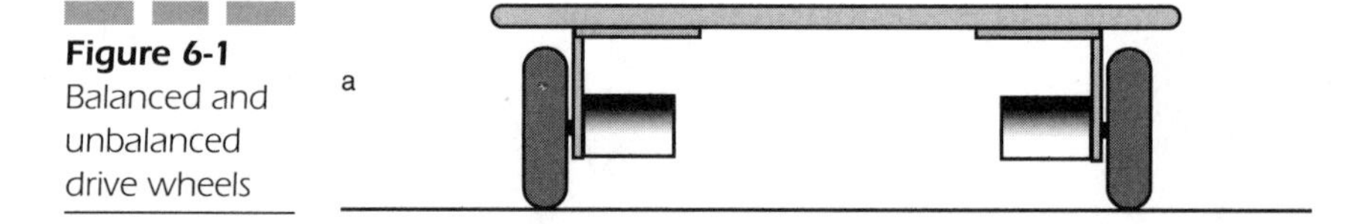

Figure 6-1 Balanced and unbalanced drive wheels

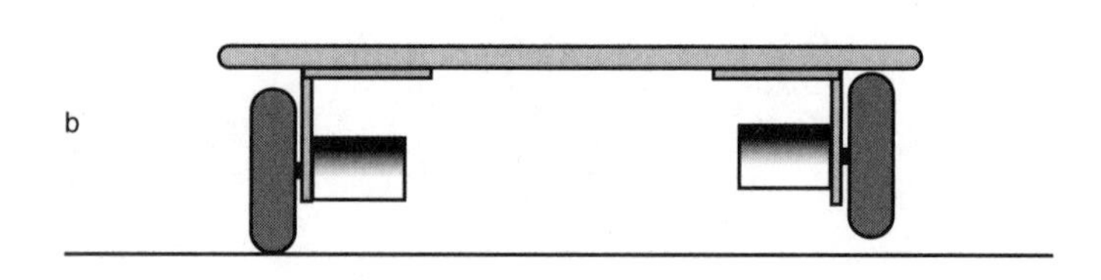

It is surprisingly easy to get this part wrong, especially when you make your own mounting brackets. To check for the problem, place the robot on a smooth, flat surface and have a look. Look along the front of the robot to verify that neither wheel is off the surface and that the robot is not cocked at an angle. You might wish to apply pressure above each drive wheel to make sure there is no movement.

If you do find that the wheels are unbalanced, you'll want to correct the problem; failure to do so will result in a robot that tends to curve in the direction of the shorter wheel when commanded to go straight. For motors mounted via brackets bolted to the chassis, you can generally use washers between the bracket and the chassis to raise the shorter mount as required.

Even if your robot uses only a single drive wheel and two wheels for steering or balance, you will want to run this same check against the nondriven wheels as well to avoid steering difficulties later.

Symmetry For dual-drive systems, be sure that your motors are mounted along a common centerline, as in Figure 6-2a. This is easy enough to do: Simply mark the desired centerline prior to drilling holes for the mount. Single-drive systems should

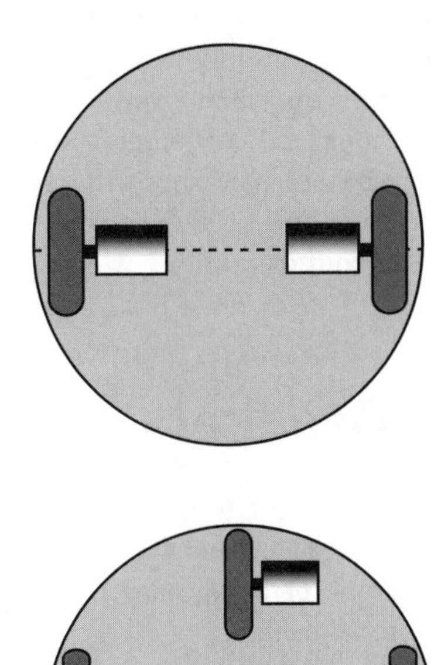

Figure 6-2 Symmetrical motor placement

Figure 6-3 Proper and improper alignment

generally be mounted so that drive wheel is centered between the support wheels, as in Figure 6-2b.

Alignment When drive wheels are connected directly to a motor, you must take care that the motor and shaft are aligned so that the wheel (or wheels) is facing directly forward, as in Figure 6-3a. A robot with wheels set as in Figure 6-3b will not be able to follow a true path without some type of software correction.

Even when your motors are not connected directly to wheels, you'll need to make sure you plan ahead when deciding where to mount the motors. Where motor output is to be coupled to wheels via chains or timing belts and pulleys, make sure you take into account the length and required tension to keep the chains or belts from slipping under load. Although flexible couplings such as vinyl tubing or manufactured couplers will tolerate some degree of misalignment, you will still need to be sure any misalignment stays within tolerable bounds.

Motor Brackets

Most of us will mount our motors to the robot chassis via some form of mounting bracket. This is certainly not the only approach possible, but it has the distinct advantage of being easy to remove later for use in another project or a quick redesign.

Many approaches to bracket design are possible, depending upon the motor form factor and the materials on hand. We will not attempt to cover all of them here, but we *will* cover some of the most commonly used designs, especially those types that can be fabricated with inexpensive materials and relatively little labor.

Metal Mounting Brackets

Sheet aluminum and galvanized steel are both good choices as materials from which to fabricate mounting brackets (see Figure 6-4). These can be cut, bent, and drilled with relative ease and are reasonably priced in small quantities.

Aluminum sheets from 1/16 (1.5 millimeters) to 1/8 of an inch (3 millimeters) thick are widely available from hobby shops in small sizes at reasonable prices. If you don't need a very wide piece, the larger hardware depots often carry "flats" of aluminum bar in suitable thicknesses.

Galvanized steel suitable for our purposes can also be had from well-stocked hardware stores, either as flats or in the form of construction braces, such as mending T's or joist holders, from which suitable pieces can be cut with a hacksaw or jigsaw. You may be able to find construction braces that are already suitably formed for your motor size. Note that steel stock that is much thicker than 1/16 of an inch (1.5 millimeters) can be hard to both cut and bend.

Cutting and Drilling Sheet Aluminum and Steel Both of these materials can be cut with a hacksaw, but the relatively thin pieces we'll be working with can be safely cut with a jigsaw. Be sure to use a blade

Figure 6-4 Commonly available metal stock

with a high tooth density—24 *teeth per inch* (TPI) or better—and select a blade created specifically to cut sheet metal. An aluminum sheet up to 1/8 of an inch (3 millimeters) and galvanized steel to 1/16 of an inch (1.5 millimeters) or even a little thicker can be cut easily and accurately with a jigsaw equipped with a suitable blade. Here are a few things to keep in mind when working with metal:

- When cutting, be sure to clamp the work piece securely. Not only will this be safer, but if not well clamped, metal stock can vibrate to the point that it becomes difficult to see any guidelines you might have drawn, leading to sloppy work.
- Use a flat "bastard" file to remove any rough edges. Both aluminum and steel can be left with nasty, jagged edges after cutting.
- If it's wide enough, the bar or flat stock can often save you a little cutting—just cut off the length you need to fashion your bracket.
- Either material can be drilled with a standard hand drill. You may want to use cutting fluid when drilling to extend the lifetime of the drill bit and give a cleaner hole.

A Bent Sheet Metal Bracket One common type of motor bracket is simply a strip of sheet metal (of the appropriate width) bent into an L shape. One face of the L is drilled to accommodate the motor shaft and motor mounting holes, which are expected to be on the face of the motor, and the other face is drilled with holes used to mount the bracket to the underside of a chassis.

Bending Sheet Metal The tricky part of making an L bracket is getting a good bend in the metal strip, particularly as stock gets thicker. If you have a metal brake, of course, bending metal is no problem. For those of us without access to a machine shop, it's still pretty easy to get a decent bend in the stock of up to 1/8 of an inch thick by doing the following:

1. Mark the point on your strip of metal where you want the bend to go. In the case of aluminum, it is easier to make your bend parallel to the milling marks (fine lines created as the stock is manufactured).
2. If you have a vise of sufficient width and depth, clamp the strip so that the bend mark runs along the edge of the vise jaw. If not, clamp the strip firmly to a sturdy workbench, making sure the bend mark runs parallel to the edge of the workbench, as in Figure 6-5.

Figure 6-5
Secured strip for L bracket

3. If the strip is of relatively pliable stock, you may be able to simply apply slow and even pressure by hand to get the bend. This is usually fine for a 1/16-inch aluminum strip.
4. If the strip is of a thicker or less pliable material, use a clamp to secure a sturdy length of lumber such as a 2×4 to the free end, and use the lumber as a lever to bend the strip of metal (see Figure 6-6). If your clamp fails to provide adequate clearance so that you can bend the strip a full 90 degrees, you can drill 4 or so holes into the free end of the stock and secure it to the lever arm via wood screws and washers instead of a clamp. Then bend the strip using the attached lever. The holes drilled for the wood screws will become your chassis mounting holes. We have used the lever approach for aluminum stock up to 1/8 of an inch (3 millimeters) thick with good results.

Making a Template Once you have bent the strip, you can drill out the chassis mounting holes if you haven't already done so. Next, you can drill out the mounting holes for the motor face and a through-hole for the motor shaft. Make a template to show you where the motor mounting holes should go. You will always get better results if you make a template or guide before drilling the mounting holes. Although you can use cardboard for your template material, it's a lot easier to use a stiff piece of plastic cellophane (such as the kind toys are packaged with) or other transparency. Carefully use a hobby knife (an Xacto, for example) to cut

Figure 6-6 Using a lever to bend tough stock

an opening in the cellophane for the motor shaft (and shaft bearing if it protrudes) so that the cellophane can be placed flush against the motor face, and mark out the locations of the motor mounting holes.

Making a Good Impression One good alternative to drawing a template for mounting holes exists: Use wax to make an impression on the surface to be drilled. To do this, first take the thing you'll be mounting, such as a motor, and screw your mounting screws into the holes. You can put the screws in partially or all the way, but in either case, the screws should be set at the same depth.

Next, mark the drilling surface with a pen or pencil at the approximate location of the holes you'll be drilling. Then drip hot candle wax over and around the marks.

Once the wax has cooled but is still somewhat warm to the touch, place the motor or other object into the wax, screw-side down, in the desired position. Note that in the case of a motor, you will want to have previously drilled a hole for the shaft to pass through. The screws will make an impression in the wax at the exact location where the mounting holes should be drilled. Phillips head screws work well in this regard; they leave little cross-hair impressions in the wax that make nice drilling guides.

If you make a placement error, just scrape off the wax and try again. When using this technique on a metal surface, be sure that the metal is free of oils or grease before proceeding. You can use a degreaser to clean the work piece up before starting. Finally, you should obviously proceed with caution when using this technique on painted surfaces, and by all means, do not allow wax to get into any mechanisms lest you gum something up beyond repair.

Drilling Using your template or wax impression as a guide, drill out the mounting holes. You can use a hole saw or fly cutter to drill out larger holes for the motor shaft and bearing. This is easiest to do in a drill press. Since most hole saws are made for use with wood, exercise patience when using these tools. Back off frequently and give the work piece time to cool a bit before proceeding. The finished bracket should look like Figure 6-7.

Using L Brackets for Servo Mounting You can use L brackets for mounting servos as well as regular DC motors. Simply drill mounting holes to fit the servo mounting holes and drill a single hole to accommodate the servo shaft. You may want to make this hole wide enough to accommodate the swelling around the base of the shaft.

Rather than attempting to cut out a section to accommodate the entire servo face, we simply use spacers to keep the servo flush against the mount. You can use brass tubing or just even a couple of hex nuts as in Figure 6-8.

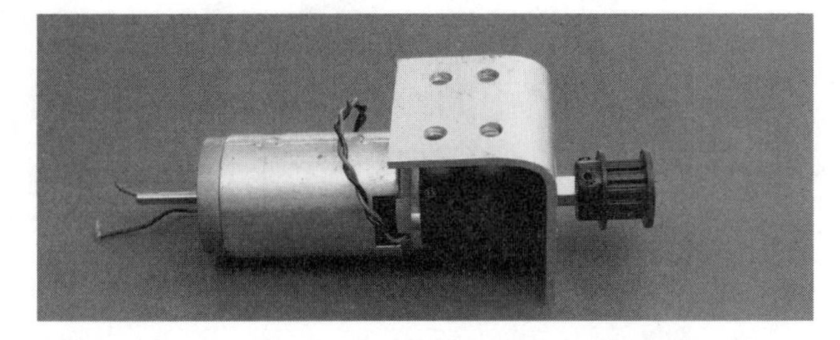

Figure 6-7 Completed L mount with motor attached

Figure 6-8 Servo mounted on homemade L bracket using spacers

Hardware Store Shelf Braces as Brackets If you can find one in an agreeable size, steel corner or shelf braces from the hardware store make ideal mounting braces for gear motors with square gearboxes and corner mounting holes (one of the most common configurations). They also require little effort to utilize (see Figure 6-9).

Usually, these come predrilled. You can use one set of holes for mounting to the chassis, but the holes for gearbox mounting will almost certainly need to be redrilled. As previously done, make a template with stiff, clear plastic for the mounting holes, making sure the holes are parallel to the sides of the bracket and that the holes on each bracket are at the same height with respect to the other (see Figure 6-10).

If you aren't happy with the selection of braces at your local hardware store, you can bend your own from suitably sized steel flats (also carried by most hardware stores) using the techniques described above for building your own L mount.

Figure 6-9 Corner braces make ideal motor mounts for many gearhead motors.

Figure 6-10 Motor with corner brace mounts

A Form-Fitting Metal Bracket This bracket can be employed when mounting cylindrical form-factor motors, and it does not require that the motor be equipped with mounting holes.

Starting with a relatively thin strip of metal—no thicker than 1/16 of an inch (1.5 millimeters)—form the strap around approximately 3/4 of the circumference of the motor housing, as in Figure 6-11a. Next, bend the free ends out 90 degrees to form "wings" that will carry the holes used for mounting to the chassis, as in Figure 6-11b. Be sure that when bending the wings, you use a vise or workbench edge as a guide.

The finished mount should look like Figure 6-12. Note that since the motor mounts are flush with the chassis, there will not be much clearance for a wheel attached to the shaft. You can prevent slippage under heavy loads by lining the inside of the curved part of the bracket with double-sided sticky foam tape.

Sometimes steel stock is coated with a light coating of oil to protect it from moisture. This can cause slippage problems, even when using foam tape, which will adhere poorly to the oil-coated metal. If you plan to use

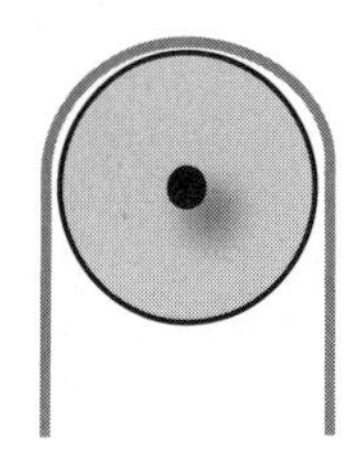

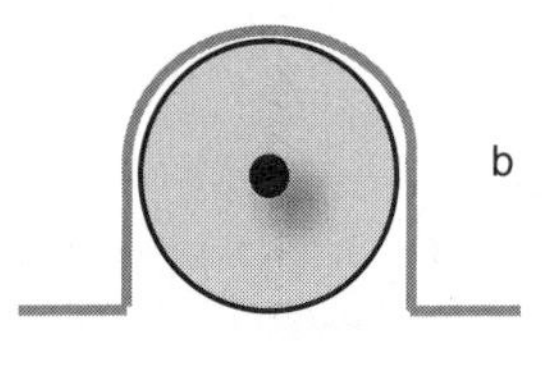

Figure 6-11 Building the form-fitting bracket

Figure 6-12 Finished form-fitting bracket (with motor)

steel straps for construction, you should remove this coating with a solvent. A number of readily available spray-on compounds can be used for cleaning automotive components suitable for this purpose.

A Wooden L Mounting Bracket

Surprisingly strong mounting brackets can be built from wood, which has the distinct advantage of being easy to cut and drill. The bracket in Figure 6-13 is built from 1/4-inch-thick (6 millimeters) birch plywood scraps; the original piece was purchased at a home improvement store.

The two pieces of the L bracket in Figure 6-13 are simply glued together with a good quality wood adhesive. For heavier-duty use, you can reinforce the bracket with a pair of small corner braces, also available at most hardware stores.

Other Ideas: U-Bolts and Hose Clamps

A couple other items from the hardware store can be used to quickly construct motor mounts. The U-bolts shown in Figure 6-14 can sometimes be used to fit a cylindrical form, if you are lucky enough to find one in the right size.

A better solution can be fashioned from large, screw-driven hose clamps (see Figure 6-15). These can be purchased at virtually all hardware stores and make great adjustable permanent or temporary motor mounts with just a little minor surgery.

Figure 6-13 A wooden mounting bracket

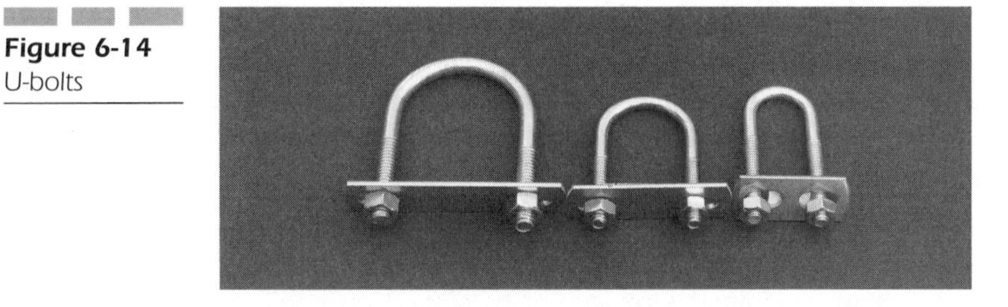

Figure 6-14
U-bolts

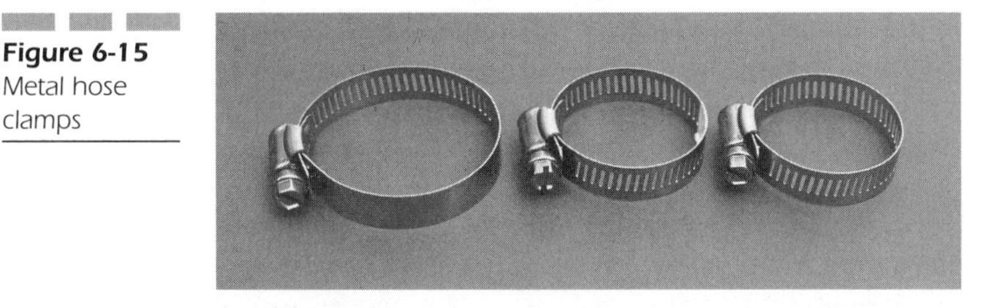

Figure 6-15
Metal hose clamps

Figure 6-16
A motor mounted with a modified hose clamp

Cut the clamp at the bottom, on the opposite side from the screw, and bend the ends out to form mounting tabs. Drill each tab for a bolt. Figure 6-16 shows a completed mount.

You can adjust the size of the clamp via the screw drive at the top. Using shims, you are not limited to cylindrical motors. Motors with square gear boxes can also be secured with these clamps as long as the clamps are used on the body of the motor itself and not the gearbox. You may need to use a shim if the top of the gearbox rises above the cylindrical motor body. For longer motors, you should use two clamps in parallel for extra support under heavy loads.

Fasteners for Motor Mounts

In general, we just use screws or bolts and nuts to attach mounts to a chassis. Even a humble 4-40 steel screw can stand up to immense loads. For most applications, the size of the screw or bolt is not critical.

If the bracket is mounted to the underside of the chassis and you'd like to be sure that the screw heads are set flush to the top of the chassis, you can use a countersink drill bit like those shown in Figure 6-17 to create a recess around the mounting hole into which a flat-head screw head will fit neatly.

Be sure to use washers to help keep any bolted assemblies tight. Unlike static assemblies, anything on your robot will be subject to continuous vibration, which will have a tendency to loosen your motor mounts over time. Helical spring washers, available from hardware stores, do a great job of keeping things tight.

Minimizing Shock and Noise As mentioned in Chapter 3, "Using DC Motors," you should at least consider shock mounting drive motors, especially if you intend to use your robot outdoors. Even asphalt roads can be surprisingly bumpy, jarring delicate components to the point that connectors loosen unexpectedly. Shock mounting can also minimize vibration from motors transmitted to the robot chassis, which without some intervening damping material, can actually amplify noise from the gearbox by acting as a soundboard.

Figure 6-17 Countersink bits

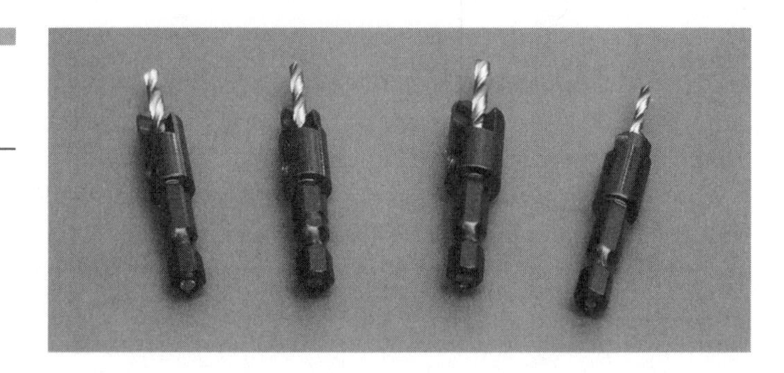

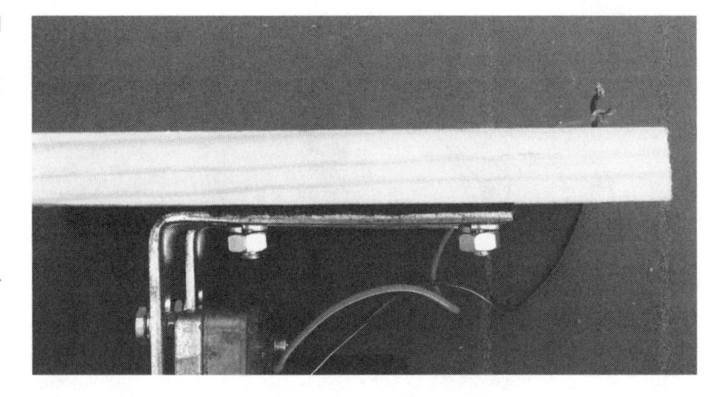

Figure 6-18 A motor mount with a damping pad sandwiched between the mount and the motor

You can cut down on noise and vibration by using a rubber washer or two on the bolts that connect the motor mount to the chassis. These can be purchased from hardware stores or plumbing supply outlets.

A more effective solution is to place a rubber sheet between the top of the motor mount and the bottom of the chassis, as in Figure 6-18. Although you can get sheets specifically meant to damp vibration, we have had luck using old mouse pads made of suitably stiff foam rubber.

Mounting Without a Bracket

Finally, it is possible (in some situations) to mount motors without using a bracket at all. Although adhesive compounds should be avoided, for light-duty or temporary applications you have a couple of options.

Double-Sided Foam Tape

Two varieties of foam tape are available. The first type can be bought at the hardware store and is often used when laying carpet. The 3-M corporation (the makers of Scotch Tape™) make a popular brand (see Figure 6-19). This tape is of medium strength, inexpensive, and easy to get.

Servo tape is a stronger version of the previous foam tape and is widely used in the RC car arena as a fastener for radios, spoilers, and other accessories. It can be purchased from any decent hobby shop.

Figure 6-19
Foam tape

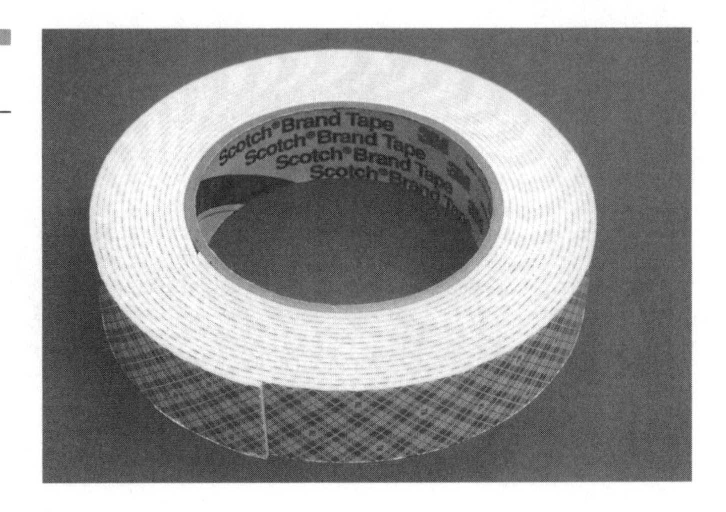

Both of these fasteners will work well when mounting modified servos to a robot chassis, and they can easily withstand loads of a pound or more without shearing off. Both types of tape are also useful for mounting sensors and other parts. Be sure that that the surfaces to be adhered are free of grease or oils before mounting. Use a little degreaser if necessary.

Temporary Mounting with Hook and Loop (Velcro®)

Finally, it is possible to use hook and loop as a motor mount, albeit a temporary one. Like foam tape, it's best to use hook and loop on items with flat surfaces to provide a decent contact area. Note that hook and loop doesn't tend to be very sturdy and is best confined to use as a temporary mount only. You can shore up items mounted via hook and loop by reinforcing cable ties.

Hook and loop *does* come in handy for mounting items to your robot that you would like to be easily removable. We use this material extensively for mounting everything from breadboards to radio data transceivers.

Most hardware stores carry hook and loop under the Velcro brand name in spools of adhesive-backed strips. One modest-sized spool lasts a surprisingly long time; stifle the urge to buy hundreds of feet of the stuff.

CHAPTER 7

Motor Control 101, The Basics

By now, you've seen all the DC motors we're going to talk about. You've seen some basic circuits and become familiar with the strengths and weaknesses of each of these motors. In this chapter, we'll show you how to control DC and stepper motors. There isn't anything more to say about the hobby servo motor when it comes to electronic interfacing, so we'll concentrate on DC and stepper motors.

Some Electronics Conventions We Use

To avoid confusion, certain conventions are assumed throughout this book, but in this chapter they will be especially important. We follow the convention that current flows from negative (the region of excess electrons) to positive (the region with a dearth of electrons.) We also use the convention that ground is 0.0V and is common throughout the circuit. We'll use the term V+ to denote the positive power line. For those of you new to electronics, we must warn you now, we'll be using some jargon that is unique to electronics and especially motor controllers. It's time you started learning it too if you want to read the data sheets. We'll introduce the jargon gently, we promise. There is more math here too, but you need only pay attention to it if you are doing sophisticated motor control design. It's strictly for the tech geeks out there that *really* want to know what is going on.

DC H-bridge Drivers: How They Work

The basic circuit for driving DC motors in both directions is the *H-bridge*. This circuit enables the motor to spin in either direction from a single supply (+ something and ground). Before we show you how to make one from scratch, let's see just how this circuit works. We'll assign *clockwise* (CW) and *counterclockwise* (CCW) arbitrarily to denote the motor going in each direction.

Figure 7-1 shows a common four-transistor H-bridge in its most basic form (don't build this; it's just for discussion!). To get our motor to rotate CW, we will turn transistors A and D on. Figure 7-1 shows the direction of current flow (from negative to positive) that will occur. To the motor,

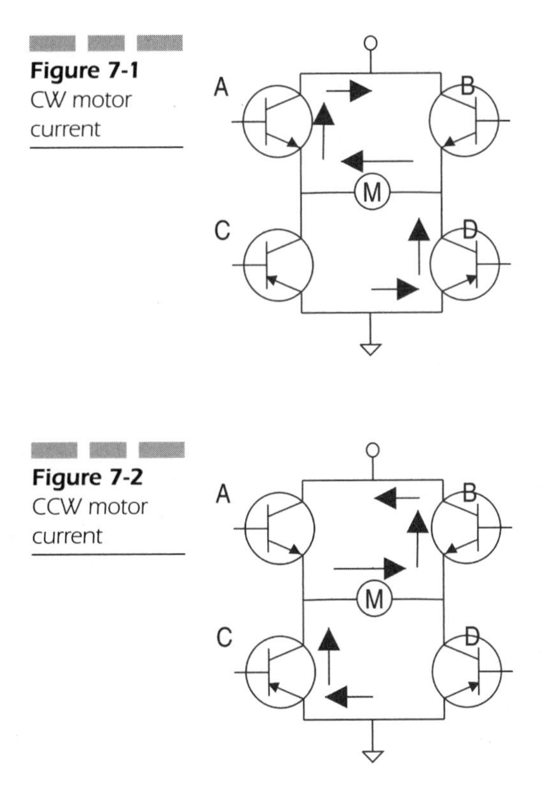

Figure 7-1 CW motor current

Figure 7-2 CCW motor current

the voltage on its right side is negative and the voltage on its left side is positive. Note the H pattern formed by the transistors and the motor. This is how the circuit derives its name. This H is also how the motor can appear to "see" both positive and negative voltages on its inputs by alternating diagonal transistors.

In Figure 7-2, we show that transistors B and C are turned on to make our motor spin CCW. Again, the arrows show the direction of current flow through the transistors and the motor. H-bridge functionality is as simple as that.

CEMF and Clamp (and Recirculating) Diodes

One of the most troublesome aspects of DC motors is that they are basically powerful inductors. It's time for another short electronics lesson.

One of the properties of an inductor is that it *resists* any change in current. This means that when you turn off power to an inductor, it tries to keep that current flowing. The faster the current is stopped (and the resultant electromagnetic field collapses), the harder it tries. It is not impossible for an amount that is 20 or more times the original voltage to appear across an inductor when its power is removed. The voltage that appears across the inductor (motor) is of opposite polarity from the original voltage and is called *Counter Electro Motive Force* (CEMF). CEMF is caused when the magnetic field generated by the original current collapses suddenly across the windings of the inductor (motor). We won't get any more involved with the physics of this phenomenon than this explanation. Usually, our transistors will not tolerate this extreme CEMF voltage. They are broken down, release their magic smoke, and often pop like firecrackers.

Fortunately, your circuits can be protected from CEMF. The most commonly used protection is the *clamp diode* or *recirculating diode*. Diodes used in this manner shunt the current caused by CEMF to ground or V+ so your electronics are protected.

Figures 7-3 and 7-4 show our original hypothetical H-bridge with clamp diodes added. Figure 7-3 shows the current path through the diodes when the switches for CW rotation are turned off. Note that current is flowing in the opposite direction during our CW rotation motor while it was powered on. Figure 7-4 shows the direction of current flow for our troublesome CEMF-generated current when we remove power from our CCW rotation motor. The clamp diodes protect the transistors from negative transients caused by CEMF by *clamping* that voltage to a safe 0.6V. It is important that these diodes be high-speed switching diodes; we'll tell you why in the next section.

Another Problem with CEMF: Transients on the Power Bus

A potential problem exists with the clamping diodes across the transistors. Current shunted to the power rail can cause voltage transients or *noise* to be generated on your circuit board. Further, if you are using *Pulse Width Modulation* (PWM) to control your motors, you get this CEMF current every time you cycle on to off. If the conducting transistors turn back on while the CEMF current is still flowing, there will be a brief period where the transistors create a direct path for current from

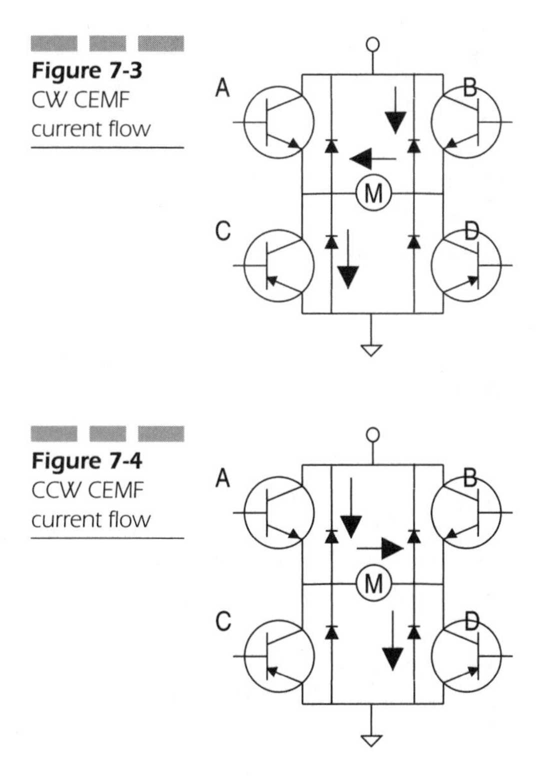

Figure 7-3 CW CEMF current flow

Figure 7-4 CCW CEMF current flow

V+ to ground. The length of time this occurs depends upon how fast your clamp diodes can shut off when they are reverse biased. This current is called the *shoot-through* current and is another source of noise. For these reasons, it is important that high-speed bypass caps (ceramic or tantalum) be located close to your H-bridge circuits to protect the rest of your electronics and even your battery. A bypass capacitor is just a capacitor intended to smooth out ripples (variations) and transients on the V+ power line. These bypass capacitors are usually .1 to .22 uF.

It's time for another short electronics lesson, this time on capacitors. A capacitor is a device that *resists* a change in voltage. A bypass capacitor attached to your power bus will sit there at the usual Vcc voltage level. When a voltage spike appears, the capacitor will quickly absorb the energy, trying to keep the voltage constant. When the voltage drops, the capacitor will discharge energy back into the circuit trying to keep the voltage up. So when your H-bridge spikes the V+ line, the bypass caps sitting there (usually .1 to .22 uF) will greatly reduce the effects of that current shunt.

A Simple Low-Current H-bridge Design

Here's a simple H-bridge design that will work for DC motors of 100 ma or less. It uses all the features we've explained earlier. The resistor values were chosen based on worst-case gain for these transistors and are quite conservative. You can create much higher current H-bridges by substituting higher-current transistors and diodes and using their correct biasing resistor values. We leave this as an exercise for the student.[1]

This H-bridge uses 2N3904 NPN transistors on the high sides and 2N3906 PNP transistors on the low sides. These transistors have identical switching speeds as well as maximum currents and gains. They are known as a *complementary pair* of transistors because their specifications are nearly identical. One hundred ma is half the rated maximum current for both transistors. This leaves us with plenty of margin so they won't overheat. The 1N5817 diodes have high voltage tolerances and can carry 1 amp of current, much more than we need, so again, there is a very high tolerance margin. Resistors of 1/4 watt or more and 5 percent tolerance will function perfectly. The end result should be a reliable H-bridge.

This H-bridge is quite simply the best, the most bulletproof and simple transistor H-bridge design you can build. It's simple to interface with because it has only two inputs (A and B) and you won't short power to ground if you turn on both sides because it's not possible to turn on the top and bottom transistors on either side at the same time. It's more efficient than any design that puts the PNP transistor on the top because it can't self-bias. Take our word on this. If you want, build this circuit, but swap the PNP and NPN transistors.

To use this H-bridge, raise side A to V+ and ground side B to spin in one direction (see Figure 7-5). Connect side A and B oppositely to spin the other direction. If you tie both A and B to V+ or ground, nothing happens and the motor freely spins unpowered. The functional truth table for the design is shown in Table 7-1. 0 means grounded; 1 means connected to motor V+.

---

[1]Hint: You need to know the *hfe*, or current gain of your transistor. Your base-emitter current multiplied by the hfe equals your collector-emitter current. Work backwards from your desired Ice.

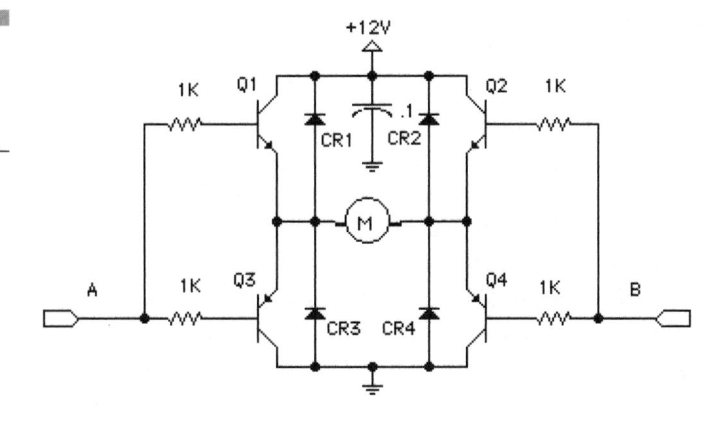

Figure 7-5 Simple transistor H-bridge

Table 7-1 H-bridge functionality

Result	A	B
Nothing	0	0
Spin CW	0	1
Spin CCW	1	0
Nothing	1	1

A Slightly Better Low-Current H-bridge Design

The main problem with the H-bridge in Figure 7-5 is that it can't be interfaced with a microcontroller or TTL or CMOS logic. It's voltage levels and current requirements are too high for a direct connection if voltages higher than 5V are used. We've solved these issues with a new circuit, designed to be controllable by a microcontroller. This circuit uses a 74LS06 to buffer the outputs of the microcontroller from the effects of a voltage and a transistor current that are too high.

Note also that this design requires a single direction input to change motor direction. However, you will notice that a motor is always turning; a "Nothing" setting isn't used. If you use a *locked antiphase PWM* from

1 kHz to 10 kHz to drive this circuit, you can have both directions and zero. An interesting aspect of locked antiphase PWM and this design is that the current through the motor is constant, no matter what the speed. This means that the torque is constant, and only the speed changes. We'll discuss PWM and what locked antiphase means later on. You'll just have to remember those concepts if you build this H-bridge.

For those of you that understand electronics a bit, you can add an active-low enable line to the circuit shown in Figure 7-6a. You have three inverting buffers left unused in the 74LS06 IC. If you take two of them, wire-or them with buffers A and C and tie these two new buffers' inputs together, then you have a disable line. (Wire-or refers to logically ORing two signals by use of open collector logic gates. Basically if any output is low, the output of all of the OR'd gates is low. This ORing of outputs will *only* function with open collector output devices, such as the ones referred to in this text.) When you pull the inputs high on this disable line, both Q3 and Q4 turn on, Q1 and Q2 turn off, and the motor will free spin. This will enable you to turn a motor full on in both directions and full off. You can also use sign-magnitude PWM by toggling the enable line. We'll show you how to do this; see Figure 7-6b for the full details.

Parts Lists for These Two Simple H-bridges

Table 7-2 shows the parts lists for these two H-bridges.

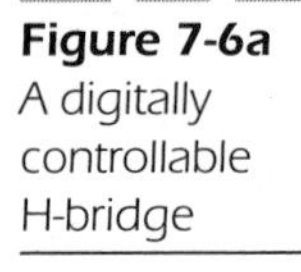

Figure 7-6a A digitally controllable H-bridge

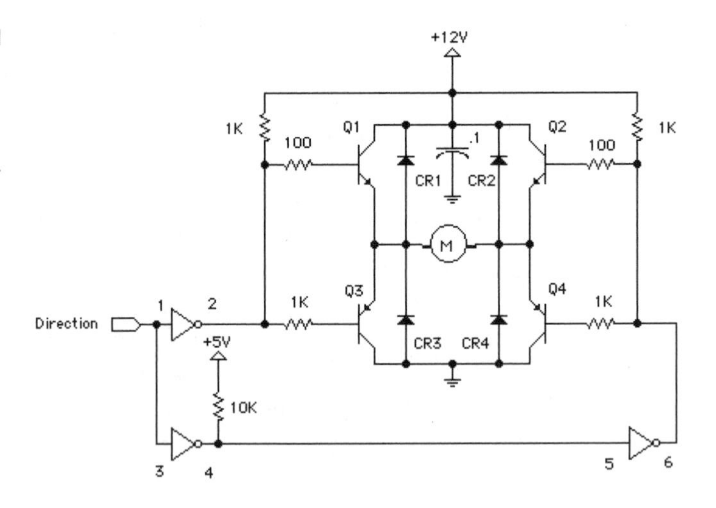

Figure 7-6b
An H-bridge with an enable line

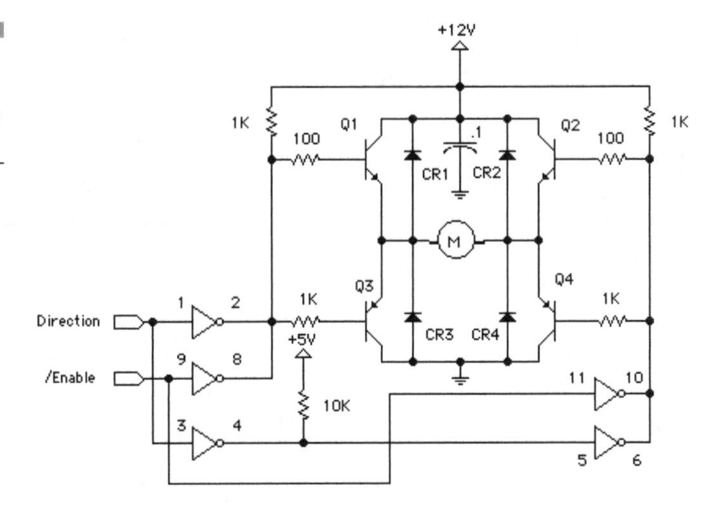

Table 7-2
Parts lists

Designation	Part number
U1	74LS06 OC buffer
Q1, Q2	2N3904 transistor
Q3, Q4	2N3906 transistor
CR1–CR4	1N5817 diode

Single-Chip H-bridge Integrated Solutions

It's time consuming to design and build an H-bridge, but several single-chip H-bridge designs can be found out there. In this section, we'll show you how to properlyconfigure and use them. Many of these driver ICs are called *quad push-pull* drivers. We'll show you how to control one or two motors with each of them and not worry about that "half-bridge" thing. All these drivers will support PWM control. Get ready for Show and Tell!

754410 and L293B or L293D Dual DC Motor Drivers

The L293B is a 4.5V to 36V quad push-pull (or dual H-bridge) driver (see Figure 7-7). It shuts down when it overheats and can supply 1 amp per channel (or per motor) current. The L293D is the same controller but has built-in clamp diodes. The 754410 is the superior replacement for the obsolete L293. The 754410 has built-in clamp diodes and a better power dissipation capability; it is otherwise functionally identical to the L293B (see Figure 7-8).

With any of these chips, if you are planning on more than 300 ma, it would be a good idea to stick on a heat sink. On both chips, the voltage on pin 16 (Vss-L293B/D, Vcc1-754410) can be 5V to minimize power usage for the logic side of the chip that doesn't need high voltage. You can simply connect the higher voltage to these pins if you wish. These chips consist of four push-pull circuits that are paired to control a DC motor. Each pair of outputs shares a single enable line. The operation of each pair of inputs is exactly as shown in Table 7-1. These are excellent chips to use for small robots using DC or stepper motors. We suggest that you use 1N5817 diodes for the clamp diodes with the L293B.

Figure 7-7 L293B typical use

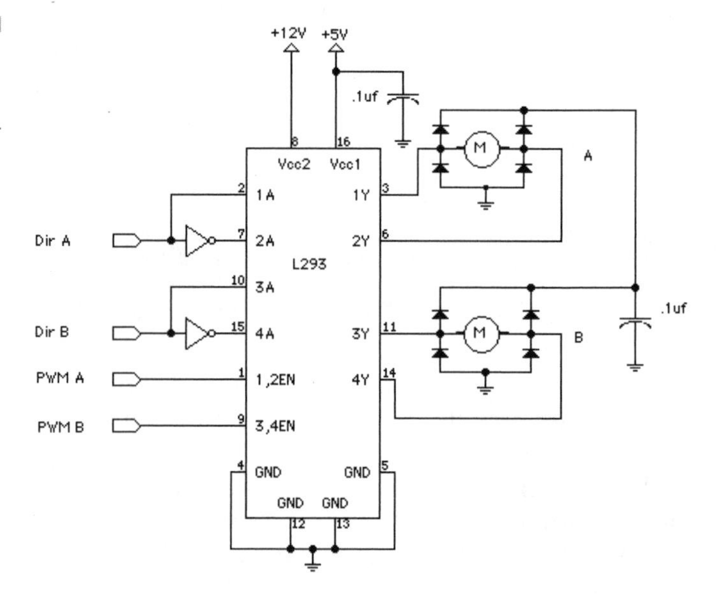

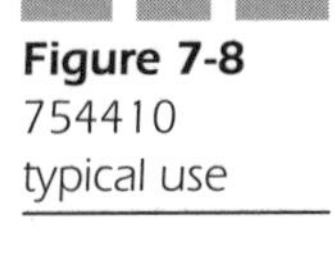

Figure 7-8
754410
typical use

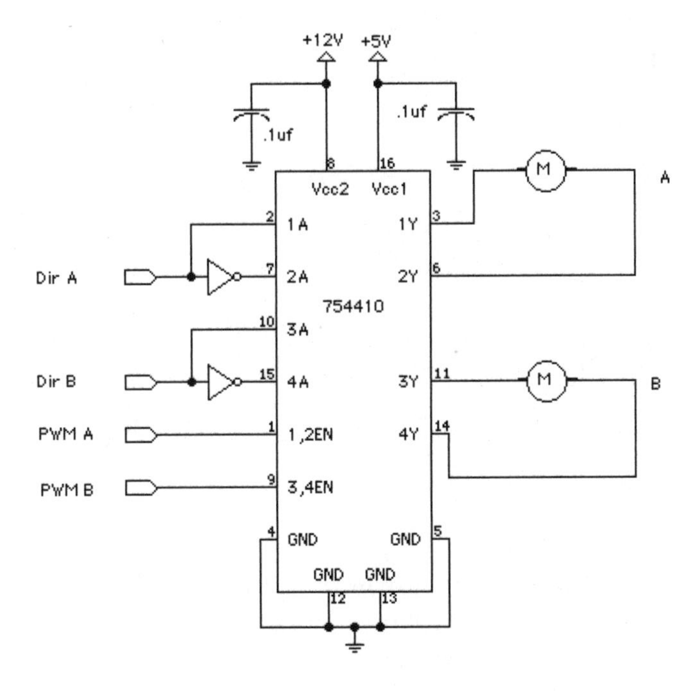

L298 Dual DC Motor Driver

This is one of the most versatile of the small motor driver chips. It's designed to drive either two DC motors or one bipolar stepper motor (see Figure 7-9). Voltages of about 6V to 46V are supported at up to a 2 amp steady current (3 amp maximum spike). The L298 will shut down when it overheats and it has a current-sensing capability, which our configuration does not use. The data sheet for the L298 will show how that is done if needed. The L298 requires clamp diodes mounted externally for proper operation. The L298 functions, looks, and acts exactly as the L293 and 754410 with respect to its inputs and how they operate the motor drivers. We suggest that you use 1N5822 diodes for clamp diodes with the L298.

LMD18200 Single DC Motor Driver

This is a very popular DC motor driver. The LMD18200 can drive a single DC motor or half of a bipolar stepper motor at voltages from 12V to

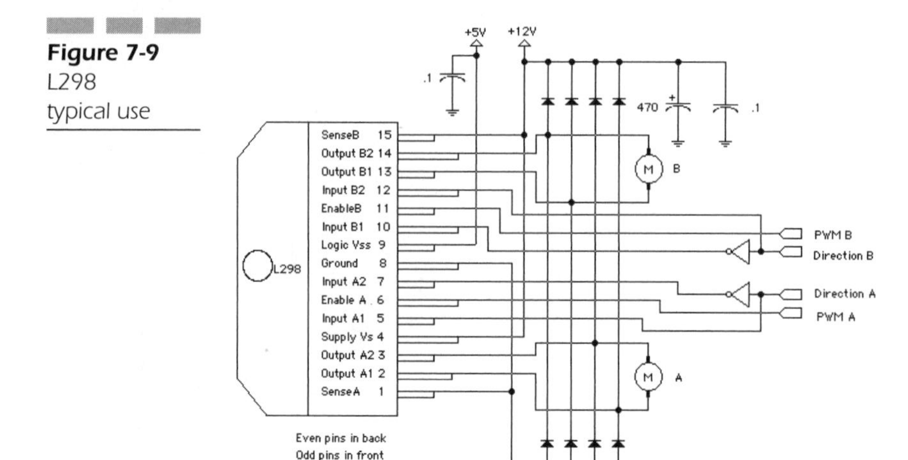

Figure 7-9
L298
typical use

60V and currents up to 3 amps continuous. It has internal clamp diodes and a current-sensing output and will shut down automatically if it overheats. The LMD18200 has a single direction pin and a PWM pin; regenerative braking is supported through a brake input pin. As long as the PWM used on this chip is 1 kHz or less, the internal capacitors used in the charge pump that creates the higher voltage for the high-side MOSFETs will work fine. At higher PWM frequencies, .01 uF capacitors need to be added between pins 1 and 2 and between pins 10 and 11. Since it doesn't hurt to have these bootstrap capacitors there at any PWM frequency, we show them connected in Figure 7-10.

UCN2998 Dual DC Motor Driver

This is a compact and simple H-bridge driver (see Figure 7-11). Few external components are required because the UCN2998 has internal clamp diodes. This chip will provide 2 amps of continuous current and 3 amps of peak current at voltages from 10V to 50V. It has an active-low (which means that a logic 0 turns it on and a logic 1 turns it off). enable and phase (direction) input for each motor driver for controls. The UCN2998 will automatically shut down if it overheats and uses a single voltage supply. This is the simplest dual-motor controller to use that we detail. No external components are required and each H-bridge has a single direction and PWM input pin. Each H-bridge has a current sense output, but we'll not use it in our designs.

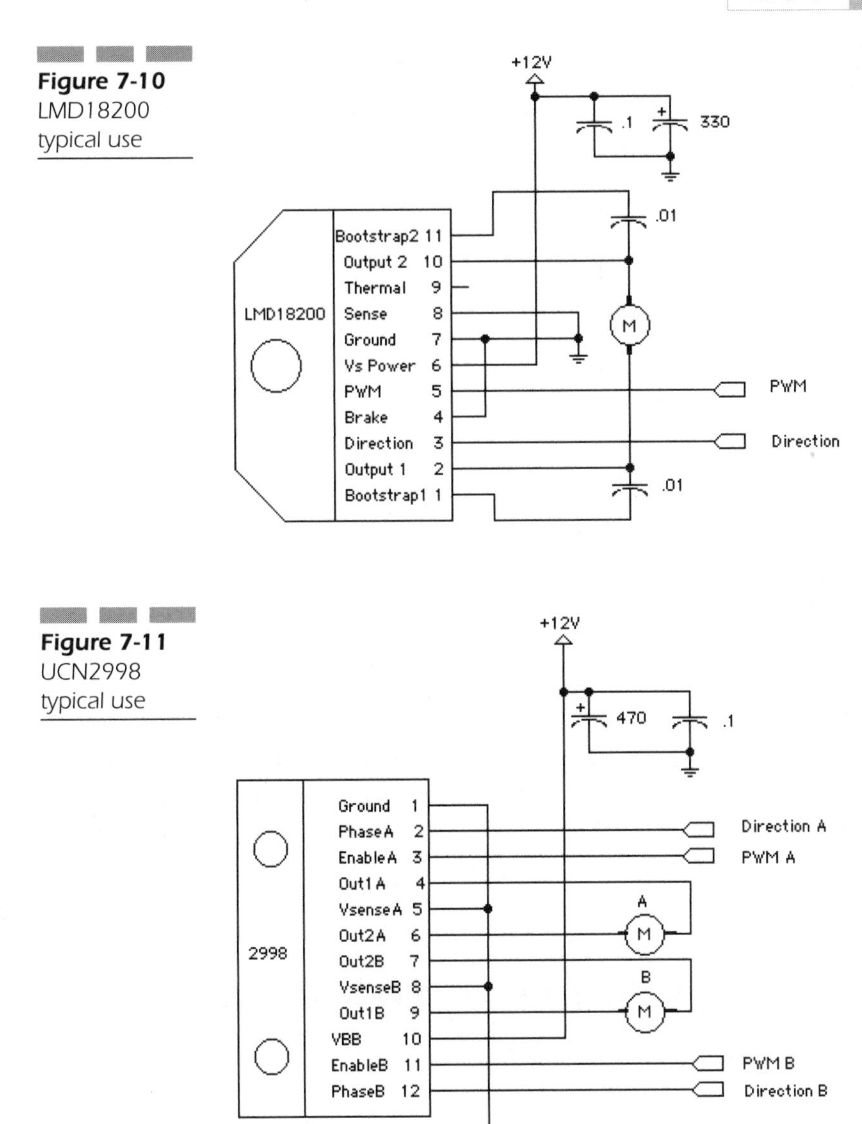

Figure 7-10 LMD18200 typical use

Figure 7-11 UCN2998 typical use

UCN3951 DC Motor Driver

The UCN3951 comes in two forms: the power DIP SW series and the 16-pin DIP SB series. This is currently an obsolete chip, but it is commonly found in surplus outlets. The UCN3951 has the capability to limit current with an internal PWM circuit; this capability is commonly called

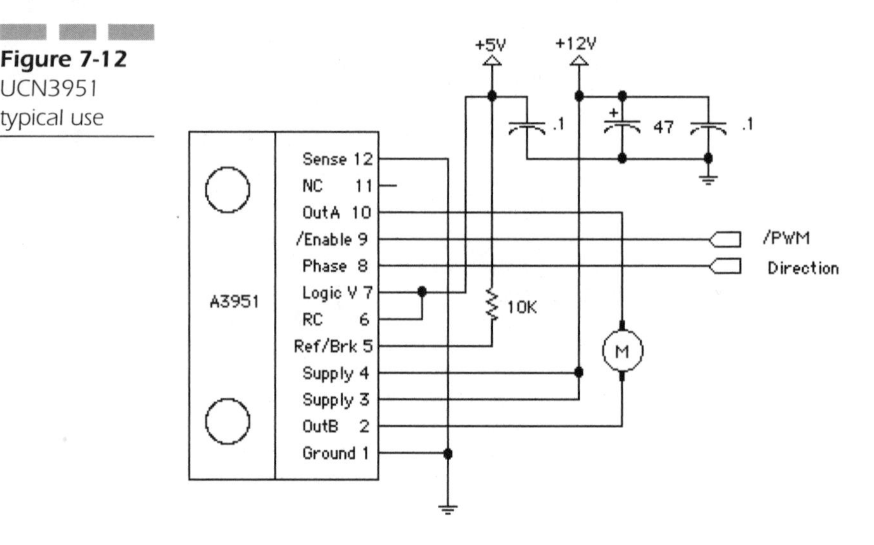

Figure 7-12 UCN3951 typical use

a *chopper / stepper driver*. You don't need to use this functionality, and our example circuit does not use it. The UCN3951 can drive two DC motors at 2 amps of continuous current and 3 amps of peak current at a 10V to 50V supply voltage. It has internal clamp diodes and will shut down if it overheats. It requires a logic supply (5V) and motor supply, and uses an active-low enable signal (see Figure 7-12).

High-Current DC Motor Controllers

Sometimes you want more than 2 or 3 amps of current. At these times, you need a high-current controller. High-current H-bridges are complex and much more costly than the devices we have discussed so far. You have two choices: purchase a costly device or build it yourself. We examine each choice.

In most cases, you simply cannot build your own high-current DC motor driver for less money than it costs to buy one. There are exceptions, but only if you are a competent electrical engineer with a nicely stocked junkbox.

Commercial High-Current H-bridges

A number of options are available for really high current DC motor controllers, but most of these are *way* out of the price range of the hobbyist. A few options are at least reasonably affordable (meaning under $500 each). The *radio-controlled* (RC) car *electronic speed control* (ESC) and ESC devices built with the robotics community in mind whose manufacturers focus on motor controllers.

RC Car ESCs A great variety of RC car ESC devices exist (see Figure 7-13). Most RC radio manufacturers have one or more models, and certain companies specialize in them. Typically, an RC motor controller will enable voltages from 6V to 8.4V and up to 60 amps peak. Some specialized RC ESCs will enable up to 14.4V in a bidirectional controller. Many RC ESCs are forward- and brake-only because they are for racing, which does not allow a reverse; these are often rated at up to 100 amps peak.

RC ESCs can be had for as little as $40 for basic units limited to 10 amps or so. The best bidirectional controllers are up to $90 and can handle up to 60 amps. Before you get excited though, remember that these motor controllers are rated in the number of cells supported. Each cell is standardized as 1.2V, which is the voltage of a single *Nickel-Cadmium* (NiCad) rechargeable battery. Most support 5 or 6 cells (6V to

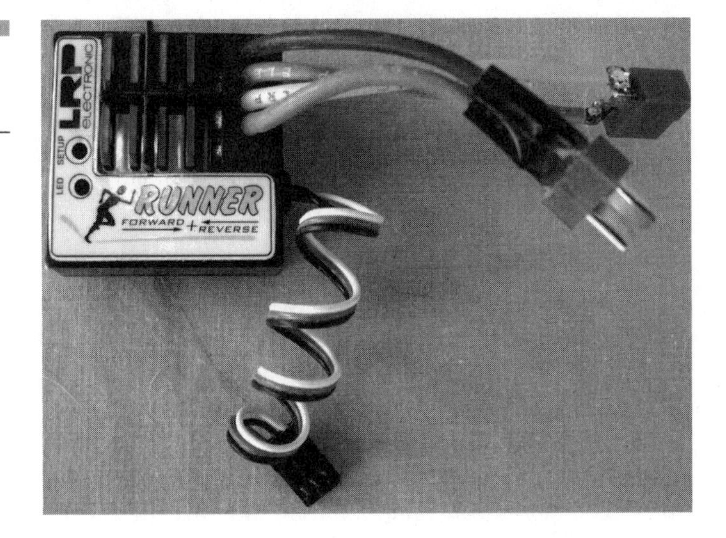

Figure 7-13 A hobby RC ESC

7.2V) with a rare seventh cell and a *very* rare 12-cell dual motor model made by Novak®.

If these voltage limits are acceptable, then RC ESCs are a great motor controller for the robot hobbyist. The RC motor controller is controlled exactly the same way that an RC servo is controlled (refer to Chapter 4, "Using RC Servo Motors") with a 1 to 2 millisecond PPM coded signal. With one *input/output* (I/O) line, you get direction and speed to the resolution of your PPM encoding controller.

Be careful which model you buy, however. Some RC motor controllers have a safety feature that does not allow you to switch from forward to reverse immediately. This feature protects the gear train of an RC car from sudden shocks. Both high-end and low-end RC motor controllers can have this feature. The more expensive models may have a configured mode that can disable the safety reverse lockout. The least expensive models may not even have this feature in them. Read the labels carefully and ask questions of the hobby store clerk or mail-order sales department if you're not sure. Going to the RC car newsgroup is a good idea as well; they can answer almost any question you have to ask. Check out the USENET group rec.models.rc.land for your RC controller questions.

Vantec® DC Motor Controllers Sometimes you want high voltage and high current, but you don't want the limitations and oddities of the hobby RC motor controller. In that case, you want a commercial motor controller, and the 800-pound gorilla in that corner is Vantec (see Figure 7-14). Vantec motor controllers run the gamut from 14 to 75 amps, but all share one similarity: They're not cheap. They do have great features, however. In general, the following features are in all their DC motor controllers:

- Highly configurable dual-motor mixing modes for boats, subs, tanks, and so on
- High current and rugged construction
- Good instructions and wire gauge selection charts

These controllers are popular with the BattleBot® and RobotWars® enthusiasts for their small size, high power, and good durability under somewhat difficult conditions (like being hurled into walls, falling into pits, running into things that don't get out of the way—that sort of thing).

Solutions3® Motor Mind C and ICON H-bridge Controllers Solutions3 has motor controller boards that are in the medium- to high-current range, as we define them for robotics.

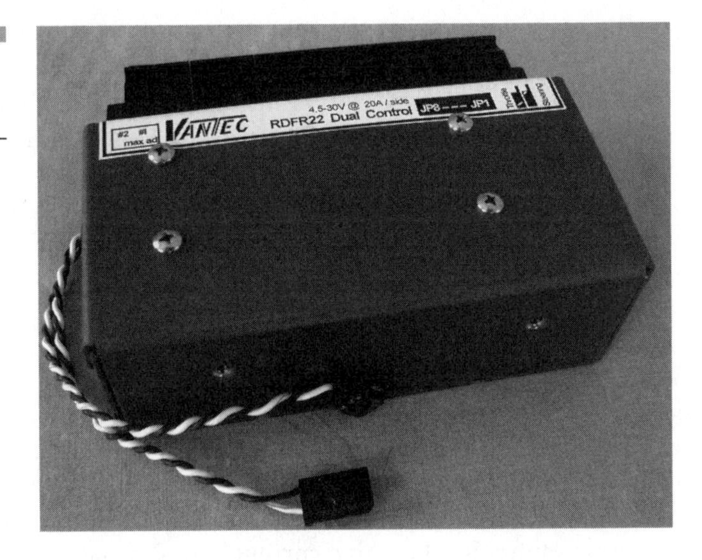

Figure 7-14 Vantec speed control

The *Motor Mind C* can handle from 10V to 24V and up to 4 amps continuously for two motors. It can be configured for control via PWM, hobby RC PPM, or a serial data link. Up to 2.5 amps, motors can be used without cooling, but a fan is required at the 4-amp mark.

The ICON H-bridge is a MOSFET motor controller that will drive a single DC motor up to 40V and 12 amps with active (fan) cooling. It is controlled via a serial data link to an onboard microprocessor. You can monitor the current draw and even set overcurrent protection fuses in the software.

Other commercial DC motor controller manufacturers exist, but this selection is the most cost effective for the hobbyist. Of course, someone may pop up after we publish the book. That's the chance we take even writing about it.

Do-it-yourself (DIY) High-Current DC Motor Controllers

For those of you out there that like a project that's challenging or that don't have the disposable income required to buy a high-current controller, there is another option: Do it yourself. In this section, we've got a couple of projects you can build for 5-amp to 20-amp motor controllers that are simple and inexpensive to build.

15V, 5-amp Transistor H-bridge This circuit should look very familiar; it's the same one used in Figure 7-6, but now we're using TIP120/TIP125 TO220 complementary Darlington transistor pairs with far higher current capabilities (see Figure 7-15). This design will work very nicely with 6V to 24V systems. It will be pushing these transistors to have them work at 5 amps at 24V; 4 amps will be safer at that voltage. At any current above 1 amp, you should use individual heat sinks on each transistor—the more current, the bigger the heat sink. We suggest you use SB550 diodes as clamp diodes.

Relay and MOSFET 20-Amp Motor Controller In reality, this design can be for any current you can find a relay and MOSFET rated to handle. These circuits can be pushed very hard current-wise if you make sure that you have the MOSFET turned off when you toggle the relay—any easy enough task to accomplish in software. This eliminates relay contact arcing when high currents are flowing.

The basic idea behind this controller is that a MOSFET is used to switch power to the motors on and off (thus providing the PWM signal), and a DPDT relay controls the *polarity* of the current, which enables us to do away with the complexity and high part count of a high-current H-bridge design. In our design, we use an opto-isolator to completely isolate the logic from the potentially noisy motor control circuitry. This is not required, but recommended, especially for controlling larger motors.

These circuits work extremely well and are among the simplest high-current controllers you can construct. Relay/MOSFET controllers have a

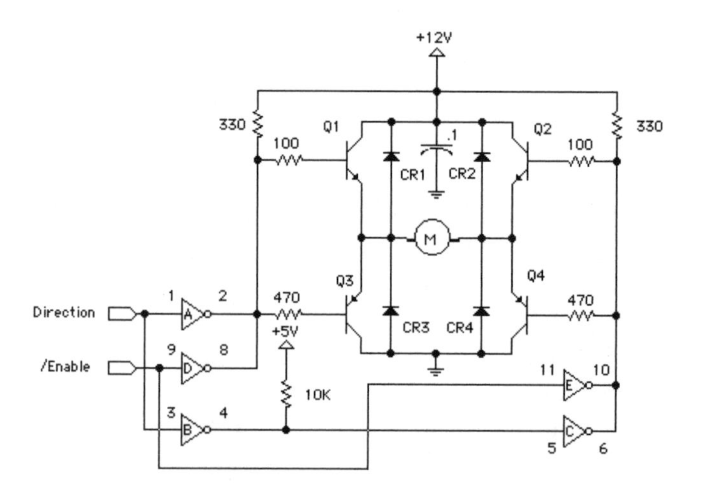

Figure 7-15 A 5-amp transistor H-bridge

couple of minor disadvantages, however, which you should bear in mind when deciding whether this design is suitable for your robot. These disadvantages, as might be expected, revolve around the relay:

- Any circuit incorporating a relay can be presumed to have a shorter working lifetime than its solid state equivalent. In practice, however, you will find that you needn't be bothered much about this aspect of our design. If you're careful to avoid arcing as outlined earlier, you should see *years* of use with no problem. Still concerned? Incorporate a socket in your implementation to ease the replacement of the relay.
- Being mechanical, the relay can potentially be affected by excessive vibration and shock. Although not ordinarily a problem in most environments, it might be cause for concern in a BattleBot or other similar scenarios where mechanical trauma is the norm.
- The relays will click when they are activated to change polarity. Depending on the relay, the click can be surprisingly loud. This can be a minor annoyance for those of you with fragile nervous systems or for those who prefer stealthier robots.

As shown, the circuit in Figure 7-16 uses an International Rectifier IRLZ34N MOSFET, with an absolute maximum current capacity of

Figure 7-16 Relay and MOSFET motor controller

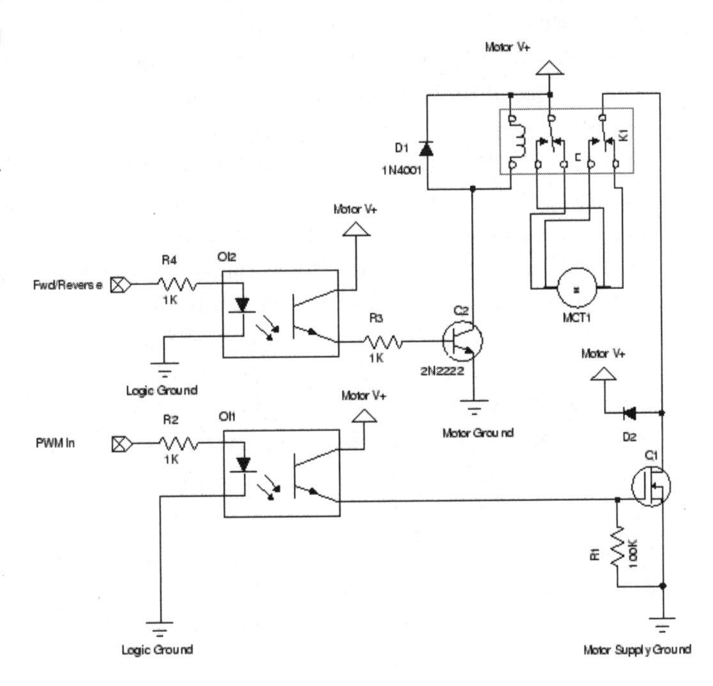

30 amps. You may substitute any N-channel MOSFET with a sufficient current rating. Use a device with a low Rds(on), which is the resistance from the drain to the source of the MOSFET. Low Rds(on) devices will carry far more current and barely even get warm while doing it. The relay can be any DPDT-type capable of carrying adequate current.

A couple of these components are fairly critical:

- **D2** The job of clamping diode D2 is to protect the MOSFET from the CEMF generated by the motor as the power is switched on and off via PWM. Because the CEMF voltage can be substantial—potentially hundreds of volts—and because most PWM implementations switch at relatively high frequencies, a heavy-duty Schottky-type fast recovery diode is a requirement here. Suitable devices usually come in a TO-220-style case (to which a heat sink can be attached) and have two or three leads. One suitable candidate includes the International Rectifier 20L15T (Digi-Key part number 20L15T-ND), but any number of parts will fit the bill. Do *not* attempt to use a garden variety 1N4001 or something similar. Not only can these devices not take the current levels they are likely to encounter, but they are not made to operate much beyond 60 Hz, which will be too slow for most PWM implementations. Such diodes will almost always fail rather quickly (and sometimes rather spectacularly).
- **OI1** The opto-isolator used to switch the gate of the MOSFET is a second critical part. Because it turns the gate of the MOSFET on and off, we want to choose a part with very fast rise and fall times. This ensures that the MOSFET is "snapped" on and off, rather than "ramped" on and off. When a MOSFET is not fully turned on, the drain-source resistance is relatively high, which causes the MOSFET to heat substantially. Look for parts with a rise/fall time of well under 3 to 5 microseconds. Suitable candidates include the NEC PS2501-1 (single channel) and PS2501-2 (dual channel). These inexpensive parts are available from Digi-Key as part numbers PS2501-1-ND and PS2501-2-ND, respectively. Note that OI2 is less critical, but you can save board space by using a double-channel version for both relay and MOSFET switching.

How Much Current Can a Wire Carry?

Wow, that's a good question, and one that very few folks remember to ask. The fully detailed and qualified answer can be quite complex and lengthy—boring too. So we'll give you the rule-of-thumb-answer and some qualifiers in the form of Table 7-3.

This data is for single-stranded copper wire at an ambient room temperature (77° F, 25° C) in the open air. The table uses the *American Wire Gauge* (AWG) standard for wire size. Every 3 AWG doubles/halves the cross-sectional diameter of the wire, hence its resistance to current, so you can gauge above and below this chart with that rule of thumb. This information has been gathered from a variety of sources, rounded off, and cross-checked between them. Hopefully, everyone isn't wrong. If your wire is stranded, that is, has many smaller wires in it, then those smaller wires clumped together have a higher current rating than a single strand of the same size would. According to one source, that derating factor is 0.8 for 2 to 5 strands, 0.7 for 6 to 15 strands, and 0.5 for 16 to 30 strands.

Table 7-3
Wire gauge and current

AWG	Amps
10	30
12	20
14	15
16	10
18	5
20	3.3
22	2.1
24	1.3
26	0.8
28	0.5

Pulse Width Modulation (PWM): What It Is and How to Use It

PWM is the most efficient means by which a DC motor's speed can be controlled. The only other way to control a DC motor's speed is by controlling its current, which is highly wasteful of energy and incidentally the device that is controlling the current gives off a lot of heat. PWM works on the principle that a motor that is on 20 percent of the time has a lower average voltage and current than a motor that is on 40 percent of the time. When the motor is not being powered, it is not using any energy from its battery; this is more efficient. This is all there is to PWM; it's simple and there is no magic. It's a good way to reduce the power required to drive DC motors. The driver chips run cooler, the motors run cooler, and your battery lasts longer. It's all good news. All you have to do is generate the PWM signal, which is explained in Chapter 9, "Electronics and Microcontroller Interfacing."

A PWM signal can be generated in two ways: one is via a fixed frequency with a variable duty cycle and the other is through a fixed pulse width variable frequency, often called *Pulse Frequency Modulation* (PFM). The former is simple to visualize. Imagine a fixed time period that consists of the time between when the waveform goes from low to high to the next time it goes from low to high. Now change how long the waveform stays high. The longer its high, the higher the duty cycle (see Figure 7-17). The latter simply adds in more fixed-width high pulses in

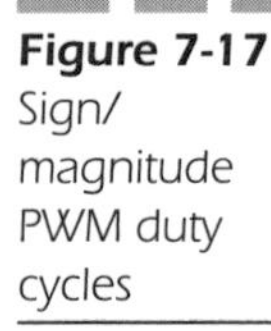

Figure 7-17 Sign/magnitude PWM duty cycles

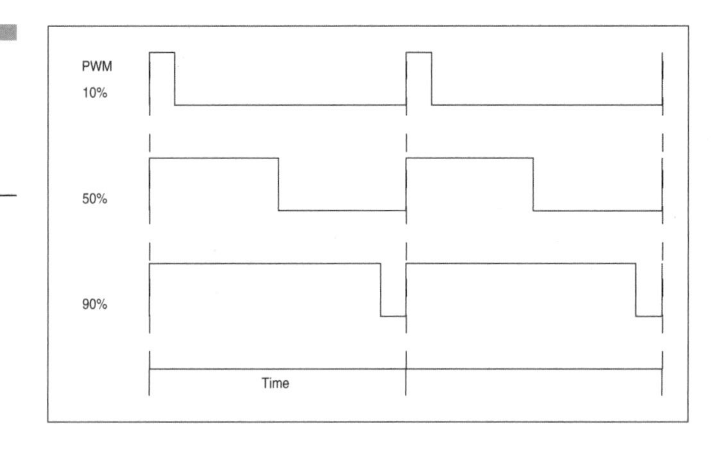

a given time period. This latter technique is somewhat frowned upon because it generates many different frequencies and may cause some motors to resonate, which causes noise and vibration. A fixed-frequency variable duty cycle is the mode used most frequently.

Two types of PWM are used with DC motors: sign-magnitude and locked antiphase. Each of these techniques has its pros and cons.

Sign-magnitude PWM

Sign-magnitude PWM gets its name from the control lines used to make the motor go. One line is the direction control (the sign, positive or negative); the other control line is the magnitude, which is the percentage of the on time during the time period of the PWM signal. This on time is called the *duty cycle*. Figure 7-17 shows the relationship between various duty cycles for PWM. Some of our motor driver chips have an active-low enable line. Active-low means that something is turned on when that input is pulled to ground and not V+. In these cases, the PWM signal is reversed so that on is defined as 0V.

This is a simple PWM to implement; all you need are one or two direction lines and a PWM enable line. To disable a motor you simply stop sending PWM and leave this line disabled. You can use the current sense lines of your H-bridges to control them with sign-magnitude PWM.

Locked Antiphase PWM

Locked antiphase PWM uses a single I/O line to generate forward/reverse and amplitude information for an H-bridge driver. Locked antiphase works with any H-bridge that can be configured with a single direction line and a PWM enable line. You enable the H-bridge and connect your PWM signal to the direction line to operate in locked antiphase mode.

You need to look at the output waveforms to understand locked antiphase PWM control. When the duty cycle is at 50 percent, the motor receives the same current to turn CW as it does to turn CCW. The net sum is 0; the motor doesn't move. If a 25 percent duty cycle PWM is sent, then side A (for instance) is on for 25 percent of the time and side B (for instance) is on for 75 percent of the time. The sum of the currents will cause the motor to turn in that direction. If a 75 percent duty cycle is sent, then side A is on for 75 percent of the time and side B is on for only

Figure 7-18 Locked antiphase PWM

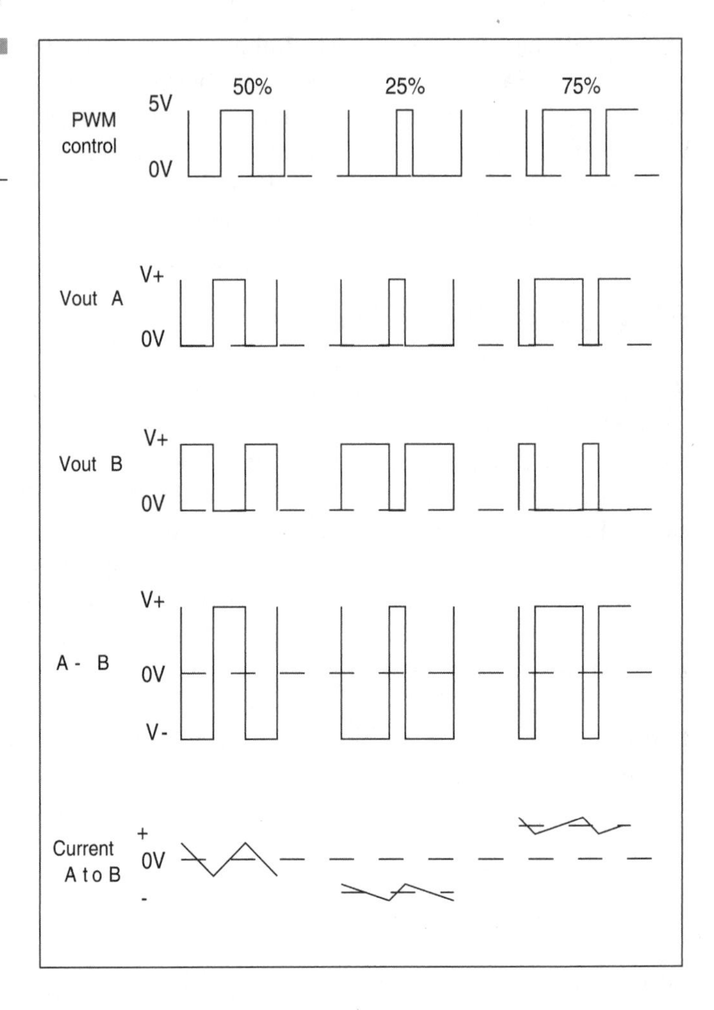

25 percent of the time. The net sum in this case causes the motor to turn the other direction. See Figure 7-18 for a visual explanation of this control mode.

Locked antiphase enables swift motor changes and minimizes shoot-through because two transistors are *always* on at any given time, so there is always a path for current to flow through. You only need one control line for direction and magnitude control. The downside to locked antiphase is that you can't use current sensing because no polarity is associated with the current flow from that measurement point.

Which PWM Frequency Is Best?

This is a question that is difficult to answer. Higher-frequency PWM is less likely to cause a mechanical resonance in motors. A resonance is a vibration that a mechanical structure, either through design or chance, is tuned to vibrate at. Lower frequencies may resonate and cause the motor to vibrate or even "sing" (you can hear a whine when the noise of the motor is quiet). The characteristics of your motor will determine how high a PWM frequency you can use. If the resistance in the motor winding is high compared to the inductance (a unit of magnetic resistance to current) of your motor winding, then your motor will not generate its maximum current during the time that the PWM signal is on and you won't reach the maximum speed or efficiency. Most motors do not come with a inductance specification so you just have to guess.

Another limiting factor to which PWM you use is the software or hardware you are using to generate the PWM. In fact, your software is usually the limiting factor; it will run out of gas long before your motor's specification limits will. Regardless, we don't recommend any PWM frequency lower than 1 kHz; your motor will be very choppy and noisy at that low of a PWM value. Unfortunately, at low PWM frequencies your motors will have a tendency to sing near the 50 percent PWM duty cycle. We've found that higher frequencies (8 kHz and up) eliminate this singing. In fact, most of the small DC motors that we tested ran quite smoothly at 10 to 20 kHz with no annoying noise (that we could hear) being generated.

If you really want to calculate the highest PWM frequency you can use for your motor, and you actually *have* your motor's specifications to that level of detail, the following is the formula that will give you what you desire. You want the left side of this formula to be much greater (>>) than the right side, which is to say 10 times more. If it isn't, then your motor current will not peak quickly enough and the motor won't generate its maximum torque.

$$2\pi f L >> R$$

Here f is the switching frequency, L is the armature inductance, and R is the armature resistance. Armature resistance can be measured with your *digital volt-ohmmeter* (DVM). Without an inductance bridge (a neat device designed to make this kind of measurement), you just can't find L if it's not given to you.

Snubber Network: What Is It and Why Use It?

A snubber network can be used in place of the clamp diodes or in addition to it. A snubber limits the rate of change of voltage across the motor to dampen (or snub) the CEMF generated by the motor's inductance between switches of the PWM motor control. A snubber network consists of an RC circuit where the capacitor limits the rate of change of voltage and the resistor simply limits the peak current flowing through the transistor when it turns back on.

Initially after the transistors shut off, the voltage across the snubber will jump to the value determined by the current flow times the value of the resistor. Then it will increase at a rate determined by the charging time of the capacitor. To properly size a snubber network, you need to choose a resistor value whose maximum motor current will produce a voltage less than the power supply voltage. If the resistor is too large, then the snubber network won't limit the voltage rise until the voltage is already higher than the power supply, which defeats the purpose. Choose your minimum voltage Vs_{min} carefully; your peak current I_{peak} is determined by the motor and any current limiting that your H-bridge circuit has implemented. Here is how you choose the resistor formula:

$$R_{max} = \frac{Vs_{min}}{I_{peak}}$$

The snubber capacitor is calculated from the peak current and the rise time of your switching circuit. The rise time, d_t, is given as a specification of your H-bridge chip or switching transistors and MOSFETs. The voltage change, d_v, is known by your choice of motor voltage minus any voltage drops in your driver transistors. These drops are also given in your specifications documentation for your H-bridge and transistors. I_{peak} can be estimated by dividing your motor max voltage (remember voltage drops in the H-bridge) by the winding resistance, which you can measure with your DVM. Here is the formula to use:

$$C = I_{peak} \frac{d_t}{d_v}$$

A properly implemented snubber network will reduce the heating caused by recirculating CEMF currents in your driver transistors and it can smooth the motor responses to speed changes.

Additional Stepper Motor Driver Chips

In Chapter 5, "Using Stepper Motors," we gave many circuits to control and drive stepper motors. Here we will add a couple more that are handy when you are using controller chips that don't have built-in drivers. These are quite useful as single-chip solutions for stepper drivers. They're much easier to wire up than a bunch of individual transistors, diodes, and resistors.

A Unipolar Stepper Driver Using the ULN2003 and ULN2004

The ULN2003/ULN2004 is a 16-pin package with seven 500 ma drivers, each having clamp diodes internally connected to their outputs. The 2003 is TTL- and 5V CMOS compatible; the ULN2004 is 6V to 15V CMOS compatible. Both can handle up to 50V for the coil voltage.

As you can see in Figure 7-19, this chip is easy for interfacing a stepper motor with a controller that cannot directly drive a stepper motor

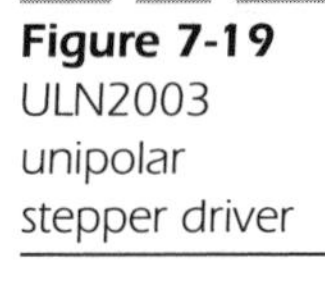

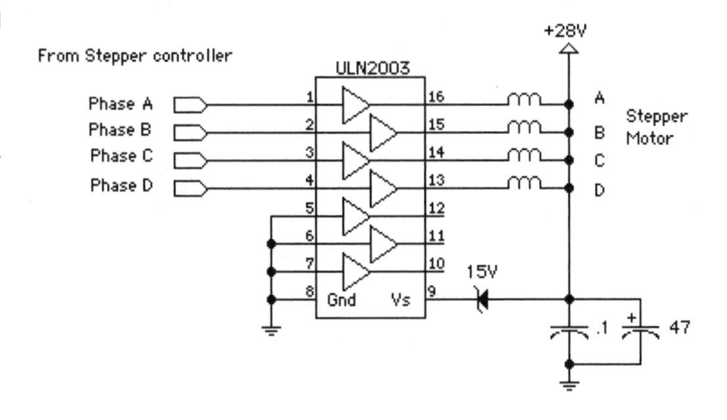

Figure 7-19 ULN2003 unipolar stepper driver

(such as an OOPic, for example.) The Zener diode in the schematic boosts the flyback voltage, causing a faster current decay. This is supposed to improve the stepper's performance according to Allegro Micro®. We recommend a 1N3024 Zener diode.

A Unipolar Stepper Driver Using the UDN2540 and UDN2544

These drivers are high voltage and high current inductive load drivers. They are perfect for driving the four unipolar stepper motor windings. Both of these chips (the UDN2544 is now obsolete but still commonly available) are rated at 50V and have up to a 2.5 amps current peak per winding (1.8 amps constant current). Both chips have internal clamp diodes. Figure 7-20 shows the simple hookup between a stepper motor and a controller that cannot drive a stepper directly. The FerretTronics FT609 serial stepper controller (the schematic is shown in Chapter 9) has a special mode just for the UDN2544 driver. As with the ULN2003, a 15V Zener diode is used to improve stepper performance. The UDN2544 has inputs 2 and 3 inverted for some reason that is unknown to us.

A Bipolar Stepper Driver Using the L298

The L298 can also be used to drive bipolar stepper motors. With a 2-amp peak current and up to a 46V rating, it's a strong driver in a small package. Current sense outputs are available and the L298 has a thermal

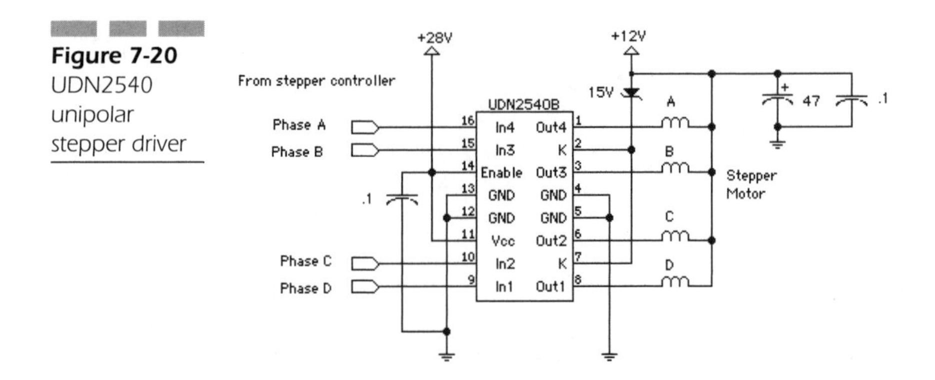

Figure 7-20 UDN2540 unipolar stepper driver

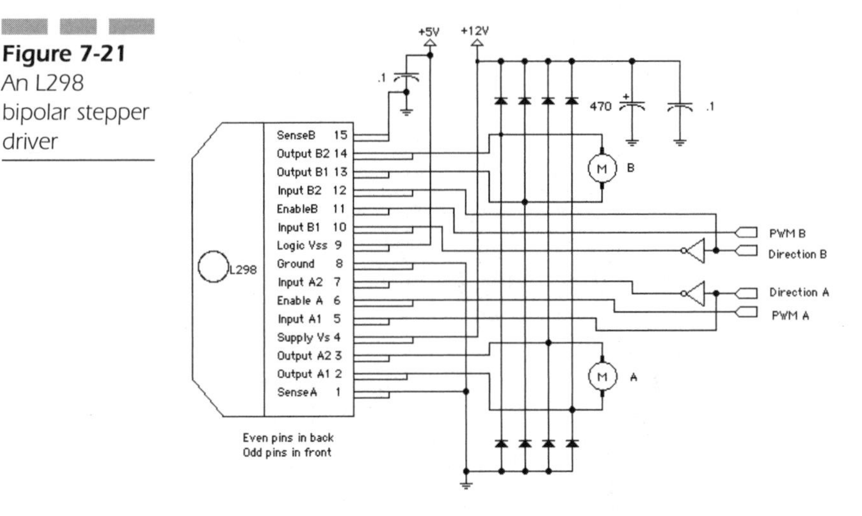

Figure 7-21 An L298 bipolar stepper driver

shutdown capability. External clamp diodes are required for proper operation. The L298 has a peer chip called the L297; together these chips provide a full chopper/stepper drive circuit for the high-current stepper motors often used with CNC mills and lathes—but that isn't our topic for this book. Figure 7-21 shows how to connect the L298 to a stepper and a stepper controller that cannot directly drive a stepper motor.

Current Sensing and Overcurrent Protection

We strongly recommend adding a current-sensing capability to your motor controller, if your system is not already equipped with it. You might want to consider over-current protection and sensing for a few important reasons:

- **Motor life** A motor that runs at maximum current levels for long periods of time is likely to have a much shorter working life than a motor that is shut off when a large, sustained current draw is detected.
- **Overheating protection** Hopefully, you've been conscientious about choosing the correct wire gauges and heat sink protection for the maximum current levels your motor is capable of drawing. If you've missed something, however, an unchecked sustained current

draw could cause real damage to your controller board or wiring. Even *fire* is a possibility. Overheating protection will go a long way toward avoiding robot *flambé*.

- **Obstacle detection** The primary cause of sustained high current draw is a motor stall condition. In this case, the motor is providing maximum torque at zero *revolutions per minute* (RPM). Although a stalled motor can be caused by something being caught in the wheels or another mechanical failure, it most often happens when the robot runs up against an obstacle that was not detected by other obstacle sensors. You can use overcurrent detection to allow your robot to sense this condition and respond accordingly. Note that not all obstacles will cause the motor to completely stall; the wheel may instead spin in place. However, you should see a substantial increase in the current draw as the motor works harder to overcome the friction between the wheel and the floor. By detecting this current, your robot can "feel" obstacles missed by other sensors.

Many of the single-chip solutions discussed earlier in this chapter have current sense outputs; some even have automatic overcurrent avoidance. If your controller lacks current sensing, or the current sensing provided fails to meet your needs, here are a few methods of creating your own.

The Easiest Method of All—The Reed Switch and Coil Method

We wish we could take credit for this elegant idea ourselves. It's about as simple and cheap to implement as it gets. The idea is to wrap a length of enameled wire around a subminiature reed switch. The coil thus created is then placed inline with one side of the motor power feed. One end of the reed switch is connected to the logic ground, and the other end to a free input port of the robot controller. When the current drawn by the motor exceeds a certain threshold—which will be determined by the number of turns in the coil—the resultant magnetic field produced in the coil will cause the reed switch to close, pulling the controller pin to logic ground. The more turns in the coil, the more sensitive the switch will be. Figure 7-22 shows a simple schematic. Note that the coil is actually wrapped around the reed switch, not sitting above it, as shown in the figure. You may also not need the pull-up resistor shown in the figure if your controller input port is equipped with an integral pull-up.

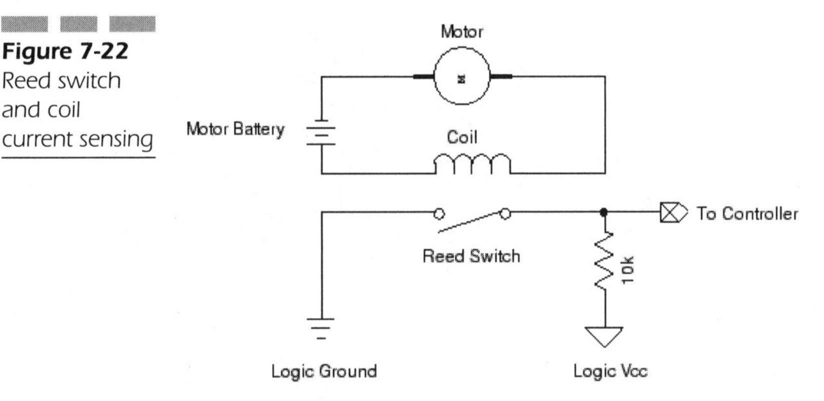

Figure 7-22 Reed switch and coil current sensing

The exact number of turns required in your sensing coil will depend on the current threshold at which you want the switch to close and the sensitivity of the reed switch. You will need to experiment. Start with about a dozen turns and work up or down. It's fine to have multiple layers of turns if needed. Also, note that if you set the sensitivity relatively high (meaning less current will be required to trip the switch), you may need to adjust the number of turns on the coil if you add substantial additional weight (such as larger batteries) to your robot, since the running current could end up closing the switch on the heavier robot.

Since the coil will be placed directly in your motor feed line, ensure that the wire gauge used is adequate to support the amount of current you'll be drawing. We recommend using enameled *magnet wire* (usually available from your local Radio Shack in gauges up to 24), which has a thin insulation and is well suited for making coils. You can solder your motor/power leads directly to the wire; the heat from soldering will remove the enamel insulation, giving you a solid connection. Once you've settled on the number of turns, use a couple of drops of glue to keep everything in place.

You will have better results with regular, miniature-glass reed switches. Avoid the types encased in plastic or epoxy blocks and sold as drop-in alarm replacement parts, since the extra casing will increase the distance from the coil to the switch, thus lowering the unit's sensitivity.

This arrangement works fine for both a continuous DC feed as well as a PWM feed. Be aware, however, that it is normal for a motor to briefly draw its stall current when first started or thrown into reverse. You will need to ignore any reading from the sensor for a certain amount of milliseconds after either of these conditions occur. Experiment a little to find

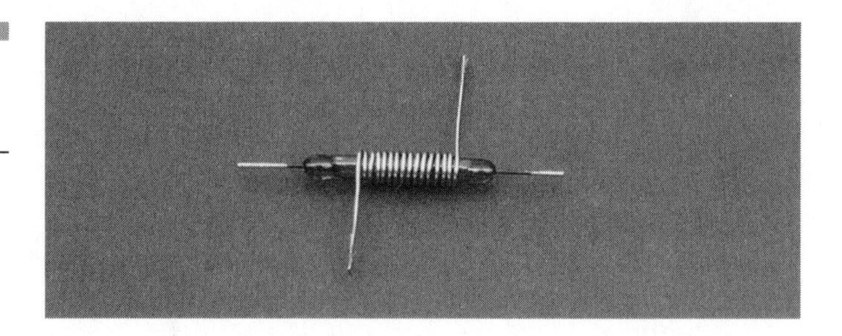

Figure 7-23 A reed switch stall sensor

how long this will actually be. We've managed good results with values as low or lower than 250 milliseconds.

Figure 7-23 shows a completed unit of about 15 turns. The reed switch was purchased surplus; the entire sensor, in fact, costs only a few cents and a couple of minutes to construct.

Current-sensing Transformers

These items can often be found on the surplus market and output a varying AC voltage that increases directly in proportion to the current being sensed. The Falco CS-200 units shown in Figure 7-24 are typical ($4 apiece).

These two-pin devices are essentially transformers. Instead of a primary coil, however, the current sensors have a hole through which the power lead to be monitored is passed. This lead acts as the primary. As current passes through the power lead, it induces a voltage in the secondary that varies in precise proportion to the amount of current passing through the lead. As you may have guessed by now, this *requires* that the voltage passing through the lead be modulated; a straight DC voltage will not work. For most robots, which are driven via PWM, this will not be a problem until the duty cycle gets close to 100 percent, at which point the secondary will cease producing a voltage. Because these sensors are

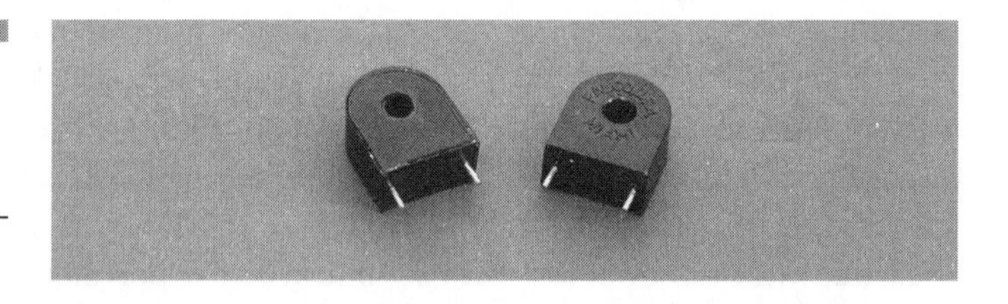

Figure 7-24 Current-sensing transformers

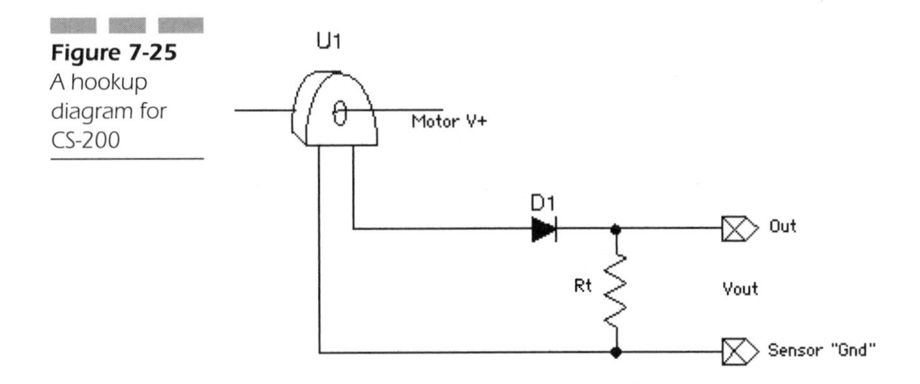

Figure 7-25 A hookup diagram for CS-200

really transformers, they output an AC voltage. One typical hookup diagram is shown in Figure 7-25.

Note that the voltage at the output has to be measured with reference to the voltage at the other pin of the sensor, *not* your logic ground. Your A/D conversion hardware must be capable of measuring this type of voltage. Also, the output (with the diode attached) will be a sort of half-wave at the frequency of your PWM implementation. In the case of this sensor, that frequency is ideally between 20 and 100 kHz for most uses. Diode D1 can be any small signal diode in your parts bin.

Resistor Rt controls the sensitivity of the sensor. As the value of Rt increases, so does the sensitivity. A value of 2K yields a voltage output of around 1 v/100 ma, which is a pretty good value for most medium-sized robots. A value of 20K yields a sensitivity of 1 v/10 ma, which is good for smaller machines. For really large robots, a 200-ohm resistor gives a sensitivity of 1 v/amp, up to a whopping 35 amps.

By adding the capacitor across the output as in Figure 7-26, you can get a steady voltage at the output, rather than the half-wave described earlier, which will be easier to process with most A/D hardware. The capacitor, however, will change the sensitivity of the circuit, requiring you to raise the value of Rt accordingly. The exact value of these components will vary with the desired sensitivity and the PWM frequency. We have used a 22 uF capacitor along with a 20K value for Rt to yield a sensitivity of 1 v/100 ma using a PWM frequency of 20 kHz. You will need to experiment with your own setup.

These devices are quite sensitive, able to get down to 1 v/10 ma with a maximum current-measuring capacity of 35 amps. They are a bit more trouble and expense than the reed switch sensors described previously, but are better suited to situations where you might want to actually

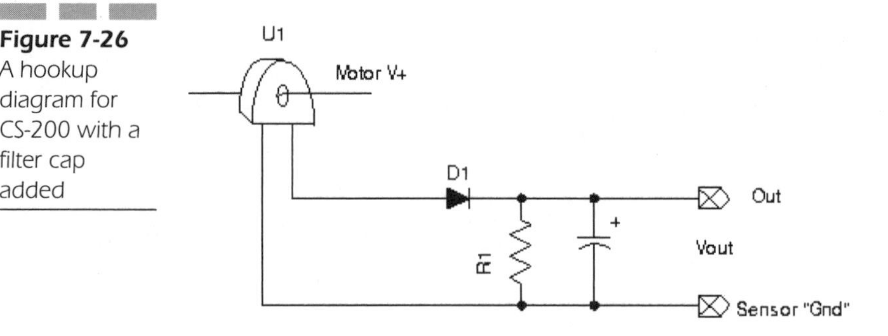

Figure 7-26 A hookup diagram for CS-200 with a filter cap added

measure current usage. Obviously, you *must* have A/D hardware to use this type of sensor.

Finally, these sensors are *sensitive to the polarity of the feed line*. For this reason, you should be *sure* to put them *upstream* of any circuitry that reverses the polarity of the motor power feed. If you don't do this, you will get odd results whenever you reverse a motor, although you won't damage anything.

In-line (Shunt) Resistors

One popular method for measuring the motor current draw is to place a resistor in-line with the motor power supply, and measure the voltage drop across the resistor. This inline resistor is often termed a *shunt resistor*. The voltage drop across the shunt resistor will increase with the flow to the motor in line with Ohm's Law:

$$V = IR$$

For instance, if we put a 1-ohm resistor in line with a 12-volt motor drawing 2 amps, we should see a voltage of 2 volts ($V = 1 \times 2$)across the leads of the resistor. Put another way, we can get the exact current draw using the formula $I = V/R$, which is really just Ohm's Law restated in terms of current. To take another example, let's say we measure a voltage drop of .5 volts across the shunt resistor. This tells us that .5 amps (.5V/1 ohm) of current are flowing into the motor (see Figure 7-27).

Unfortunately (referring back to the first example), we can count on the motor seeing exactly 2 volts less (10 volts) than the supply voltage (12 volts). Even worse, the lost power (2 volts $\times$ 2 amps $=$ 4 watts) will

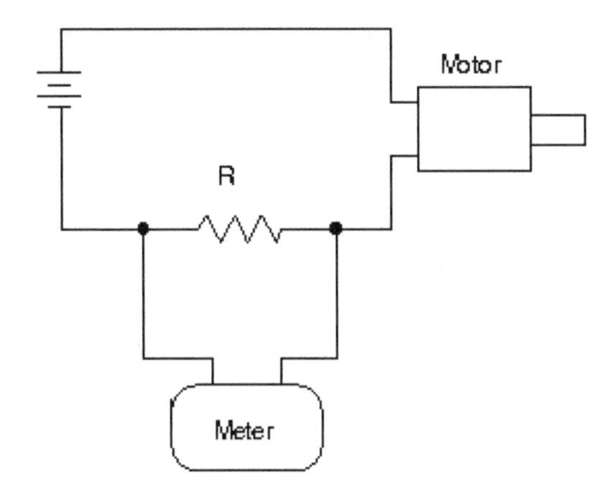

Figure 7-27 Deriving current from the voltage drop across a shunt resistor

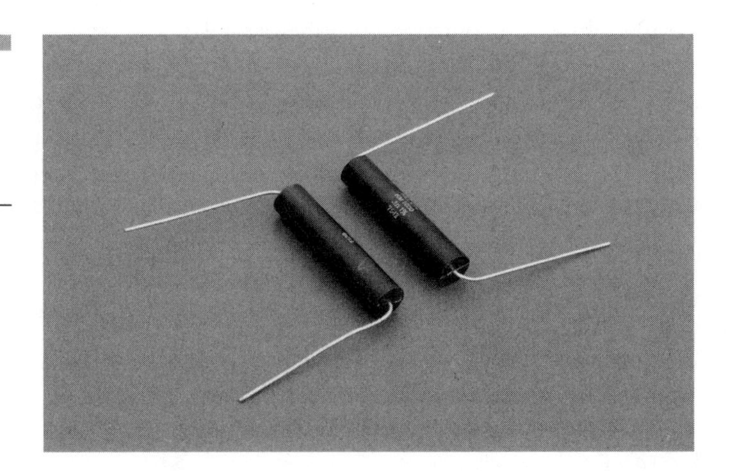

Figure 7-28 Low-value resistors, 0.1 ohm and 0.3 ohm

be dissipated by the resistor as heat. So how do we get around starving our motor and heating the garage?

The idea is to pick a very low value for resistor R. A good value to start with is 0.1 ohm, but resistors can be had in a variety of values well below 1 ohm (see Figure 7-28). Many electronics surplus houses carry these resistors, though not your local Radio Shack, and they can be had quite cheaply, often for less than $1. In our first scenario earlier, a resistance of 0.1 ohm, across which 2 amps are being drawn, would cause a voltage drop of a mere 0.2 volts and dissipate only 0.4 watts of heat—a great improvement. Our motor will still see 11.8 volts. We could improve our situation by going with an even smaller value for R, such as 0.01 ohm. *The more current drawn by the motor, the smaller R needs to be.*

Remember that the voltage dropped across the shunt is not seen by the motor, and the dropped voltage increases with current and shunt resistance.

Without incredibly accurate A/D hardware, however, our control logic may have a hard time reading the relatively small voltage drop across the shunt resistor R, especially for smaller resistance values. We'll need to amplify this voltage somehow, and the easiest way to do this is via a simple *operational amplifier* (opamp)-based circuit.

The LM324 forms the heart of this circuit (see Figure 7-29). The chip itself actually has 4 opamps on a single chip, so you can add more amplifiers easily. This chip costs about $1 and as of this writing is widely available; it's even stocked by Radio Shack.

As configured, the circuit is set up as a differential amplifier with a gain of around 10. In other words, it amplifies the voltage across shunt Rs (0.1 ohm, in this case) by a factor of 10. The exact amplification is determined by dividing R1/R2. Thus, to increase the amplification to 100 —which you might do for a higher current controller that requires a shunt of 0.01 ohm—you would change R1 to 1 megohm. Note that R4 must match R1, and R3 must match R2. So if you decide to go with an amplification of 100 and change R1 to 1 megohm, you'll have to be sure to change R4 to 1 megohm as well.

Rs should be placed between the battery and the controller, upstream of any circuitry that might change the polarity of the motor feed, since the circuit is sensitive to the polarity of the current flowing through Rs.

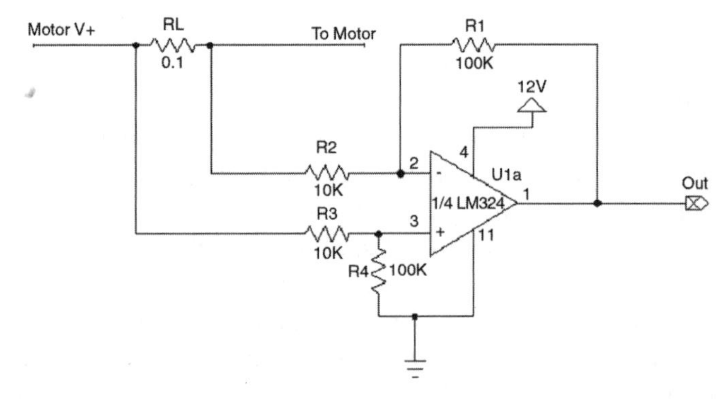

Figure 7-29 A LM324-based amplifier

CHAPTER 8

Motor Control 201—Closing the Loop with Feedback

Although simple *Pulse Width Modulation* (PWM) control, as detailed in Chapter 7, "Motor Control 101, The Basics," is certainly a great improvement over simple on-off control, PWM alone still has significant shortcomings. To really get the most out of a motor controller, we need to have some way of knowing how fast each wheel is turning. We can feed this information back to a microcontroller or other mechanism, which can then adjust the PWM output so that the motor speed stays at some commanded input value. Motor output might stray from the commanded value for a few reasons:

- When negotiating an incline, the motor speed will naturally drop as the robot climbs. Even worse, the speed drop may be different for each wheel on a dual, differentially steered robot when the incline is approached at an angle. This can cause the robot to veer along a curved path, rather than a straight one. To maintain a steady speed and heading, we'll need to increase the power feed to the motor as the wheels slow.
- Uneven terrain can cause one wheel to slow relative to the other. If more power isn't applied to the slowing wheel, the robot will stray from its expected path.
- Even matched motors sometimes run at different speeds given the same input voltages. Without feedback, you can compensate for this somewhat by applying a fixed power reduction to the faster motor. The difference in speed, however, may vary with the amount of power applied, making this method only marginally effective. By utilizing feedback, a motor controller can dynamically adjust power to achieve a reasonably precise output speed without having to resort to unreliable "fudging."
- As battery power drops, a given PWM duty cycle will yield increasingly slow motor speeds because of the reduced voltage. By using feedback, a smart controller can compensate for this effect by increasing power applied to the motors as the battery depletes with use.

Open-Loop Control The term *open-loop* control is somewhat misleading; there is, in fact, no *loop* at all. Instead, motor control is strictly a linear affair, as illustrated in Figure 8-1.

Information flow in an open-loop system is simple. A microcontroller or other logic generates a PWM signal that is fed to an H-bridge or other controller, which in turn applies power to the motor. In an open loop system, the controlling logic must know *a priori* which PWM signal will

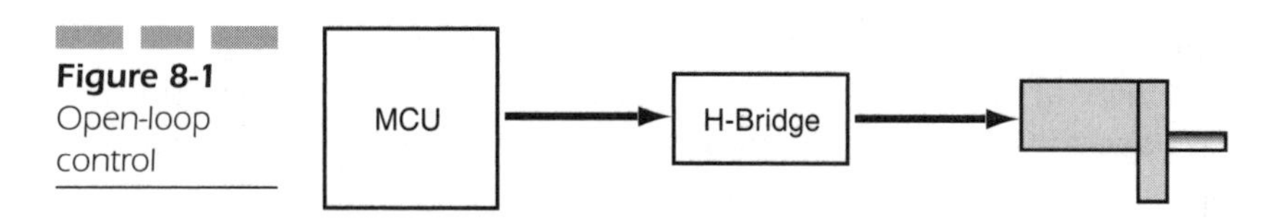

Figure 8-1 Open-loop control

result in a specific motor output speed. Unfortunately, the amount of speed a motor attains given a specific level of power varies with motor load conditions, and in the case of a mobile robot, with the degree of battery depletion.

Closed-Loop Control Closed-loop control adds feedback to the picture, as in Figure 8-2. In a closed-loop system, the controller starts with a commanded velocity and applies power to the motor. A sensor measures the actual velocity and feeds it back to the controller. Based on the difference between the actual velocity and the desired velocity, termed the *error*, the controller adjusts the power to the motor up or down, usually proportional to the amount of error detected. The process then repeats until the desired speed is reached.

Two common sensor types are used to determine motor speed:

- *Shaft encoders* measure the position or velocity of a rotating shaft. An *absolute position encoder* measures the actual position of the shaft; these are commonly used in servos to maintain a specific shaft position. *Incremental encoders* are used to measure shaft velocity (speed and direction). Incremental encoders produce a pulse train that corresponds directly to shaft rotation speed and, in the case of quadrature encoders, direction. Note that an encoder that measures only speed is technically a *tachometer*, but like most references, we'll use the terms *tachometer* and *encoder* interchangeably. Incremental encoders are often implemented using

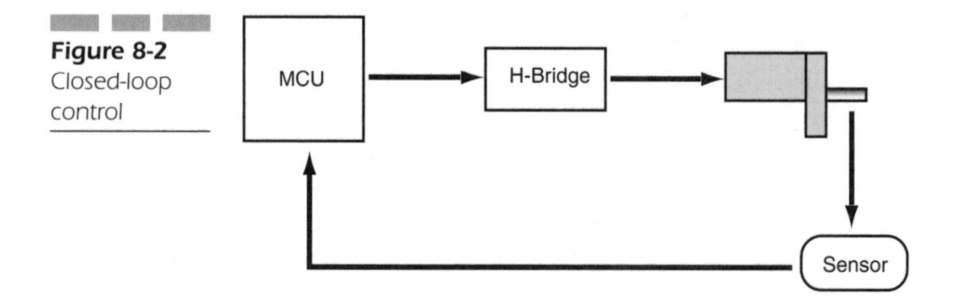

Figure 8-2 Closed-loop control

optical techniques, but other methods, such as magnetic field sensing with Hall-Effect devices, are also used. We will cover the use and construction of shaft encoders extensively in this chapter.

- *Analog tachometers* are devices coupled mechanically to a motor output shaft that output a voltage proportional to motor speed. Some motors can be purchased with integral tachometers. It is also possible to purchase them separately, although such motors can be expensive and hard to find in smaller sizes on the surplus market.

Encoders and Odometry

For each fractional rotation of the attached shaft, an incremental encoder changes its output from low to high or vice versa. We can count the number of pulses produced every second to get the shaft velocity, but we can also count the total pulses produced to get the distance traveled. The latter function is termed *odometry*. Within limits, odometry data and simple trig can be used to estimate a robot's position in relation to some starting point. This is termed *dead reckoning*. Although we will not be covering this topic here, you should be aware that because incremental encoders can be used to measure velocity as well as speed, they can be used for both motor control *and* navigation.

Building Incremental Shaft Encoders

This section will cover the construction of incremental shaft encoders. For the most part, encoders are relatively cheap and easy to build; however, it *is* possible to buy encoders ready made. Some of these units have a shaft that must be coupled to the motor shaft via a belt and pulley or similar arrangement. Others can be coupled directly to a shaft. Prices range from a few dollars for mechanical models up to hundreds of dollars for high-quality optical models (see Figure 8-3).

The mechanical models can have poor resolution, however, as low as 6 to 24 pulses per shaft resolution, and in general they are not made for continuous duty. The higher-cost optical models are not always made for continuous duty either (often these units are manufactured for use as high-performance rotary controls, not for use with motors). Still, we have been able to use them in this manner without problems, although your

Figure 8-3 Commercial optical encoder

results may vary. Commercial optical encoders offer much higher resolution than their mechanical counterparts, with 128, 500, or more pulses per revolution being typical. If you decide to use a commercial encoder, you may be able to find optical encoders on the surplus market for $15 or less.

A Simple Wheel-Mounted Reflective Optical Encoder

Reflective encoders are some of the simplest types to build. One typical approach involves mounting a white disk with black stripes on to the motor shaft or on the inside face of the wheel. The disk is illuminated by an *infrared* (IR) *light-emitting diode* (LED). As the disk spins, the light from the LED is reflected onto a phototransistor as a white segment passes by, and is shut off as a dark segment passes by, causing the phototransistor to produce a corresponding train of output pulses. Typical disk patterns are shown in Figure 8-4.

The more stripes on the encoder disk, the more pulses are produced as the disk rotates. The more pulses produced, the greater the *resolution* of

Figure 8-4 Encoder disk patterns

the encoder. Resolution is often stated in terms of *cycles per revolution* (CPR). Higher resolution is generally better, but if the stripes on the encoder disk become too small, it will be difficult for the photodetector to differentiate them as they pass by and errors will occur. As a rule, the smaller the photodetector, the smaller the stripes can be.

High resolutions improve the accuracy of the encoder, both for odometry and velocity feedback. For example, assume your motor drives a 5-inch diameter wheel, and your encoder disk has a total of 16 stripes. The circumference of your wheel is $\pi \times 5$ inches, or around 15.7 inches. Dividing the number of stripes into the wheel circumference gives us 0.98 (15.7/16) inches per encoder pulse. We can double the resolution to 0.49 inches per pulse by doubling the number of stripes to 32 (15.7/32).

Good resolution is important for responsive motor control as well. This is because a motor control unit typically samples the pulse train from an encoder at fixed intervals to determine the robot speed. The more pulses produced per rotation, the easier it will be to determine velocity accurately, especially at lower speeds. Higher resolution means that the encoder can be sampled more frequently as well, which will make the control system able to respond more quickly and accurately.

For example, consider the case of a motor that is capable of speeds from 0 to 45 *revolutions per minute* (RPM). Using an encoder with a resolution of 32 CPR and a relatively slow sample time of 1 second, we can only expect to see about 0.533 pulses per RPM per sample. This is not terribly accurate. Since we can't detect partial pulses, we can expect sampled velocity estimates to be accurate only to within 2 RPM, or about 4 percent. Doubling the resolution to 64 CPR doubles our accuracy to 1 RPM, or about 2 percent. We could also shorten our sample interval, if we were willing to accept a 4 percent error increase, to make our control loop doubly responsive.

Making the Encoder Disk Using a compass, ruler, marker, and a little patience, you could draw your own encoder disk, but it's easier to use a computer graphics program to draw the disk pattern. Even the freebie graphics program that comes with most personal computers will be adequate for our purposes. Simply draw a circle, subdivide it using straight lines, and then "flood fill" every other segment with black, as illustrated in Figure 8-5. This becomes somewhat tedious after 64 segments, however.

You can also print a nice pattern using a Postscript program and a Postscript viewer, such as GhostView. A number of programs for printing encoder wheel patterns can be found on the Internet. Try a Google®

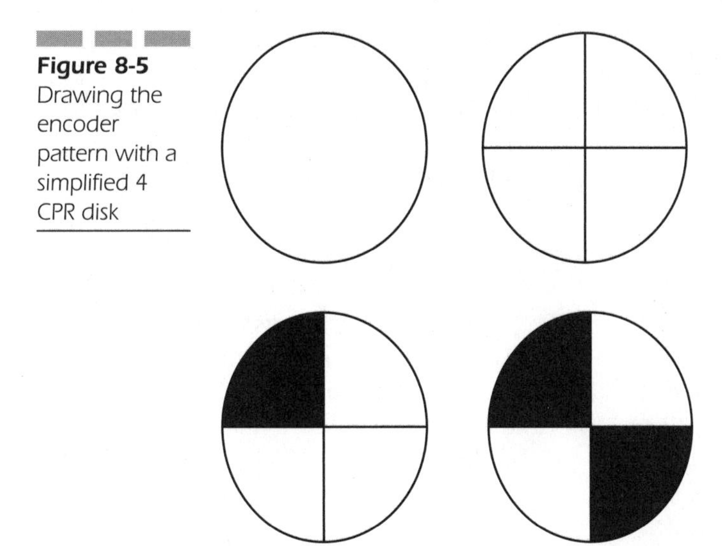

Figure 8-5 Drawing the encoder pattern with a simplified 4 CPR disk

search (www.google.com) for "postscript" and "encoder disk" or "shaft encoder." This should get you a number of suitable programs. A Postscript viewer/printer can be found at www.ghostgum.com.au.

Next, you'll need to print your pattern. Here you have a few options:

- Print the disk on regular white paper. You'll probably need to glue the printed pattern on to heavy cardboard stock. *Try this first*. Often, you'll get the bests results from plain paper, especially with a sensitive photoreflector. Be sure the paper is well glued to the cardboard backing, with no waves or edges sticking out. We want the disk to be as flat and level as possible with respect to the detector.
- Print the disk on glossy printer paper made for photographs using the highest quality printer setting. The idea is to have the white stripes as reflective as possible and the dark stripes as dark as possible. The contrast will often give you a better and more consistent photoreflector response. Use the heaviest stock you can find; this helps ensure the surface of the disk will be as planar as possible. As mentioned earlier, waves along the surface of the disk can cause encoder output irregularities. You can also use lighter stock and glue the result onto a stiff cardboard backing.
- Print the disk on a printer transparency and glue the transparency on to a reflective backing. A piece of cardboard with some aluminum foil glued to it will work well. For larger encoder disks, you can use

an old CD-ROM. This will usually give great contrast; however, some printer inks may be less opaque to infrared light than others, making this approach less effective under some circumstances. One other potential problem with this method is the relative expense of printer transparency—over $45 per box for some types.

NOTE: *Although it is possible to copy the patterns from Figure 8-4 directly, the black stripes, as reproduced in this book, may be insufficiently dark to give an adequate contrast. If you want to use these patterns, take the book into a printer and have them create a high-contrast negative from the printed pattern. They can then reproduce the pattern in any size you need with full contrast.*

Encoder disks are frequently mounted to the inside of the robot drive wheels. Glue usually works well. Use sticky, blue art putty to mount things temporarily for testing. A mounted encoder is shown in Figure 8-6.

However you mount the encoder disk, make sure it is level. A warped disk or a disk mounted at an angle can easily cause erratic output from reflective photodetectors, which are very sensitive to the distance of the encoder disk.

Figure 8-6 Mounted encoder disk

The Reflective Photodetector The photodetector consists of two parts:

- An IR LED that illuminates the encoder disk.
- A phototransistor or photo diode that changes output state as the reflection from a white stripe, or shadow from a dark stripe, passes across the detector surface.

It is possible to construct your own photodetector from a phototransistor and an IR LED. However, matched pairs in convenient sealed plastic housings can be had cheaply and easily either new or surplus. These are sometimes termed *reflective photosensors* (see Figure 8-7). Our recommendation is that you go with these units, rather than build your own. The packaging is more convenient and the phototransistor is selected to have an optimum frequency response at the paired LED wavelength.

Some of these units contain active circuitry to condition the output signal. The cheapest and most commonly available units are passive, however, consisting only of an LED and a phototransistor. Typically, these will be four-lead devices. One pair of leads is used to power the LED. The other pair is connected to the collector and emitter of the phototransistor. If you have no datasheet on these units, don't despair; usually the device will have the schematic symbol for an LED printed on the LED side, and the symbol for a phototransistor printed on the phototransistor side. Figure 8-8 illustrates the minimum supporting circuitry you'll need to use these devices.

Figure 8-8(A) shows the circuit you'll need to drive the LED side of the detector. This is a typical LED driver circuit, consisting of a power supply and a current-limiting series resistor. The value of R1 can run anywhere from 220 ohms to about 1K. The lower values will give better responses but use a good bit more power. Although 220 ohms is a lower resistance than you might normally use as a current limiter with a

Figure 8-7 Reflective photosensors

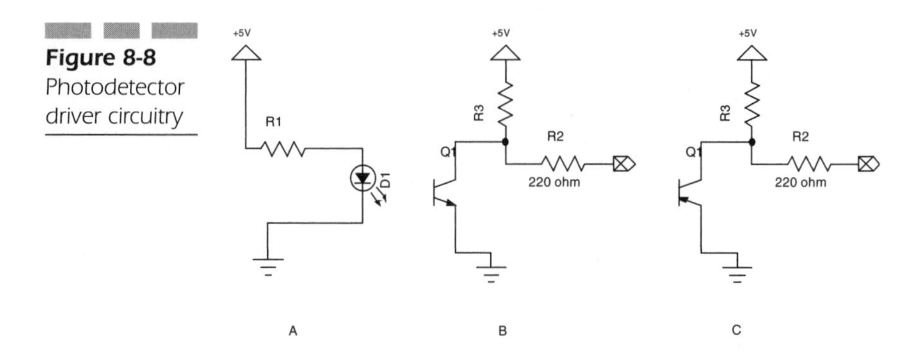

Figure 8-8 Photodetector driver circuitry

visible LED, this is perfectly safe, as IR LEDs are usually capable of handling significantly higher currents than visible LEDs.

Figure 8-8(B) shows the support circuitry for the phototransistor side of the circuit. R3 controls the sensitivity of the detector and may range from 10K up to 2 megohms. A higher resistance for R3 results in a more sensitive circuit. The exact value will depend on the sensor. If you have spec sheets for the sensor, there should be a recommended value. If not, you can experiment. The cheap surplus sensors used by our larger robots seem to perform best with a value of 1 megohm for R3.

Note that Figure 8-8(B) assumes you'll be using a sensor utilizing an NPN phototransistor. If your sensor uses a PNP phototransistor, you'll need to reverse the ground and output circuitry as in 8-8(C).

Ideally, when the sensor encounters a black stripe, V_{out} will hover within at least 1.5 volts of the supply voltage. When a white (reflective) stripe is encountered, the transistor will conduct, pulling V_{out} down to about 0.5 volts or less. V_{out} can usually be treated as a digital output and be connected directly to a microcontroller's input pin, although as we'll see, the output we get from this arrangement is a long way from being a true digital signal. If V_{out} is read directly, bear in mind that the input must be of high impedance. Fortunately, most microcontroller input pins fit this description.

We strongly suggest that when using an unknown sensor, you first build the driver circuitry on a breadboard before committing to a soldered assembly. This will allow you to experiment with the resistor values to determine the optimal response of the circuit.

Mounting the Photodetector Closer is usually better, but if mounted too close, the phototransistor will be unable to see the reflected IR from the emitter. In any case, start by mounting the detector in such a way that it can be easily moved. Double-sided foam tape (available from most hardware stores) can work well in this regard. For a less firm hold, use the blue, sticky art putty mentioned earlier to temporarily secure the sensor in position, as in Figure 8-9.

You may need to shield the photosensor from ambient light, which usually means mounting the sensor on the underside of the robot. You can also construct a housing to block out ambient light (to some extent) from a bit of stiff cardboard.

For indoor use, mounting the photosensor beneath the robot usually works well; additional light shielding will almost certainly not be required. This arrangement may not work as well outdoors; in direct sun, encoder disk black-to-white transitions can occasionally get lost, but much will depend on the construction and position of your sensor. Experiment before going to the trouble of providing additional light shielding.

Often, you can vastly improve the detector's response by *changing* the sensor angle with respect to the encoder disk surface. This is particularly true when using a glossy or otherwise reflective encoder disk. You may also find that an angled sensor does a better job reading higher-resolution disks. Try about 10 to 30 degrees off perpendicular. For higher-resolution disks, you might also orient the sensor so that the emitter and detector are parallel to the stripes, rather than perpendicular.

Figure 8-9
A test-mounted photodetector using sticky putty. Notice the angular orientation; this is intentional.

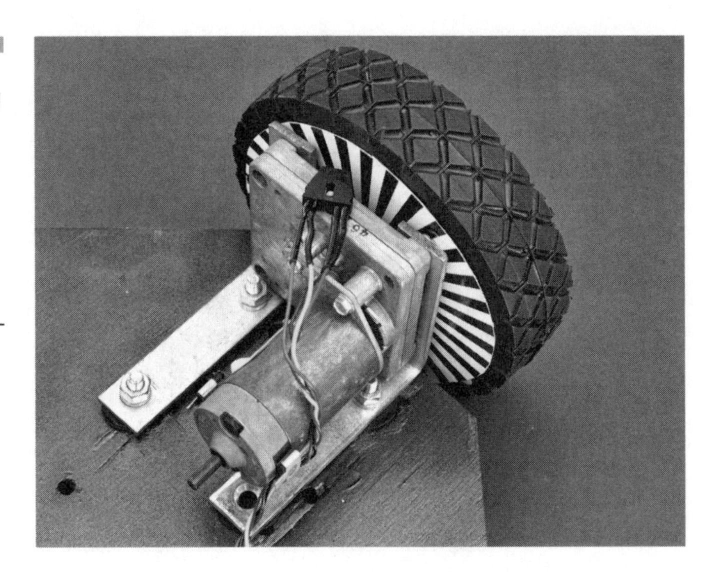

Crisper Output with Schmitt Triggers

The output from our photodetector is not particularly crisp. That is to say, rather than swinging from a microcontroller-friendly 0 to 5 volts (logical 0 to 1) as the encoder disk stripes pass by, the actual output seen at V_{out} looks more like Figure 8-10.

It would be great if we could digitize the output of the photosensors so that our controller could more easily distinguish between transitions. For example, many microcontrollers are not guaranteed to see a logical 0 at an input unless the input voltage is below some very low value, often a couple of tenths of a volt. A logical 1 may not be seen unless the input voltage exceeds 3 volts or so. Photosensors, however, produce continuously varying voltages in response to the passing of encoder stripes, and these voltages will often not fall consistently into the ideal range required by microcontroller I/O pins, particularly when ambient light is a problem. The bottom line is that the microcontroller may miss enough encoder counts to significantly affect both odometry and motor control accuracy.

One way to improve our sensor signal would be to condition V_{out} with a *comparator*. A comparator is a special kind of circuit that compares an input voltage against a reference voltage, and it changes state when input voltage exceeds the reference, and again when the input voltage falls below the reference. Assume a reference (or threshold) voltage of 2.5 volts, a comparator that outputs +5 volts when the signal is over the threshold, and 0 volts when the signal goes below the threshold. This cir-

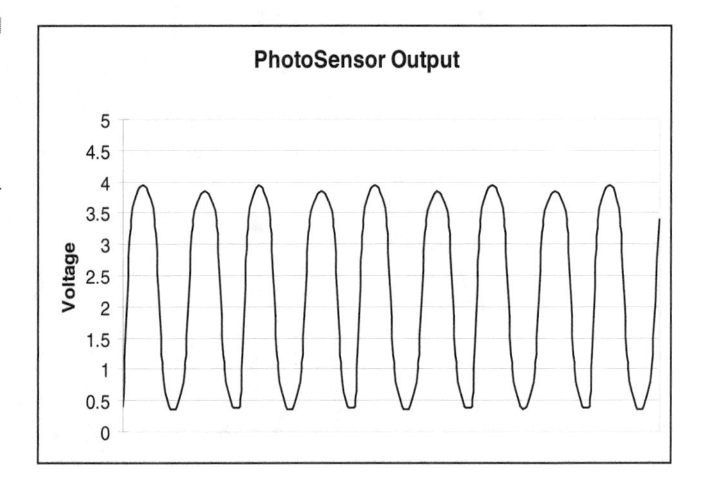

Figure 8-10 Unconditioned reflective photosensor output

cuit might exhibit a response to a moderately noisy input, as shown in Figure 8-11.

The output from the comparator does indeed "square up" the input signal nicely, but it lacks noise immunity when the input signal briefly crosses the threshold voltage and back again. We can improve on this by using a *Schmitt trigger* in place of a simple comparator. A Schmitt trigger is a special type of comparator that implements *dual* thresholds. When the input voltage exceeds the upper threshold (sometimes termed the *positive-going threshold* and abbreviated V_{t+}), the output of the trigger goes high. When the input voltage falls below the lower threshold (the *negative-going threshold*, abbreviated V_{t-}), the trigger output goes low. The trigger does not change state at all when the input voltage falls within the "deadband" between V_{t+} and V_{t-}. This behavior is termed *hysteresis* and is useful because it provides us with a certain amount of

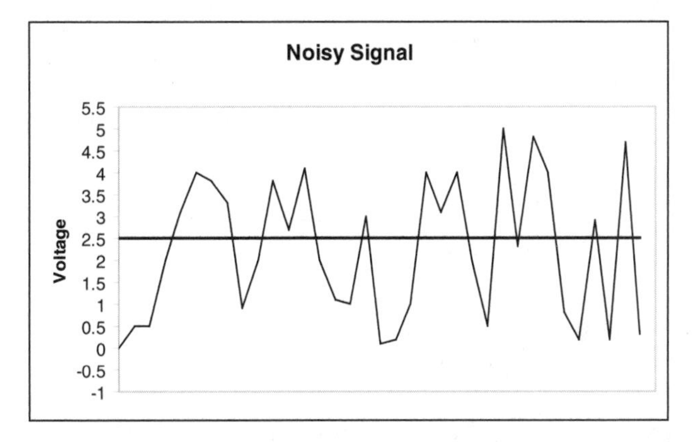

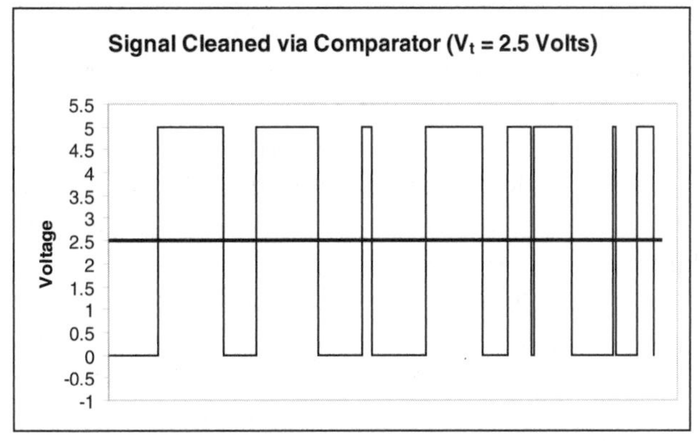

Figure 8-11 Using a comparator to clean up a signal

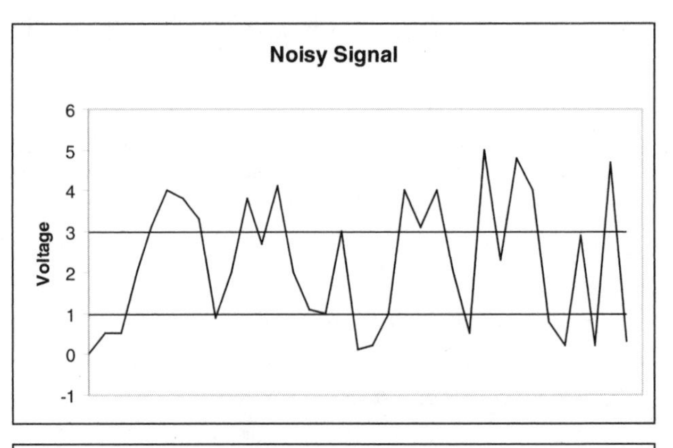

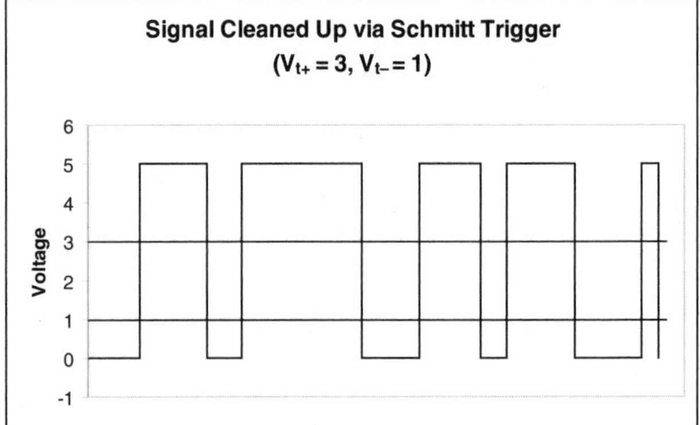

Figure 8-12 Processing an input signal with a Schmitt Trigger

noise immunity. Figure 8-12 shows how our noisy input signal might look if processed using a Schmitt trigger instead of a comparator.

Notice that the output signal is now missing the brief noise pulses we saw when we used a simple comparator for signal conditioning. This noise immunity makes Schmitts desirable for use in many signal-conditioning applications; in fact, they are used frequently enough that they have acquired their own schematic symbol, as shown in Figure 8-13. The Schmitt on the left is *noninverting*; that is, its output goes high when V_{t+} is exceeded and low when the input drops below V_{t-}. The Schmitt on the right is *inverting*; its output goes *low* when V_{t+} is exceeded and *high* when the input signal drops below V_{t-}.

The venerable 7414 is one of the most common Schmitt triggers around. We recommend the CMOS 74C14 for its low power consumption and high input impedance. You can also use its high-speed cousin, the 74HC14, although at the slow signal speeds we'll be encountering, the HC chip speed provides no real advantage. It should be noted, however,

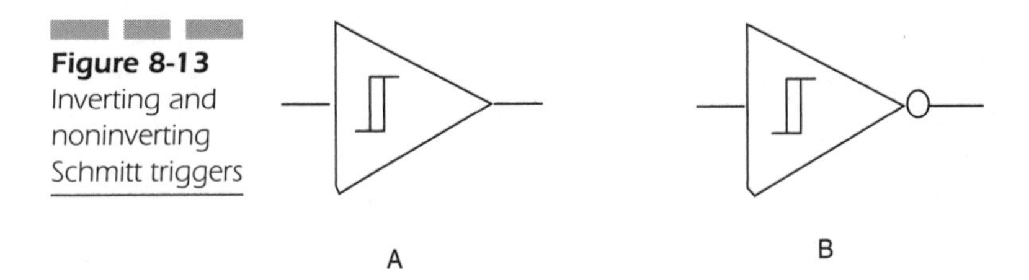

Figure 8-13 Inverting and noninverting Schmitt triggers

that V_{t+} and V_{t-} differ between the C and HC families. Assuming a power supply of 5 volts

Part	V_{t+}	V_{t-}
74C14	3.6V	1.4V
74HC14	2.9V	1.9V

The 74(H)C14 contains six inverting Schmitt triggers and costs about $0.45 to $1 new in single-unit quantities, but it is much less when bought surplus or in bulk. Adding this chip into our encoder circuitry gives us the following simple schematic in Figure 8-14.

The Hamamatsu Reflective Photosensor As it turns out, you can purchase reflective photosensors with the Schmitt trigger and other conditioning circuitry already built in. The most popular units are the Hamamatsu reflective photosensors. Not only is the output of this sensor microcontroller-ready, but its tiny size means that the Hamamatsu

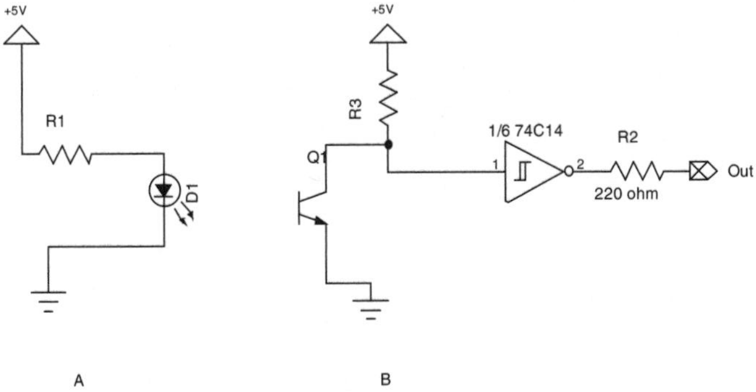

Figure 8-14 Encoder support circuitry with Schmitt trigger output

can be used in much tighter spaces and can read encoder wheels of a much finer resolution than its larger cousins. These units cost a bit more, around $3.25 per unit, but are worth the money, especially when space is an issue.

The primary drawback of the Hamamatsu sensors is the size of the sensors themselves, especially the leads, which are quite thin and closely spaced (see Figure 8-15). Leads 3 through 5 won't fit on a standard perf or prototyping board without being spread apart. This can make them somewhat difficult to work with, although both the chip and its leads are really less fragile than they appear at first glance.

The Hamamatsu sensors commonly come in two models, the 5587 and the 5588. Both are functionally identical with the sole exception that the output of the former goes high in the presence of light, and the latter goes low. For purposes of incremental encoder construction, either unit will work equally well, since our interest lies primarily in sensor output *transitions* rather than the actual output itself.

These sensors require some minimum supporting circuitry, including a current-limiting resistor for the onboard LED, and a 0.01 to 0.1 uF filter cap. Because the sensor output is an open collector, a 6.8 to 10K pull-up resistor is required. The circuit shown in Figure 8-16 is a typical example.

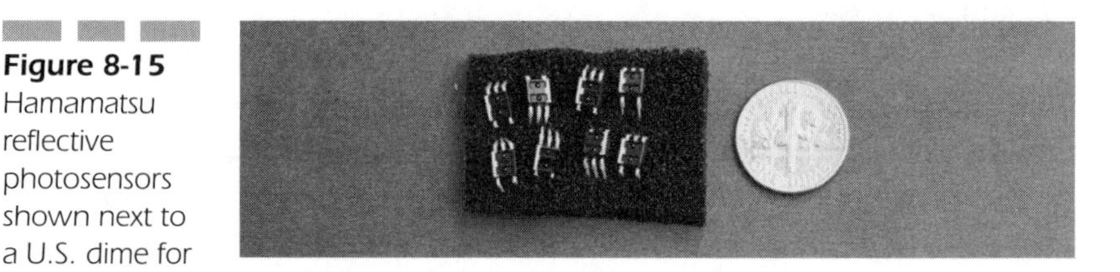

Figure 8-15 Hamamatsu reflective photosensors shown next to a U.S. dime for scale

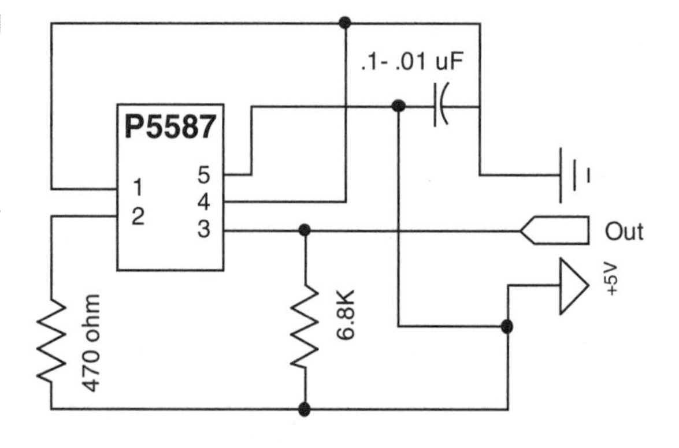

Figure 8-16 Hamamatsu sensor-supporting circuitry

We have found that it is often convenient (but certainly not required) to build the sensor and supporting circuitry on a single bit of PC board, strip board, or perf board and mount the board and sensor as a single unit. The example in Figure 8-17 was constructed on a small piece cut from a Radio Shack General Purpose IC Board (Cat 276-150a). Notice that leads 3 through 5 have been spread apart so that they can be inserted into the holes on the board, which are spaced for standard DIP parts.

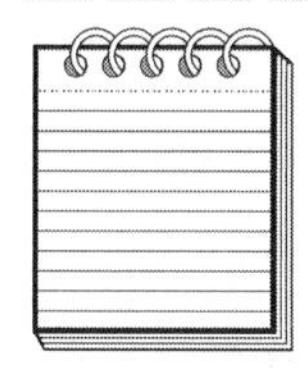

NOTE: *You can easily cut small (or large) segments of PC board by using the following method:*

1. Clamp the board to be sectioned securely to a work surface.
2. Clamp a guide piece, such as a flat piece of board or metal, along the top of the board about where you would like to cut it.
3. Using a sturdy and sharp utility knife and with the guide board secured according to the previous step, carefully and firmly score the board. Try *not* to remove any of your fingers as you do so. The deeper you score the material, the better.
4. Optionally, turn the board over, reclamp your guide, and score the board from the other side. Make sure your score is exactly opposite (or as close as possible to) the score you made on the other side.
5. Clamp the board to your workbench or other surface so that the section to be removed hangs over the side, with the score line just at the bench edge.

Figure 8-17 Assembled Hamamatsu sensor board

6. You should be able to apply pressure and snap the section off at the score line. Just give it a good whack with the flat of your hand. Use a file to neaten up the break if neatness is a concern.

We have had great luck with plain fiberglass or phenolic PCB, photosensitive *printed circuit board* (PCB), and perfboard using this technique. Although in theory it is possible to minutely damage or weaken the foil on the PCB, we have yet to see a problem attributable to the "score and stomp" method described here. As always, your mileage may vary.

The finished sensor should be mounted in a location that gives the optimal response. In Figure 8-18, the sensor is mounted between the robot chassis and wheel. If you don't have a convenient mounting point, the board can be glued to a stiff length of coat-hanger wire or brass strip, which can be fastened to the chassis and used as a mounting bracket.

How steady are your hands? In cramped quarters, the sensor itself (without the supporting components) can be glued directly to a stiff piece of wire (a paper clip works well) that serves as a flexible mount. Power and signal wires can be soldered directly to the chip leads (you'll need to carefully spread them a bit), insulated with a bit of heat-shrinkable tubing, and run to a remotely placed board containing the support circuitry. Figure 8-19 illustrates this approach.

Figure 8-18 Mounted Hamamatsu sensor

Figure 8-19 A standalone Hamamatsu sensor—we used 26-guage stranded wire for hookup.

Sensor Positioning

For both Hamamatsu and passive sensors, start with the sensor fairly close to the encoder wheel—about 1 to 2 millimeters—and work back. The easiest way to measure the response is with an *oscilloscope*. Just run your motor and watch the scope output. If you are using the Hamamatsu sensors or Schmitt trigger conditioning, you should see a clean train of pulses (see Figure 8-20). If not, play with the sensor position until the pulse train is consistent. If you are not conditioning the sensor output, try to achieve a signal that shows the maximum voltage difference between the peak and valley. Remember that any irregularities on the encoder disk surface or uneven mounting can cause inconsistent performance.

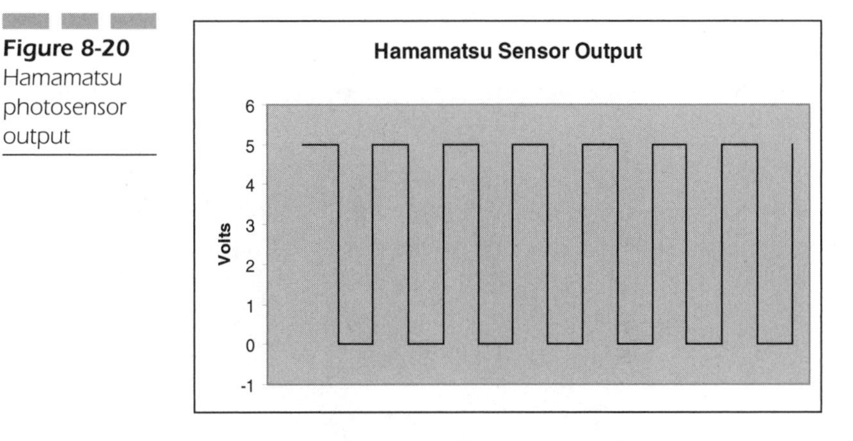

Figure 8-20 Hamamatsu photosensor output

Improving Resolution: An Armature Shaft-Mounted Encoder

Some gearhead motors have a second motor output shaft connected directly to the motor armature. This second shaft spins at a rate directly proportional to, but many times faster than, the main motor output shaft. The exact proportion is dependent on the gearing ratio in use. For example, in the case of a gearhead motor with a gearbox providing 200:1 reduction, the armature shaft can be expected to spin 200 times for each single revolution of the output shaft. Even with an encoder disk of low resolution attached to the armature shaft, say only four stripes, we can expect an output from our incremental encoder of 800 state changes per revolution of the motor output shaft, which is plenty of resolution for motor speed control.

Odometry and dead reckoning also benefit from this arrangement. In the case of a robot with 6-inch wheel diameters, each encoder pulse transition indicates, in theory, wheel movement of $(6\times\pi)/800$, which equals 0.0236 inches. Of course, in practice, wheel slippage, gear-train losses, and other factors mean you won't achieve anything near this level of accuracy—nevertheless, you will still find the increased resolution invaluable when attempting to judge the distance traveled or the robot position based on encoder counts.

We consider the availability of an armature shaft sufficiently important to strongly suggest that you look for this feature when shopping around for gear motors. The shaft need not be very long; a fraction of an inch or a few millimeters is fine. Fortunately, armature shafts are fairly common; Figure 8-21 shows a couple of examples. The motor on the left

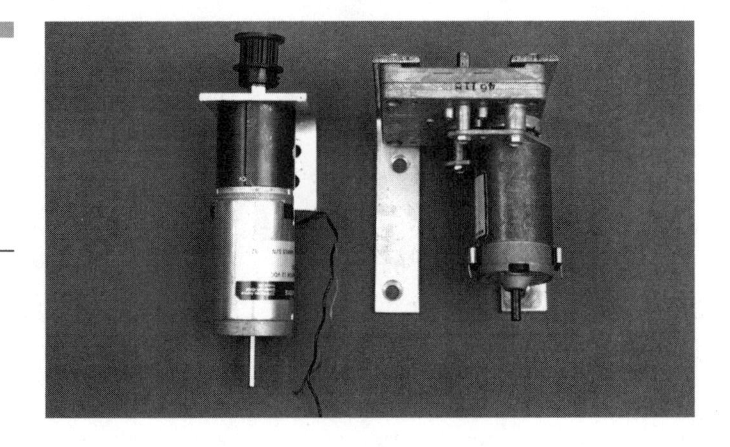

Figure 8-21 Two DC gearhead motors sporting armature output

originally came with a mechanical brake that, when removed, revealed a long armature shaft (it's easier to apply braking force to the armature output, which produces much less torque than the gearhead output).

Constructing the Encoder Disk Any pattern in Figure 8-22 should be adequate for an armature-mounted encoder disk. For the reasons discussed earlier, there is no need for lots of stripes; four- or eight-stripe disks should prove more than adequate.

Print or draw the pattern on firm card stock or heavy photo paper. Make sure the disk diameter is at least about 1/8 of an inch (3 millimeter) smaller than the outside diameter of the motor housing.

Attach the encoder disk directly to the end of the shaft with a drop of glue, but be absolutely sure that the disk is as level as possible before the glue dries and is centered on the shaft. If you need to replace the encoder disk later, simply tear it off and use a file to remove any leftover glue.

You may also glue the encoder wheel to a small shaft collar (available from most hobby shops) and insert the shaft collar into the shaft, assuming the shaft is long enough (see Figure 8-23). This allows a little more

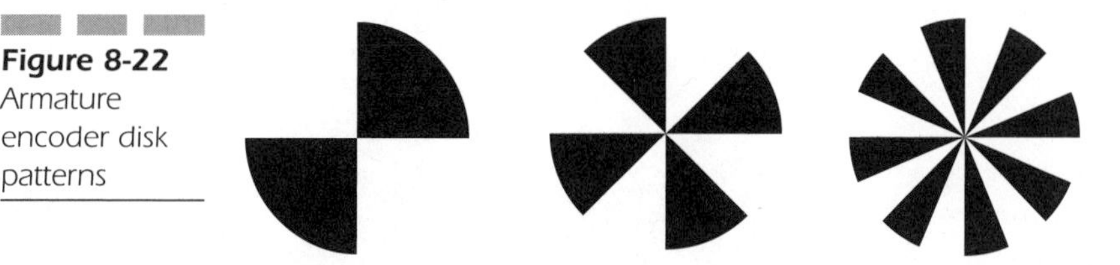

Figure 8-22 Armature encoder disk patterns

Figure 8-23 An encoder disk attached to the armature shaft

flexibility if you want to experiment with different types of encoder disks. The motor shaft will need to be flatted for the collar to fit correctly. You can do this with a file, but be sure that the shaft is well supported as you file it down.

Sensor Mounting Next, you'll need to mount the sensor and several options are available. For this application, we have chosen the Hamamatsu sensor mounted on a small piece of prototyping board along with some supporting components (see Figure 8-24).

We'd like to be able to adjust the sensor a bit with respect to the encoder disk. To make an adjustable mount, start with a length of *piano wire*. This stiff metal wire can be purchased in a variety of gauges from hobby shops, or you could even use a piece of clothes hanger wire. Bend a short section of one end of the wire 90 degrees, and secure to the board with epoxy or glue. This should yield the arrangement in Figure 8-25.

Place the whole assembly on to the back of the motor so that the Hamamatsu sensor is about 1/8 of an inch (3 millimeters) away from the encoder disk. Secure the arm to the sides of the motor housing using a hose clamp. You can adjust the sensor distance from the encoder wheel by loosening the hose clamp, adjusting the mount, and tightening the clamp back up again (see Figure 8-26).

Figure 8-24 The Hamamatsu sensor board

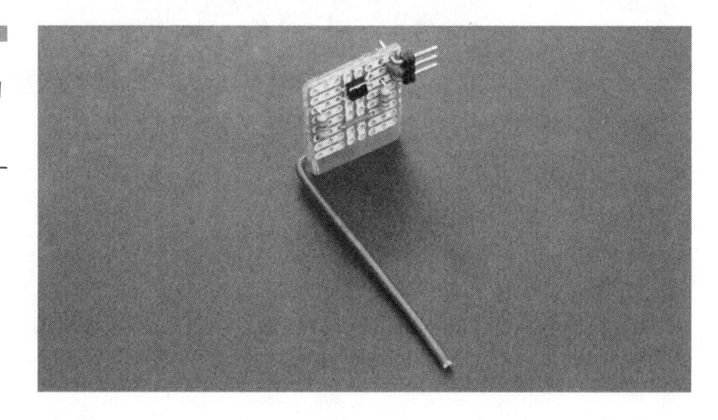

Figure 8-25 A sensor board and a wire mounting arm

Figure 8-26 A mounted sensor board

Shielding the Sensor from Ambient Light Building a light shield is easy but will probably not be necessary. We can enclose the entire assembly with just about anything cylindrical and close to the motor-housing diameter. A short length of PVC, a 35-millimeter film canister, or even a section of a paper towel roll will all work well. Of course, we recommend painting the paper towel roll metallic silver so your friends don't accuse you of making a robot out of toilet paper rolls. If your housing is a little larger than the motor-housing diameter, you can carefully wind some electrical tape around the housing to enlarge it and create a better fit.

Sensing Direction: A Quadrature Encoder

A quadrature encoder improves on our basic incremental encoder in two ways:

- Quadrature encoders provide rotational *direction* information as well as speed. If you expect your robot to find itself in situations where it may unexpectedly roll backward, this capability is invaluable. Robots that need to negotiate steep inclines or rough terrain may find this capability useful. Quadrature encoding can also be used to minimize errors during dead reckoning when brief but unexpected wheel reversals (which might occur when braking on an incline) can contribute to losses in accuracy.
- Quadrature encoding effectively doubles the resolution of an incremental encoder, since two state changes are generated for each encoder stripe pass.

Quadrature encoding is a popular technique; most commercial incremental encoders are quadrature encoders. However, as we'll soon see, the additional capabilities of quadrature encoding come with an associated increase in complexity and cost.

How Quadrature Encoding Works Quadrature encoding uses two sensors. The sensors are arranged along the encoder disk such that when one sensor encounters a black-white boundary, which will cause it to change state, the other sensor is exactly in the middle of a stripe, which will ensure that it is outputting a steady state as the first sensor is changing state. Figure 8-27 shows one such arrangement.

The resulting output from each sensor will be exactly 90 degrees out of phase with respect to the other. Figure 8-28 illustrates the waveforms from sensors A and B as the encoder wheel turns counterclockwise. A

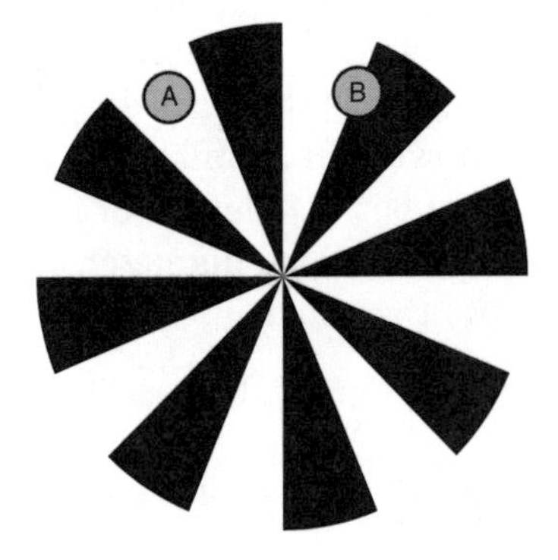

Figure 8-27 Sensor placement for quadrature encoding

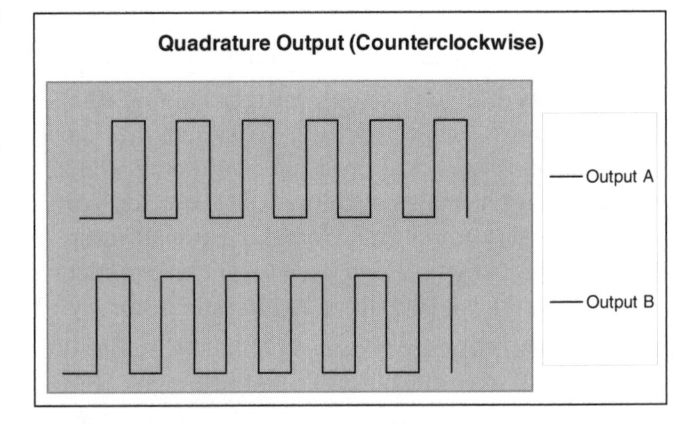

Figure 8-28 Quadrature encoder output

black stripe produces a high pulse at the sensor outputs. A light strip produces a low output.

We see an output transition on either sensor each time the encoder disk moves the distance of about half a stripe. This effectively doubles our encoder resolution, assuming accurate placement of the sensors. Moreover, by identifying the leading transition, we can tell which way the wheel is moving. For example, if B goes high to low after A goes high to low, or B goes low to high after A goes low to high, the disk is moving clockwise (see Figure 8-29).

If, on the other hand, B goes low to high after A goes high to low, or B goes high to low after A goes low to high, then the disk must be moving counterclockwise (see Figure 8-30). We'll cover a simple algorithm for reading quadrature sensor output in the "Processing Quadrature Output" section.

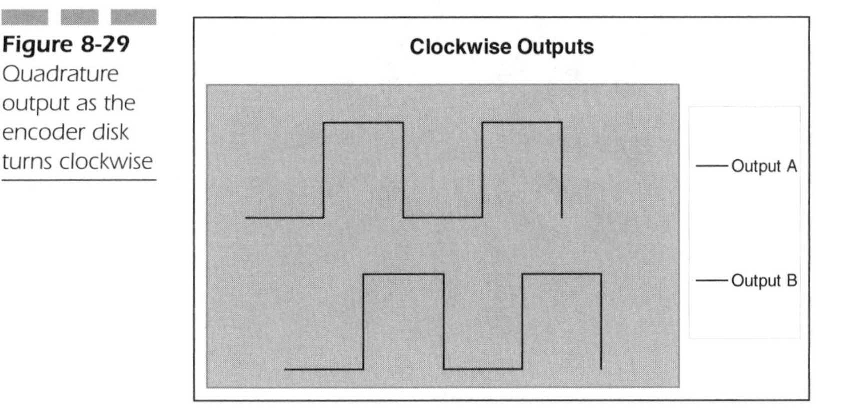

Figure 8-29 Quadrature output as the encoder disk turns clockwise

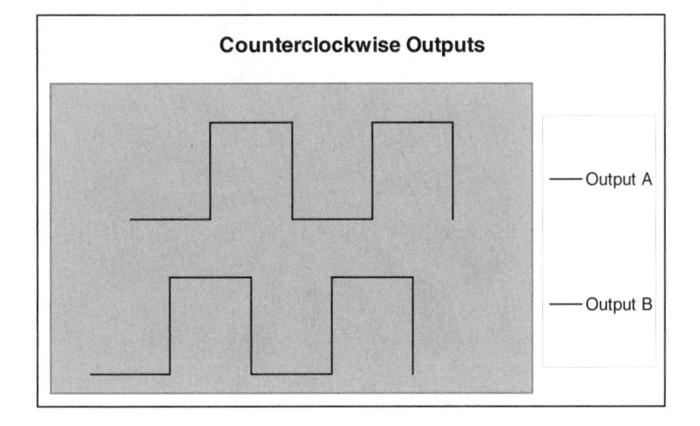

Figure 8-30 Quadrature output as the encoder disk turns counterclockwise

Constructing a Quadrature Encoder Rather than attempting to use a standard encoder wheel and struggling to get the photoreflective sensors aligned for proper quadrature output, we have found that it is simpler to just design an encoder disk with suitably offset stripes. The sensors can then be aligned vertically. The 32 segment design in Figure 8-31 is suitable for most uses; you may, of course, add segments to increase resolution.

The Hamamatsu sensors are a good choice for this project. Mount both encoders on the same board. The encoders should be aligned vertically but be separated wide enough so that one unit reads the top set of stripes and one unit reads the bottom set of offset stripes (see Figure 8-32). If

Figure 8-31 A quadrature encoder disk pattern

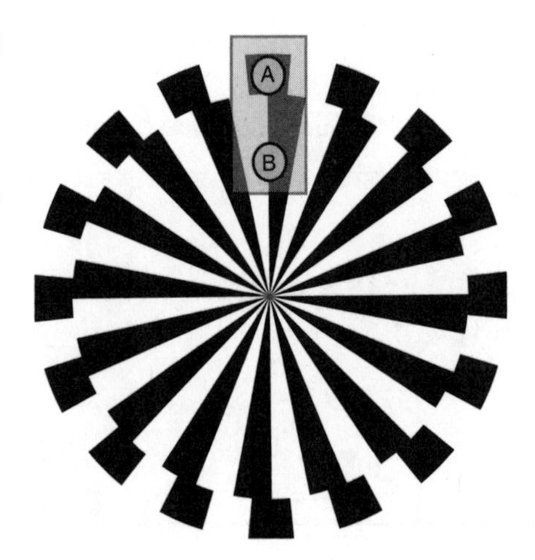

Figure 8-32 Sensor placement over disk

you have room, you can build the supporting circuitry on the same board where the sensors are mounted.

Mount the board and test just as you would a standard encoder board. A dual trace scope is handy here for checking the phase difference between sensor outputs.

Processing Quadrature Output As is the case with a regular encoder, we can determine the distance traveled by counting photosensor pulse transitions. The only difference is that we need to accumulate the number of transitions on *two* sensors rather than one. We determine the velocity of the wheels by sampling the accumulated pulse count at fixed intervals.

Finding the direction of rotation is not much harder. We can treat the output from the sensors as a two-bit binary number, where the top sensor output forms the high bit and the bottom sensor output forms the low bit. As the disk rotates clockwise, the following repeating sequence of binary numbers is output:

00, 10, 11, 01 . . .

If the disk rotates counterclockwise, the exact reverse sequence is output:

01, 11, 10, 00 . . .

It now becomes trivial to calculate the direction of travel. We continually sample the encoder input and compare the current value to the previous value. For example, if our current reading is 10 and our previous reading is 00, the disk is moving clockwise. If the previous reading is 11, then the disk is moving counterclockwise. If our previous reading is 01 or 10, then we have most likely missed a sample, in which case we have to wait for the next sample to accurately determine the direction. This sequence of numbers is often referred to as *gray code*. Gray code comprises an ordered sequence of binary numbers, where adjacent numbers differ only by 1 bit. Our quadrature encoder outputs *2-bit gray code*.

The Index Channel One additional feature usually found in commercial quadrature encoders is the addition of a third output channel called the *index*. The index channel outputs exactly one pulse for each revolution of the shaft. An index channel enables you to not only better synchronize the data from the encoder with the rotation of the shaft, but it can also be used to estimate the *absolute* position of the shaft relative to the index.

For example, assume an encoder with a resolution of 32 counts per revolution. We see the index transition and eight more transitions on channels A and B, where A leads B. Because we are using quadrature encoding, we know that this means the wheel has turned clockwise and is now 90 degrees from the index point. Eight more clockwise transitions and the shaft will be 180 degrees relative to the index position. This is not an ideal way of keeping track of absolute position, of course, since if we somehow lose encoder transitions, we have no way of resyncing unless we encounter the index pulse again, which tells us unequivocally what the current shaft position is. We also can't even begin to estimate position until we've seen the index pulse. A better method, if knowing shaft position is important, is to use *absolute encoding*, which we'll cover briefly toward the end of this chapter.

Quadrature encoding with indexing can be implemented as illustrated in Figure 8-33. This is the same setup we might use for quadrature encoding, with the addition of a third single index segment on the encoder wheel, the passing of which is detected by a new third sensor. The output from the new sensor is our index output.

Another Approach: Beam Interruption Encoders

Beam interrupters are a second type of sensor commonly used in encoder designs. Rather than sensing light reflected from an encoder wheel, how-

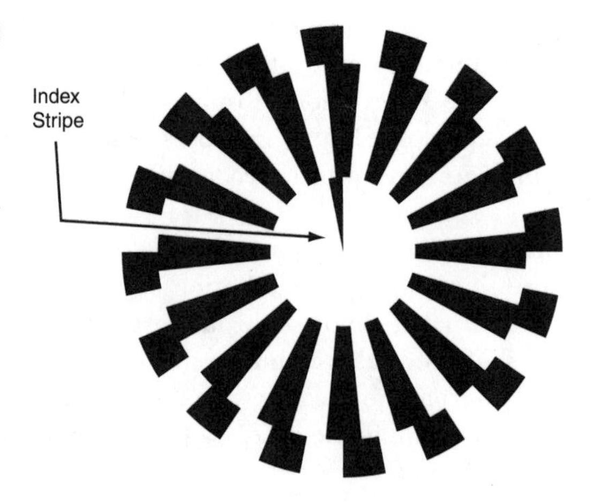

Figure 8-33 Quadrature encoding with a third index channel

ever, beam interrupters are arranged so that an LED shines directly on to a photodiode or transistor across a short space. When an object passes through this space, it interrupts the light from the LED, and the sensor changes state. Once the passing obstacle has cleared the space, light again shines directly on the phototransistor, which again causes the sensor to change state. Figure 8-34 illustrates a simple beam-interrupter arrangement.

Beam interrupters provide somewhat better reliability than reflective photosensor-based encoders, although the performance difference is usually negligible when the reflective sensors are set up correctly. Beam interrupters *may* require a little more mounting work, since the encoder disk, sometimes called a *photointerrupter disk*, must usually be mounted to a shaft with sufficient clearance for the interrupter sensor. You can't simply glue the encoder disk to the robot wheel.

Building the Photointerrupter Disk Photointerrupter disks are commonly found in a variety of commercial products, including computer mice (see Figure 8-35). These disks tend to be quite small, however, and you'll need to find a way to couple them to your motor output shaft.

Using a hole saw, you can cut a circle of plastic, wood, or even metal. Drill holes evenly around the perimeter. You may want to create a drilling template using your favorite drafting software to serve as a

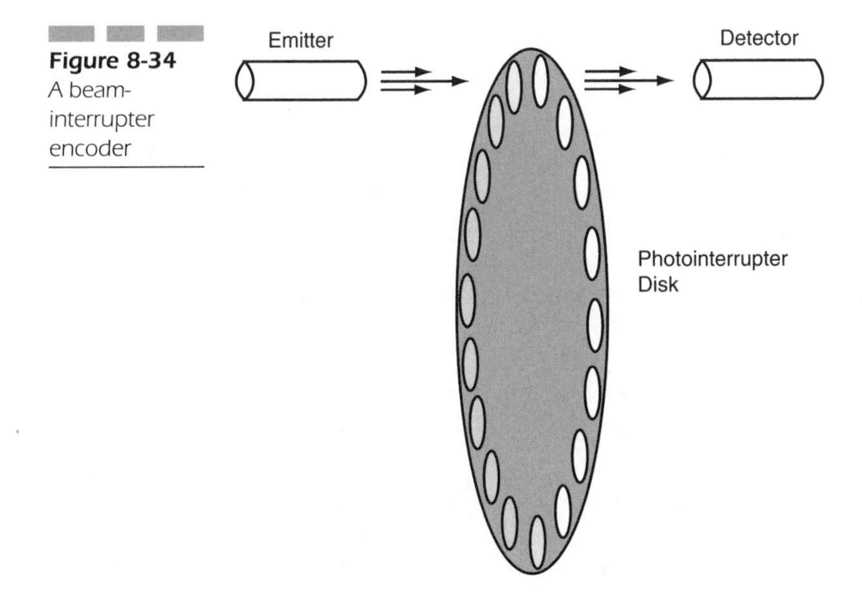

Figure 8-34 A beam-interrupter encoder

Figure 8-35 A disemboweled mouse, showing encoder parts

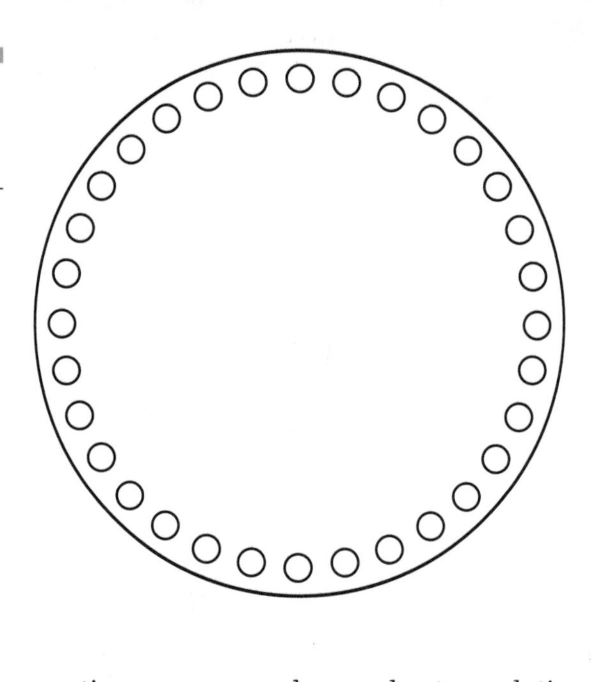

Figure 8-36 A drilled encoder disk template

guide. With some patience, you can make a moderate-resolution encoder. Figure 8-36 shows an example.

Finally, you may be able to print out an encoder onto transparency stock. The patterns used for reflective encoder disks discussed earlier are perfectly suitable. As when printing reflective encoder disks, make sure you use the printer setting that produces the darkest blacks possible. Printed interrupter disks with a small diameter (1 to 2 inches) may be sufficiently rigid so that no backing is needed. Larger disks may need to be laminated onto a stiff backing of transparent plastic or even acrylic stock.

Note that the ink or toner you use to print the wheel may not be sufficiently opaque to IR to give a clean make-break transition at the photosensor. This can be the case even if the ink appears to be completely opaque to visible light. You can test the opacity of your printer inks by printing a dark square onto a transparency and, using a video camera, viewing the transparency as you pass it over a source of IR light. You may use a TV remote as a source of IR light. Most video cameras are sensitive to IR in varying degrees. You can also purchase special cards that fluoresce in the presence of IR, but we have generally found these insufficiently sensitive to be very useful. The camera-test method can be a little flaky, since not all cameras exhibit the same sensitivity to IR; moreover, the IR source that you use may have a wavelength differing significantly from the source used in the interrupter hardware.

The Photointerrupter Sensor You can build your own sensor from an IR LED and phototransistor. When possible, try to get matched pairs, where the phototransistor is constructed for an optimal response at the wavelength of its companion LED. A separate LED and phototransistor can give you extra flexibility when it comes to mounting the sensors, although you will need to make sure everything is properly aligned.

You can also buy interrupter units ready-made with an integral LED and matching phototransistor set facing one another in a U-shaped plastic housing. These units are as cheap and readily available as the photoreflector units discussed previously. The integrated units come in a number of sizes, so when selecting a unit, be sure that the gap between the LED and phototransistor has enough clearance for the photointerrupter disk to pass unobstructed. Some surplus photointerrupter units are shown in Figure 8-37.

Whether you choose a ready-made interrupter or build your own, you'll need the additional support circuitry shown in Figure 8-38. This circuit provides the required current-limiting resistor for the LED, and biasing and pull-up resistors for the phototransistor. Does this circuit look familiar? It should—it's the same support circuitry used with photoreflective sensors.

Figure 8-37 Commercial photointerrupter units

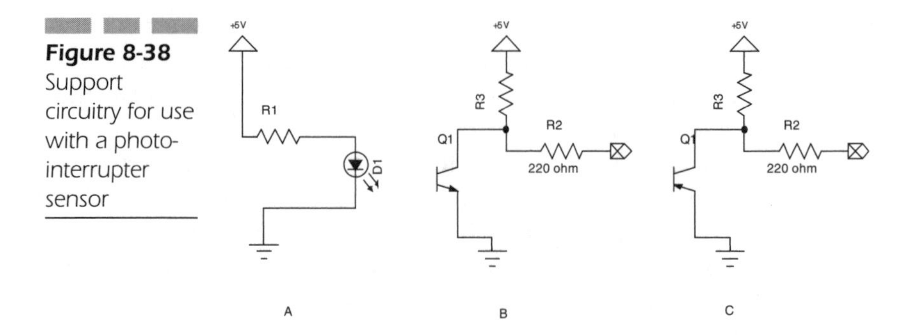

Figure 8-38 Support circuitry for use with a photo-interrupter sensor

Circuit A provides a current-limiting resistor R1 for the emitter portion of the sensor. R1 may range from 220 ohms to 1K. Unlike the case when using reflective sensors, the photointerrupter LED does not need to be very bright. In fact, if the LED is too bright, not only will the unit consume far more power than necessary, but the IR light may actually shine through otherwise opaque disk material, causing response problems (or no response at all). For this reason, we suggest you start at the high end (around 1K) for R1, and only lower the resistance if you have response problems.

Circuit B provides support for the phototransistor detector. Use Circuit C if your detector uses a PNP phototransistor. For both circuits, R3 controls sensitivity and can range from 10K up to 2 Meg, depending on the detector. You may need to experiment with this value if you do not have any datasheets for your detector. Finally, you should consider conditioning the output of the detector via a Schmitt trigger, as illustrated earlier in Figure 8-14.

Mounting the Photointerrupter Disk and Sensor You'll need some free real estate on your motor output shaft in order to mount the interrupter disk. Try gluing the disk onto a small shaft collar and then attaching the collar to the output shaft. This allows you a bit of wiggle room if you need to adjust the disk position along the shaft. In Figure 8-39, we opted for a low-resolution (4 CPR) encoder mounted to the motor armature shaft. We constructed this encoder by printing a simple 4-stripe encoder wheel on photo paper, and then cutting out evenly sized sections of the white stripes. We used a high value current-limiting resistor on the IR emitter so that it could not be detected through the black portions of the disk.

The interrupter unit should be reasonably sheltered from ambient light, although this is even less critical than when using reflective sen-

Figure 8-39 A photo-interrupter disk mounted on a motor armature shaft

Figure 8-40 A photo-interrupter assembly

sors. If using a manufactured sensor, ensure that the disk and sensor are sufficiently secure so that the disk does not loosen and "wander" into the encoder. The sensor can be epoxied to some stiff piano wire, coat hanger wire, or a metal strip. A completed assembly is shown in Figure 8-40.

A Hall-Effect Sensor-Based Encoder

Nonoptical methods of sensing shaft velocity have the advantage of being immune from "noise" created by ambient light. Although it may be possible to connect a motor shaft to a mechanical rotary switch and count the pulses generated as the switch opens and closes, a purely mechanical method is prone to failure over time as the switch mechanism degrades from continual use.

Figure 8-41 illustrates an approach sometimes used in light-duty applications. A magnet is attached to a rotating shaft in such a way as to pass close to a reed switch as the shaft rotates. When the magnet passes close to the reed switch, the switch closes. A microcontroller can count the number of times the switch opens and closes to sense shaft velocity.

The reed switch, being a mechanical beast, will only be able to open and close a finite number of times until it fails. For the same reason, the rotary switch we first considered will eventually fail. We will also need to debounce the reed switch signal, and the switch itself can only be reliably triggered at relatively low motor velocities, making it unsuitable for armature mounting. A better solution would be to incorporate something in place of the reed switch that has no moving parts. A *Hall Effect* switch will fit the bill nicely.

The Hall Effect Switch The Hall Effect switch is named for Johns Hopkins University graduate student Edwin Hall, who in 1879 discov-

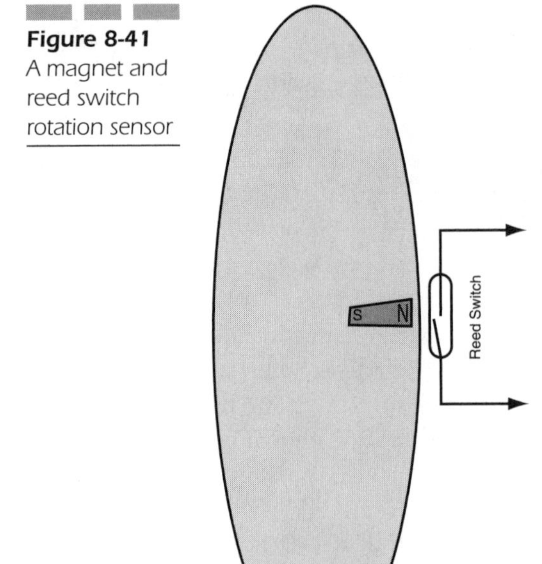

Figure 8-41 A magnet and reed switch rotation sensor

ered that when a charge-carrying conductor was placed in a magnetic field, a small voltage potential was created at right angles (transverse) to both the magnetic field and the direction of current flow. This is due to the deflection of electrons toward the side of the conductor. Figure 8-42 illustrates the phenomenon.

A Hall Effect switches are small-footprint devices, generally having three leads, that usually pull a pin low in response to the presence of a magnetic field (although the exact behavior of the output pin will depend on the device). These devices are perfect reed switch replacements. They need no debouncing, can switch at speeds in excess of 10 kHz, and don't wear out. It is for these reasons that Hall Effect devices are frequently used to sense rotation in brushless electric motors.

Figure 8-43 illustrates how such a device is used. The Hall Effect switch used is the Panasonic DN6848 available from Digi-Key (Digi-Key

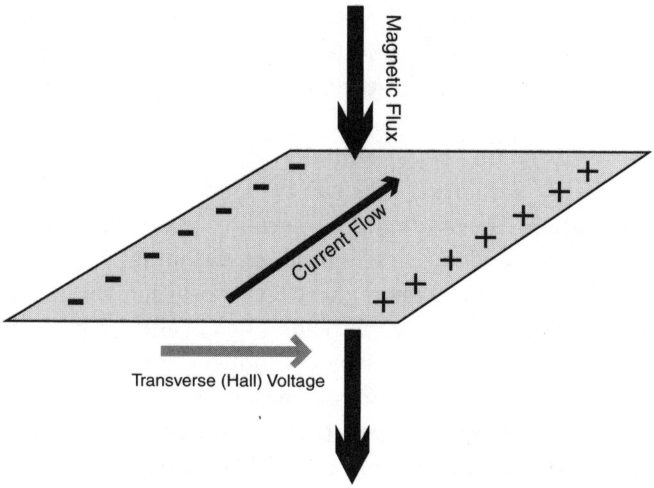

Figure 8-42 The Hall Effect

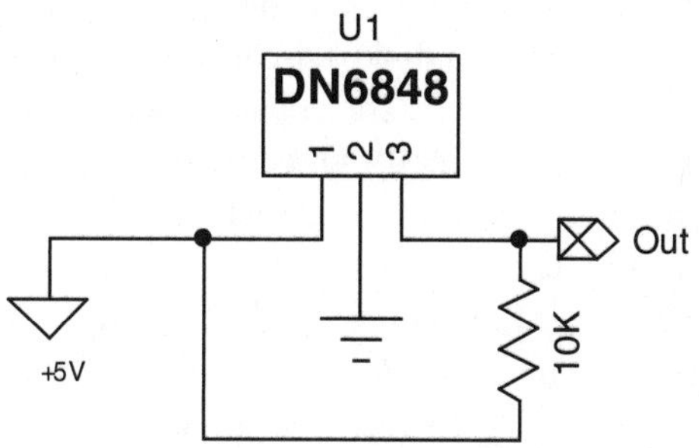

Figure 8-43 Using a Hall Effect switch

part number DN6848-ND) and costs about $1 new. Other suitable devices can be found surplus for even less. These devices tend to be made for speed control or related applications, so most types you are likely to encounter will be suitable for our purposes.

The Panasonic device is open-collector. That is, it will pull its output pin low in the presence of a magnetic field and enable the pin to float when the field is removed. A 10K pull-up resistor ensures that a microcontroller sees 5 volts when the magnetic field is removed, but this will not be necessary if the microcontroller provides internal pull-up.

Be sure that the switch you select is not the *latching* variety. A latching device does not change state when the magnetic file is removed. It instead remains on until either a reset or the power is removed.

Building the Encoder Disk The basic idea behind our encoder disk is to glue magnets around the perimeter of a nonmagnetic disk. Unfortunately, a limit exists on how close we can place adjacent magnets; the exact distance will vary with the strength of the magnets and the sensitivity of the sensor—before the sensor fails to see any break in the activating magnetic field. Although you may be able to construct larger (six-inch or wider) encoder disks with decent resolution, Hall Effect devices are really better suited to armature shaft mounting, where a relatively low-resolution disk can be used.

On other caveat should be borne in mind: Although it's easy to construct an optical encoder to produce a symmetrical square wave output, you may need to experiment a bit to position the magnets correctly so that the Hall Effect switch transitions evenly as the disk rotates. If this turns out to be too much trouble and you can live with relatively low resolution, you can choose to read only high-to-low (or, alternatively, low-to-high) transitions. Although this will halve your resolution, it will make the encoder disk somewhat easier to construct.

The disk can be cut from any nonferrous material, such as aluminum, plastic, or PC board. A small hole saw can be used to cut a disk if the guide drill bit isn't too large. You may also be able to find a suitable aluminum fender washer or other disk. Of course, there's no requirement that the encoder disk be circular; the most important thing is that the magnets be arranged symmetrically. Of course, as you glue the magnets down, remember that they *do* tend to attract one another. This can make accurate placement a little unruly at times.

Figure 8-44 is designed for armature shaft mounting. The substrate disk was cut from an old piece of PC board using a hole saw. A shaft collar is glued on to render the disk easily removable. To keep down the

Figure 8-44 A two-CPR magnetic encoder disk

weight, the encoder disk uses small chips of powerful neodymium magnets, instead of larger (and heavier) alnico or other types. The disk only provides two CPR (the encoder pulses will be quite narrow, so we only count leading edges), but this should be adequate for an armature-mounted encoder.

NOTE: *Some Hall devices are sensitive to the polarity of the magnetic field. The Panasonic device used here will only turn on if facing the south end of a magnet. Not all devices behave this way, however.*

A Complete Assembly Figure 8-45 shows a completed assembly. The sensor and pull-up resistor are built on a small strip of prototyping board and mounted via a bent coat-hanger mounting arm so that the assembly is suspended over the armature-mounted encoder disk.

Figure 8-45 A Hall Effect encoder assembly on a motor armature shaft

Nondriven Encoder Disks

All our encoder schemes to this point have been designed to measure the rotation of a shaft connected directly to the output of the robot drive train. This approach is generally adequate for motor speed control and is the route taken by most amateurs. In the case of wheeled robots, however, it is possible to improve on the accuracy of our encoders by attaching the encoder disk to a free-spinning wheel on the robot base. As the robot moves, the wheel spins along the surface of travel. Such an arrangement confers a number of advantages.

Nondriven encoders can be used to measure the velocity of the robot as opposed to the velocity of the motor output shaft. These two velocities frequently differ due to the effects of *wheel slippage*. In a perfect world, each rotation of the robot's wheel would move the robot a distance exactly equal to the circumference of the wheel. However, even on the friendliest of surfaces, the wheel will have a tendency to slip slightly, so that each wheel rotation moves the robot a distance less than the wheel circumference. The amount of error encountered depends upon a couple of factors:

- **Wheel construction** "Stickier" tires will grip smooth surfaces better and produce less error, assuming a smooth surface. On carpet, heavily treaded wheels will be less prone to slippage.
- **Travel surface** As might be deduced from the previous factor, the surface along which the robot travels will also influence the amount of slippage encountered.
- **Linkage slip** If the wheel is linked to the robot motor via a mechanism that allows slippage at high loads, such as a slip clutch, belt and pulley, or similar mechanism, then some slippage will be encountered, especially at higher loads.
- **Robot weight** Heavier robots will tend to have fewer problems with wheel slippage because of the greater amount of force exerted by the wheel against the surface of travel.

For the most part, wheel slippage over unobstructed surfaces is more of an issue when odometry and dead reckoning are used to determine current position. In these cases, wheel slippage can, over time, account for a significant amount of error in position estimation.

Wheel slippage can become a bigger problem when an obstacle is encountered. Consider Figure 8-46. When our hypothetical robot encounters an obstacle—which we'll assume was missed by any onboard obsta-

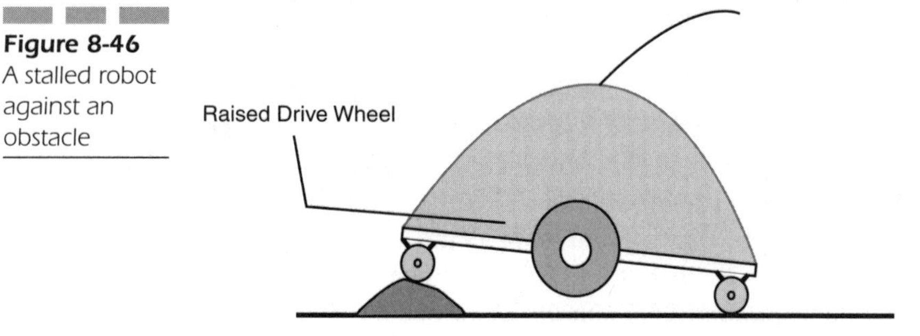

Figure 8-46 A stalled robot against an obstacle

cle sensing—the center drive wheels do not stop spinning, as we might expect. Because the obstacle has caused the robot to incline slightly, thus raising the wheels a bit off of the surface, the motor continues to spin the wheels at the rate requested by the microcontroller, with only a little extra current draw. Thus, we can use neither the controller's current-sensing capability nor shaft-mounted encoders to detect a stall since technically the motor isn't stalled, only the robot.

Variants of this situation might be encountered when a single wheel is blocked, slows unexpectedly, and causes the robot to follow an unanticipated curving path instead of stalling in place.

Even if the wheels are not raised off the ground when an obstacle is encountered, they may continue to spin in place anyway, slipping against the floor. They will do this if the motor is sufficiently powerful to overcome the friction between the surface and the wheel.

For simple motor control feedback, these scenarios may not need to be considered. However, if you plan to use a single feedback mechanism for both odometry and speed control, you may want to look at the possibility of using nondriven encoders for velocity feedback. You might also decide to use this method to augment your existing encoders as a means of detecting robot stall or other unexpected impediments.

The benefits of using nondriven encoders are offset by increased mechanical complexity, the level of which will vary with robot design. If your robot uses car-type or tricycle-type steering, you may be able to simply mount encoders on the nondriven wheels used for steering. In this case, implementing nondriven encoders is a trivial task.

A dual-differential robot, on the other hand, will require you to do more mechanical work. Although it may occur to you to simply mount encoders directly onto caster wheels, we have found this approach to be problematic. First, most manufactured caster assemblies are notoriously difficult to disassemble in such a way as to allow mounting of an encoder disk.

The larger problem, however, lies in the fact that for the caster to function properly it needs to be able to swivel a full 360 degrees. This makes supplying power to the encoder and getting back signals problematic without using expensive and difficult-to-find slip rings or similar mechanisms. Caster motion can be mechanically limited with stops, but this can interfere with the correct movement of the caster in response to direction changes. One approach that can be used on a dual-differentially steered robot is to mount a wheel and encoder that are parallel to one or both drive wheels, as shown in Figure 8-47.

Be sure that the free-running wheels are mounted so that they are in line with the centers of the drive wheels to avoid problems when the robot turns. You will also need to be sure that the wheels are level with the drive wheels.

A Simple Adjustable Free Wheel Mount Mounting a free-running wheel on your robot can be done in any number of ways, but all assume that you have enough unoccupied real estate on the chassis so that there is room for one or two wheels to be mounted along the robot centerline. Unfortunately, this is often not the case, since motor casing and the gearbox can take up a lot of space along the bottom of the robot. If batteries are slung below the chassis, you may have to get creative with wheel mounting (such as attempting to mount the free wheels to the *outside* of the drive wheels).

Assuming you find the room, one easy way to mount free-running wheels is to use the standard parts that RC airplane enthusiasts use for making landing gear assemblies. Figure 8-48 shows a completed landing gear assembly. The parts required are available from most any hobby shop (or over the Internet) that caters to RC airplane hobbyists.

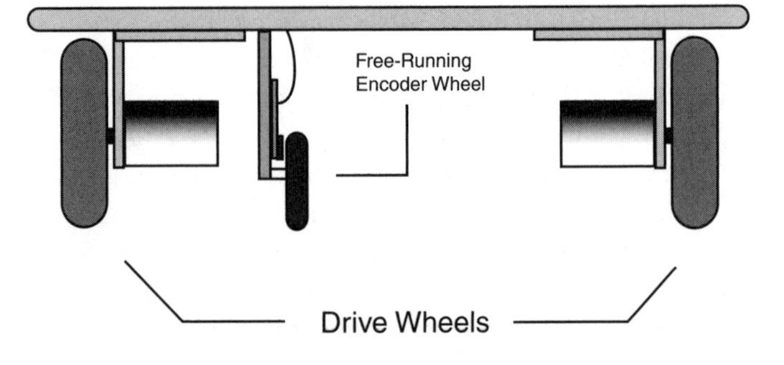

Figure 8-47 Mounted free-running wheels and encoders on a dual, differentially steered robot

Figure 8-48 An assembled, unmounted landing gear assembly

The parts you will need include the following:

- **Two model airplane foam landing wheels** The diameter will depend upon the amount of clearance available beneath the chassis.
- **Two landing gear axle/collar assemblies** The size needed will depend on the axle bore of the wheels used.
- **Two small shaft collars** The bore should match the diameter of the landing gear axle.
- **Two lengths of piano wire** The landing gear axle collars attach to the wire, so the diameter of the wire should match the landing gear collars. The length of each piece of wire should be the radius of the landing gear, plus the distance from the top of the wheel to the top of the robot chassis on level ground (this should include the chassis thickness), plus an inch and a half (38 millimeters) and about another inch (25 millimeters) extra.
- **Four small U ties** These are also sometimes sold as *landing gear straps*. They must be fitted to the diameter of the piano wire.

To build *each* wheel assembly, follow these steps:

1. Bend the piano wire into an L shape, with the short end being about 1 to 1.5 inches (25 to 38 millimeters).
2. Drill a hole in the robot chassis. The hole should lie along the same centerline as the drive wheel. The wheel mount will be inserted

into this hole, so be sure that you have enough clearance for the wheel once mounted.

3. Insert the long end of the piano wire into the hole. Secure the wire to the top of the chassis with two of the U ties. Figure 8-49 shows how the short leg of the strut is secured to the top of the chassis.
4. Next, attach the axle to the piano wire. Do not tighten the set screw all the way just yet.
5. Slide the wheel onto the axle. Adjust the whole setup on a flat surface so that the wheel firmly contacts the ground.
6. You may want to mark the axle collar position at this point. Remove the piano wire strut and file a flat into the strut to keep the axle from loosening and rotating during use.
7. Secure the wheel to the axle with the shaft collar.

Figure 8-50 shows a completed wheel and strut assembly sans an encoder. An encoder disk can be glued to the wheel and the sensor board can be attached directly to the strut arm.

Improving the Nondriven Encoder We can make at least two possible improvements to our nondriven encoder system:

- We might want to spring-load the strut/wheel assembly so that a positive force keeps the wheel in contact with the ground. This is particularly important to do when we encounter obstacles, since if a drive wheel is raised off the ground by even a small amount, the

Figure 8-49 A piano wire strut secured via landing gear straps

Figure 8-50 A completed wheel assembly

encoder wheel can be lifted along with it and will cease to rotate even as the robot continues to move. This can be a major source of encoder undercounting as the robot passes along a door threshold or carpet-floor boundary at an angle.

- Most texts recommend the use of a firm, razor-thin encoder wheel. Instead of a model airplane wheel, you can manufacture your own wheel from aluminum or a thin piece of wood. Both can be cut into an accurate disk using a hole saw, preferably in a drill press. Attach a rubber O-ring around the perimeter of the wheel to provide traction.

Commercial Shaft Encoders and Analog Tachometers

You can save a considerable amount of time and effort by acquiring motors with built-in encoders or tachometers, or by purchasing these devices as ready-made units. Deep pockets aren't necessarily required, but you will need luck if you're shopping surplus.

Remember that it is possible to purchase motors with these features already built in. You may have to do some looking, however, to find a motor with the correct attributes for your design (voltage, torque, speed, and so on) that includes a tachometer or encoder. Still, it's worth the time to search if you'd rather not create your own.

Commercial Shaft Encoders

Commercial rotary encoders are available from a number of manufacturers. Lest you purchase the wrong part, bear in mind that these encoders come in three flavors.

Low-speed Mechanical Rotary Encoders These parts are inexpensive but are really designed exclusively for use as dials on instrument panels or for other applications where relatively low-speed, intermittent rotation needs to be measured. Although these parts output 2-bit gray code, they often have fairly low resolution (as low as 16 to 23 positions per revolution), need debouncing, and will not stand up well to continuous use, often being rated for as few as 10,000 cycles (complete rotations) before failure.

Some versions can output absolute position via *Binary Coded Decimal* (BCD). Such devices, while not well suited for speed feedback, could be put to good use in other types of applications where continuous rotation does not occur. For example, these devices might be utilized to sense the current angle of a robotic arm or leg joint. One commonly available encoder is the Grayhill series 25L 36-position encoder. This device is meant specifically for control-panel applications, but its 100,000 cycle rating (long by mechanical encoder standards) and relatively low price tag of about $5 in single-unit quantities might recommend it for light-duty rotary-sensing applications.

Low-Speed Optical Rotary Encoders Also intended for use in manual control-type applications, these devices are essentially optical versions of the mechanical rotary encoders but utilize optical encoders internally rather than mechanical switches. Unlike their mechanical cousins, these devices have long life cycles, some models being guaranteed for up to 10 years of continuous operation. Resolution is much higher for optical encoders, with 128 to 500 pulses per second being typical, depending on the model. Most optical models are built to withstand rotary speeds of up to 300 RPM, making them good candidates for shaft velocity feedback. You will still, of course, need to find a way to couple a drive shaft to the encoder input shaft if you want to use these encoder types.

Unfortunately, the cost of rotary optical encoders is commensurate with their increased capabilities. Prices range from $20 to $50 when purchased new. Fortunately, these items occasionally turn up on the surplus market, where they can be had at a considerable discount. The Hewlett

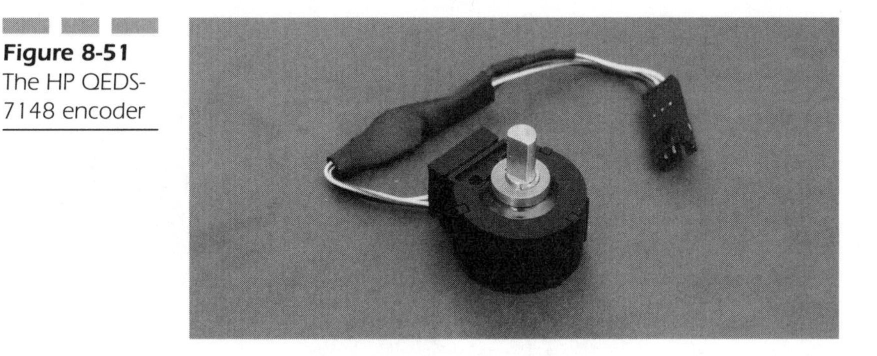

Figure 8-51 The HP QEDS-7148 encoder

Packard QEDS-7148 encoder shown in Figure 8-51 provides 500 pulses per each revolution of its input shaft.

Because the low-speed optical encoders discussed here are manufactured for use in manual electronic controls, they often have a slight built-in resistance meant to provide tactile feedback to a human operator. This means that using these units to measure motor shaft speed will put an extra bit of load on your motor. In general, however, the amount of shaft resistance is very small but could become a factor on particularly small robots using low-torque motors.

As mentioned earlier, these units are not made to exceed speeds of 300 RPM. This is fine for most indoor robots, where motor speeds won't usually exceed 75 RPM, but it would not be a good idea to mount these to motor armature output shafts or to very high-speed motors.

These encoders are almost always equipped with input shafts, as opposed to shaft hubs. The shafts are meant to mate with knobs. Pay close attention to the shaft size; you'll need to couple the shaft to your motor output or wheel, which may cause you some trouble if the input shaft is of an odd size.

High-Speed Encoders These encoders are built specifically for use in high-speed applications, such as motor control. As the name implies, high-speed encoders can be run at very high speeds and are suitable for motor armature mounting. Unfortunately, getting these types of encoders in small quantities can be difficult and expensive, with prices varying from $40 on up for quadrature-type encoders.

Some companies, however, do offer high-quality encoders in single-unit quantities. One such unit is the U.S. Digital E2 Series Encoder (www.usdigital.com) shown in Figure 8-52. This unit offers quadrature

Figure 8-52 US Digital E2 encoder. (Source: U.S. Digital Corporation)

output in resolutions up to 1,024 CPR and an optional index channel. Mechanically, the unit supports maximum shaft speeds in excess of 100,000 RPM, but the usable range depends on the CPR of the encoder disk (the formula 100,000/CPR can be used to determine the maximum usable RPM). All of this comes with a price tag of around $40 in single-unit quantities, with substantial discounts on volume purchases.

The E2 devices consist of a base, an encoder disk with a hub and setscrew, optics, and a housing. The base is mounted over the shaft, the encoder disk is mounted to the shaft (over the base), and the optics and housing are mounted onto the base. The encoder disk can be ordered with a variety of collar sizes, from 1/8 inches to 3/8 inches and 2 millimeters to 10 millimeters, making these devices suitable for use with most standard shaft sizes. The cover can also be ordered with a through-hole so that the encoder module can be mounted “inline.” If you don’t want to build your own encoder and have some extra money to spend, these units offer a reasonably priced and effective alternative.

Analog Tachometers

Analog tachometers output a DC voltage that varies precisely with motor speed. For example, an analog tachometer might output 0.005V for each RPM of movement. Since the output of the tachometer is always linear, we would expect to see 2.5V output at 500 RPM, 5V output at 1,000

RPM, and so on. This can give a slight advantage to applications interested only in measuring motor speed, since there is no need to sample a pulse count at fixed intervals to derive shaft speed.

When selecting a tachometer or a motor with one already built in, consider the following:

- Clearly, you will need to be able to convert the output from the tachometer into a digital value. If your microcontroller or other control logic lacks this feature, you will need to build a circuit that is capable of doing so.
- Be sure that your AD conversion circuitry has enough resolution to accurately convert the tachometer output. This is especially important when you expect to be using the tachometer at low speeds, since the output voltages could be fairly small. Such a situation might arise where you have connected the tachometer output to the output of a relatively slow gearhead motor.
- Make sure that your AD circuitry can handle the possible range of voltages produced by the tachometer. Some AD circuitry may use the supply voltage as a reference and thus will be unable to measure (and may even be damaged by) voltages greater than the power supply voltage.

Both tachometers and motors with tachometers built in show up regularly on the surplus market, although you may have a tough time finding a tachometer-equipped motor that fits your torque, speed, and input power requirements.

Using Motor Feedback: Control Algorithms

A motor control algorithm should be able to take a commanded velocity and apply power to the motor in such a way as to bring it quickly and smoothly up to speed and maintain that speed. Once achieved, the commanded speed should be maintained, even in the face of unexpected factors such as inclines, bumps, or other factors that might work to slow down or speed up the robot.

The control algorithm (also sometimes called a *control law* by roboticists) is the heart of any closed-loop control scheme. An enormous number of control algorithms exists; in fact, control theory is a science in its own right, but we will concentrate in this section on *Proportional Integral Derivative* (PID) control laws.

An Introduction to PID Control Laws

Before proceeding, we should define a few terms commonly used to describe PID control loops:

Process Sometimes called the *plant*, this is the object being controlled by the control loop. In our case, this would be the robot drive motor (or motors).

Control variable This is the output of the control loop to the process. It is usually abbreviated CV. For a loop controlling a robotic drive, the CV might be a new value for the PWM duty cycle controlling a motor.

Process variable Usually abbreviated PV, the process variable is the *feedback* value returned by the process. In our case, this would be the current motor speed as read by a shaft encoder or tachometer.

Set point The *set point* (SP) is the value we desire from the output of our process. A correctly designed control loop continually seeks to make the PV equal the SP. The SP for a motor controller would specify a motor speed.

The operation of many error-based, single-output controllers, whether of the PID variety or not, can in general be broken up into the following steps:

1. Read the currently commanded SP. In our case, the set point will be a desired motor speed in RPM.
2. Read the PV. Our process variable will be the current speed of the motors as sensed by a tachometer or encoder. Note that we will need to be sure to convert the encoder counts or tachometer output to the same units used by the SP—in this case RPM.
3. Subtract the PV from the SP to get the *error signal*. The error signal tells us how far away the process output is from the desired output. A high value means that process output is far from the desired output (SP). A low value means we're getting close, and a value of zero means that we are dead on. Error signals can usually be positive *or* negative. For our motor controller, a positive value

means the motor is running too slow. A negative value means it's running too fast.

4. If the error signal is nonzero, transform it into a valid value for the CV that will change the process output to more closely match the SP. Put another way, we use the error signal to derive a new value for the CV (a power command to the motor) that will cause the PV (the currently sensed motor speed) to move closer to the SP (the desired motor speed). Note that some controllers may ignore small error signals within a certain range, termed the *dead band*.
5. Repeat the entire sequence.

When diagrammed, the control system described looks like Figure 8-53.

In the case of a PID controller, the error transformation box corresponding to the previous step 4 is actually broken out into three boxes, or modules, each of which transforms the error signal simultaneously. These boxes include a *proportional* module, an *integral* module, and a *derivative* module. The outputs of all three are *summed* and fed to the plant to produce a new output. Figure 8-54 shows a generic PID controller loop.

It is possible to do without the integral or derivative controller terms. For example, we might find that our controller needs no integral term. This would leave us with a *proportional-derivative* (PD) controller.

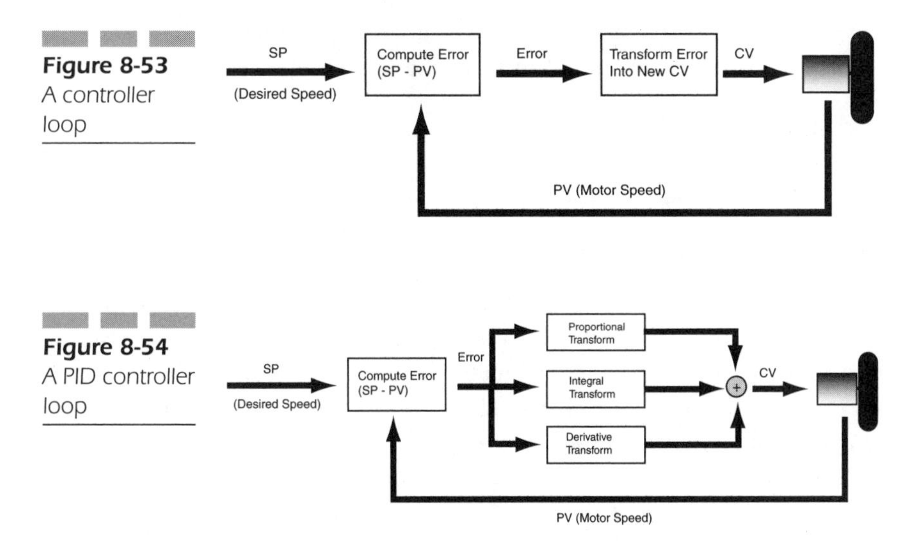

Figure 8-53 A controller loop

Figure 8-54 A PID controller loop

Alternatively, we could omit both the derivative and integral term and end up with a simple *proportional* (P) controller.

Implementing a PID Controller in Pseudocode We'll illustrate the use of the proportional, integral, and derivative terms using the example of a simple, single-motor speed controller. This controller only attempts to control the motor in one direction, although speed control in two directions is easy to implement. Our theoretical setup looks like the following:

- We have a PWM control module that takes a number between 1 and 100, indicating the duty cycle of the PWM pulse to output.
- The PWM module controls a gear motor capable of outputs of up to 45 RPM.
- Connected to the gear motor output shaft is an encoder with a resolution of 64 counts/rotation.
- The control loop accepts an SP in desired RPM, from 0 to 45.
- Our control loop runs a modest once per second. On each pass, it samples the encoder output from the motor and converts the count taken since the last sample into RPM by multiplying it by 0.9375. This value is our computed PV.

The Proportional (P) Term The first error-transformation module (or term) in our PID loop is the P module. This module simply multiplies the error signal by a constant, K_{prop}, to derive a new value for CV, which should be between −100 and 100. The algorithm for a basic proportional controller would look something like the following:

```
Loop:
      PV = ReadMotorSpeed()
      Error = SP - PV
      CV = Error * Kprop
      SetPWM(CV)
      Goto Loop
```

Note that our SetPWM does not treat CV as an absolute PWM duty cycle value. This is because decreasing values of Error would cause the output to approach 0. Since the motor requires a continuous PWM signal to operate, the loop will always tend to stabilize at low speeds, making our control loop largely ineffective.

Instead, CV is treated as a *change* to the current PWM duty cycle and is added to whatever the currently applied PWM value happens to be.

This implies that SetPWM must be sure to limit the resulting PWM signal such that the duty cycle remains between 0 and 100 percent.

Positive values of CV will cause the power to the motors to *increase*. Negative values cause it to *decrease*. If CV equals 0, there will be no change in the duty cycle. With a well-chosen value for K_{prop}, we might get the motor response curve shown in Figure 8-55.

Lower values for K_{prop} will tend to give smoother but slower responses. Higher values will give much quicker responses but may cause overshoot, where the output oscillates around the setpoint before settling. Excessively high values may even throw the loop into an unstable state where the output oscillates without ever settling at the setpoint, as in Figure 8-56.

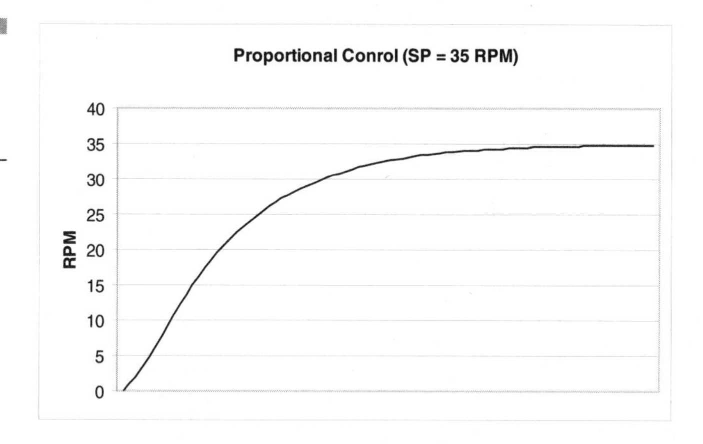

Figure 8-55 Proportional control loop response

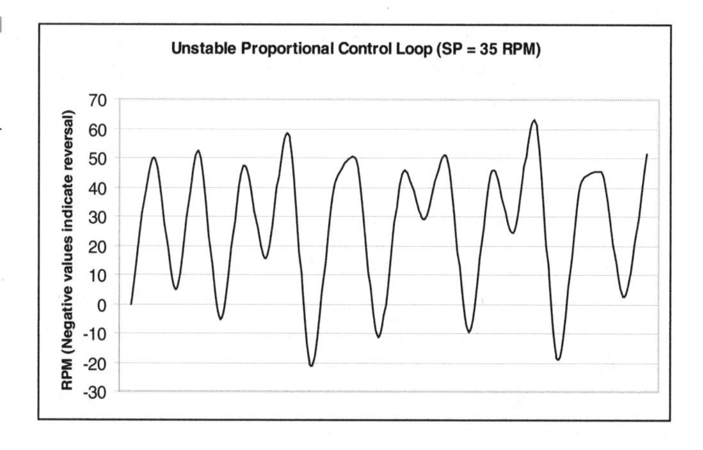

Figure 8-56 Control loop instability

Although there is a science to choosing an ideal proportional constant, we have usually just opted for the trial and error method. In general, we have found proportional control alone usually suffices for simple motor speed control.

The Derivative Term The *derivative* of any variable describes how that variable changes over time with respect to another variable, often elapsed time. In other words, the derivative of any variable is its rate of change. For example, the derivative of distance with respect to time is termed *speed*. The derivative of speed with respect to time is *acceleration*.

In the case of our PID controller, we will be interested in the derivative, or rate of change, of the error signal with respect to time. Most controllers define the derivative term of a controller, sometimes called the *rate*, as

$$\text{Rate} = (E - E_{last})/T$$

where E is the current error value, E_{last} is the last error value, and T is the elapsed time between measurements. Many controllers simply assume T equals 1 and eliminate it from the equation, leaving us with

$$\text{Rate} = (E - E_{last})$$

Negative values of Rate indicate an improvement (reduction) in the error signal. For example, if the last error measurement was 20, and the current measurement is 10, Rate will be -10. If the last error was 10 and the current error is 20, Rate will be 10. An unchanged error signal yields a rate of 0.

When used in a controller, a derivative term is multiplied by a constant, K_{rate}, and added to the output of the proportional term to give us the final value of CV:

$$CV = (K_{prop} \times E) + (K_{rate} \times Rate)$$

Because it goes negative as the error signal approaches 0, the derivative term is used to back off the corrective output as the error signal begins to improve. In some situations, this can help eliminate overshoot and allow the use of a larger value for K_{prop}, which can speed up the response of the control loop. The derivative term, in a sense, *anticipates* improvement in the error signal that may not yet be apparent from the current measurement. This being the case, derivative action is often helpful when the process tends to respond slowly to changes in controller output.

A pseudocode version of our control loop with a derivative term added in would look like the following:

```
Loop:
      PV = ReadMotorSpeed()
      LastError = Error
      Error = SP - PV
      Rate = Error - LastError
      CV = Error * Kprop + Krate * Rate
      SetPWM(CV)
      Goto Loop
```

In practical motor control applications, a derivative component should seldom be necessary, although you may want to experiment. There may be other robot control applications, such as wall following or obstacle avoidance, in which you might find a rate component to be a useful facility to have in your control loop. If you do want to incorporate a derivative term in your control loop, consider suppressing rate calculation when the set point is changed to avoid large swings in the value of the Rate variable.

As is the case with the proportionality constant, you'll need to take care when you choose K_{rate}. Excessive values can actually result in a loop that is less stable than when the term is omitted entirely. K_{rate} will usually be less than 1, but again you may experiment.

The Integral Term An *integral* is the exact opposite of a derivative. Given an expression describing a rate of change (a derivative), the integral of the expression yields the value of the thing being changed. For example, integrating acceleration yields speed. Integrating speed yields distance.

In the case of our PID control loop, the integral of the error is merely a running total of the error signals encountered since startup. This total is multiplied by a constant, K_{int}, and is added into the loop output.

So why might we want to incorporate an integral component into our control loop? Let's say that our loop, for whatever reason, stabilizes, but not at the desired set point. Instead, the loop response looks like Figure 8-57.

The space between the desired stabilized output (at 35 RPM) and the actual stabilized output (at 26.6 RPM) is called the *offset error*. The integrator in our controller is used to overcome the offset. As the offset causes errors to continue accumulating, the integral term of our PID loop grows ever larger, providing an increasing bias to the output, which will (hopefully) overcome the error offset to bring PV in line with CV.

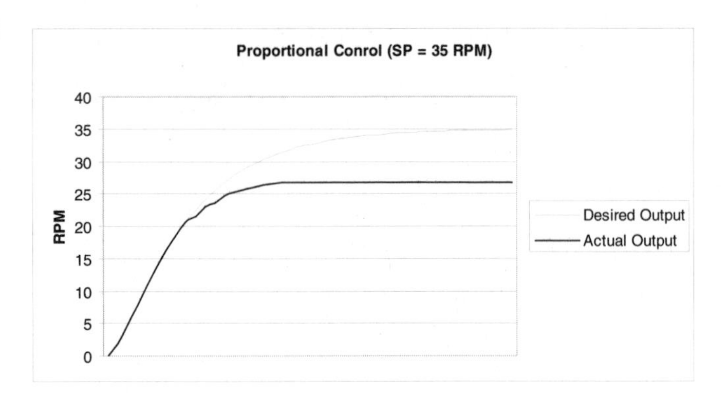

Figure 8-57 A controller response with an offset error

Pseudocode for a full (albeit simple) PID controller is provided here:

```
Loop:
      PV = ReadMotorSpeed()
      LastError = Error
      Isum = Isum + Error
      Error = SP - PV
      Rate = Error - LastError
      CV = Error * Kprop + Krate * Rate + Kint * Isum
      SetPWM(CV)
      Goto Loop
```

Because the integral term can grow quite large, making it difficult for the loop to respond quickly to set point changes, some applications stop accumulating Isum when the CV is at its maximum or minimum values (in our example, −100 or 100). Isum can also be cleared on setpoint changes.

So how might we use an integral component for speed control? For controlling a single motor, our experience has shown that an offset error is seldom a problem with decent values for K_{prop}, and thus an integral component is not usually necessary. As we shall see, however, an integral term can be a useful application in some situations.

Using Integral Control to Slave Dual-Drive Motors One possible application for integral action in a motor controller is to effectively slave two drive motors together. This slaving action can be handy in a dual-differential drive robot. For example, let's assume we want to steer a small robot along a straight path at about 50 percent speed. To do so, we command both the left and right motors to spin at 50 percent of full speed—so far, so good.

Along the way, the left wheel encounters an obstacle—say the family hamster, Lester, who has yet again escaped the confines of his hamster habitat—and the wheel begins to slow. Fortunately, the proportional control loop senses that the wheel is slowing and applies additional power to the motor to keep the output speed up to 50 percent as the wheel negotiates Lester. The robot continues to steer a more or less true path (although the exact amount of deviation depends on the responsiveness of the controller).

What would happen, however, if the same situation occurred when each wheel had been commanded to go full ahead at 100 percent of full speed? The left wheel would slow as above when it encountered Lester, but this time the control loop would not be able to apply additional power to the motor to bring the wheel back up to speed, since full power (or close to full power) is already being applied.

In their book, *Mobile Robots: Inspiration to Implementation*, Jones, Seiger, and Flynn describe a method of using integral action to mitigate the effects of just such a situation. To understand how this works, let's consider a simple bit of code that can be used to keep two motors running at identical speeds. In this loop, we issue a single set point command and expect the loop to keep both motors running at the same speed. If one motor is slowed by an external force, we expect the loop to either increase power or *slow* the other motor so that the speeds are matched:

```
Loop:
     PVLeft = ReadLeftSpeed()
     PVRight = ReadRightSpeed()
     Isum = Isum + (PVLeft - PVRight)
     LeftErr = SP - PVLeft
     RightErr = SP - PVRight
     CVLeft = Kprop * LeftErr - Kint*Isum
     CVRight = Kprop* RightErr + Kint*Isum
     SetRightPWM(CVRight)
     SetLeftPWM(CVLeft)
     Goto Loop
```

In this loop, our integral error is not accumulated based on the usual SP-PV method. Instead, we accumulate the difference between the left and right motor speeds. As long as the speeds remain the same, there will be no accumulated error signal. Should the right wheel slow, however, Isum will begin accumulating positive values. Should the left wheel slow, Isum will begin accumulating negative values.

Isum is multiplied by K_{int}, usually well below 0.5, is *added* to the right motor CV, and is *subtracted* from the left motor CV. When the motors are running at the same speed, no net effect occurs.

When the left wheel slows, more power is applied to the left wheel at the same time that *less* power is applied to the right. If the speed of the left wheel can't be increased, no problem: The right wheel will be slowed. The opposite happens if the right wheel slows: More power is applied to the right wheel, and power to the left is reduced. In this way, the loop eventually achieves the same speed for both motors, although possibly less than the commanded set point.

Of course, we would like to steer our robot as well, rather than simply making it follow a straight line. For our dual-differential robot, steering simply means changing the speed of one wheel with respect to the other. A simple steering command might simply accept two parameters, a left-wheel velocity and a right-wheel velocity (where negative values for either indicate that the wheel should move in the reverse direction). For example, to curve to the left, we might command the left motor to 10 percent and the right to 50 percent of full speed. We could rotate in place toward the left at 50 percent speed by commanding the left motor to −50 percent, and the right to 50 percent. It would be ideal if the control loop could use the same integral action described previously to maintain each motor at the desired speed and, if that proved impossible, at least maintain the desired speed ratio between left and right motors.

We can accomplish this by adding a small change to the control loop described previously:

```
Loop:
      PVLeft = ReadLeftSpeed()
      PVRight = ReadRightSpeed()
      Isum = Isum + (PVLeft - PVRight) + Bias
      LeftErr = SP - PVLeft
      RightErr = SP - PVRight
      CVLeft = Kprop * LeftErr - Kint*Isum
      CVRight = Kprop* RightErr + Kint*Isum
      SetRightPWM(CVRight)
      SetLeftPWM(CVLeft)
      Goto Loop
```

The computation of Isum has been changed so that an additional term, *bias*, is added to the accumulated value each time Isum is computed. We specify this steering bias along with the desired speed whenever we command the robot. When the bias is 0, the robot will travel straight. When the bias is positive, the robot will travel to the left; when negative, the robot travels to the right. Bias is a purely rotational attribute. If we set the speed SP to zero and the bias to 50, the robot will spin on its axis to the left at a 50 percent speed. If we set the speed to 35 and the bias to −40, the robot will curve to the right.

This approach works well in practice, but bear in mind that because the effects of bias tend to be mitigated by K_{int}, the effect of a steering command is not instantaneous but can take a few passes through the loop to attain.

Other PID Issues You should be aware that good velocity control does not guarantee an accurate path of travel. For instance, even with a well-tuned control loop supplying corrective power to our motors, our encounter with Lester the Hamster would still result in a certain amount of course deflection. Good speed control can mitigate this deflection but not eliminate it. Put another way, motor control is not a substitute for navigational control. Unfortunately, navigation is well beyond the scope of this book.

We have assumed in our pseudocode examples so far that arithmetic overflow will not be a problem when calculating PID terms; this is usually not the case. You will likely want to limit most of the terms in your loop as they are computed to minimize artificial and unexpected changes brought on by overflow.

Many platforms will not support floating point at all. In such cases, consider using clever combinations of multiplication and division with appropriate integer values for the K_{prop}, K_{int}, and K_{rate} values. Again, take care to ensure that overflow errors don't occur.

One possible problem that you may encounter when using any control loops with shaft encoders is a jittery encoder count, which causes PV to bounce around a bit. This can be caused by a variety of factors, including the motor gearbox output irregularities and encoder-sensor-mounting problems, among others. It will be difficult for any control loop to stabilize if PV jumps around unexpectedly. If this appears to be a problem, consider keeping a running average of the encoder counts and use these instead of an instantaneous value to compute PV. Your loop will respond a bit more slowly but will stabilize more readily.

Finally, we have used motor RPM as our preferred unit of measurement in the examples so far. This has been to simplify the examples a bit and eliminate wheel radius as an additional factor to consider. However, we have generally found that distance-based units, such as *inches per second* (IPS) or *centimeters per second* (CPS), are generally more useful for robotics than RPM. We would suggest that as you implement your own system, you use either IPS or CPS for both SP and PV.

Absolute Encoders

Although not required for motor control, we really can't consider this chapter complete without a short discussion of absolute position encoding. As the name implies, the output of an absolute encoder corresponds directly to a particular angular position of the attached shaft. The output is in the form of a binary number corresponding to one of n possible positions. The size of this binary number, in bits, is often used to describe the resolution of the encoder. For instance, a 4-bit shaft encoder can encode 16 unique positions. Figure 8-58 shows how we might naively design an absolute position encoder. The dark-light patterns on the disk essentially equate to binary numbers representing a particular position. A light segment represents a 0; a dark segment represents a 1. The corresponding sensor reads each bit of the number, all of which are output in parallel as a binary number from 0 to 16 representing the position of the disk.

We can improve our encoding scheme by switching to *Gray code*. As touched on earlier, Gray code is an ordered sequence of n binary numbers encoded in such a way that only a single bit is changed between adjacent numbers. Put another way, the *Hamming distance* between adjacent numbers is exactly one. Because only one bit changes for each unit of shaft rotation, the Gray code output reduces the ambiguity during a state change. For this reason, Gray code tends to be used almost exclusively by commercial encoders.

A sequence of numbers as Gray code can be encoded in more than one way. The most commonly used variant is termed *binary reflected* Gray code. Figure 8-59 shows our redesigned encoder wheel.

We would expect to see the outputs in Table 8-1 as the disk rotates from position 0 to position 15.

Figure 8-58 Naive absolute encoder

Figure 8-59 Gray code encoder wheel

Table 8-1 Gray code absolute encoder output

Position	Output
0	0000
1	0001
2	0011
3	0010
4	0110
5	0111
6	0101
7	0100
8	1100
9	1101
10	1111
11	1110
12	1010
13	1011
14	1001
15	1000

Although the construction of an absolute encoder entails significantly more work than the construction of an incremental encoder, especially as the resolution of the absolute encoder increases, you may find such a device useful for sensing angular displacements of joints, steering linkages, and other mechanisms.

CHAPTER 9

Electronics and Microcontroller Interfacing

Eventually, you will need to connect your motors to something that will control them. This chapter discusses that final stage of design, connecting your motor drivers to the controlling electronics that will tell them what to do. We'll cover all the steps from the power supply to the microcontroller and, because we like this stuff, we'll detail some schematics and software too as examples.

First Things First: The Power Supply

Several ways exist for powering your controller boards and motors, but they all eventually sort into two categories: Use one battery for everything and use separate batteries for the motors and the controller boards. Let's explore the reasons why you could choose one over the other.

Using Two Batteries, One for the Motors and One for the Controller

Here we have some isolation between power sources. One or more larger batteries will power the motors and another (usually smaller) battery will power the controller. Unless you are using opto-isolators (more on them later), you will *need* to connect the grounds between the separate power supplies and control electronics together. If you don't, the separate functions will not work properly, if they work at all.

For larger robots, multiple batteries aren't uncommon; they have the capacity to afford the extra weight and space used. Let's look at the reasons why you would use separate power supplies:

- The motor voltage is more than about 12V. This means that a lot of power would be wasted in converting down to 5V for the controller. Instead of wasted power, you may use a more expensive and elaborate voltage regulator, which complicates things.
- The current draw is more than about 2 to 3 amps. This kind of current draw could seriously affect the stability of your voltage regulator and maybe even cause *noise* (transient spiking) or *brownouts* (low voltage or sags in voltage levels) on your controller's

power bus. The symptoms of these problems are constant resetting of the board and erratic behavior for no apparent reason.

If either of the two previous reasons apply, consider using two power supplies in your robot: one for the motors and one for the controller. Typically, the battery for your controller will be much smaller and have much less capacity because your controller will not require much current for normal operation.

One caveat: This kind of depends; this may not be true over the long run for robots carrying hefty electronics. For example, robots that carry PCs or other major electronics may exert a greater demand on the controller battery over the long run. (This is one of the reasons we discourage people from building PC-powered robots.)

Where the motors only run intermittently, the electronics may actually want more heft batteries. REGIS is a good example. The controller battery has to be recharged every night. The drive battery has to be recharged about once every six weeks, even if the robot has been fairly busy.

Because this is often overlooked, we'll state it again: *Always connect all circuit and battery grounds together once and only once* (see Figure 9-1).

Don't make multiple ground connections to the same circuit; this can create what is known as a *ground loop*, which is when you have current going "down" one ground and "up" another to the same circuit (see Figure 9-2). This is a bad thing. If your grounds are not common, that is, connected, then the circuits just won't work because the voltage levels can't be determined; they "float" with respect to each other. A ground is a common reference point by which electronics determine the current flow. Again, opto-isolation circuits change the rules (more on that later).

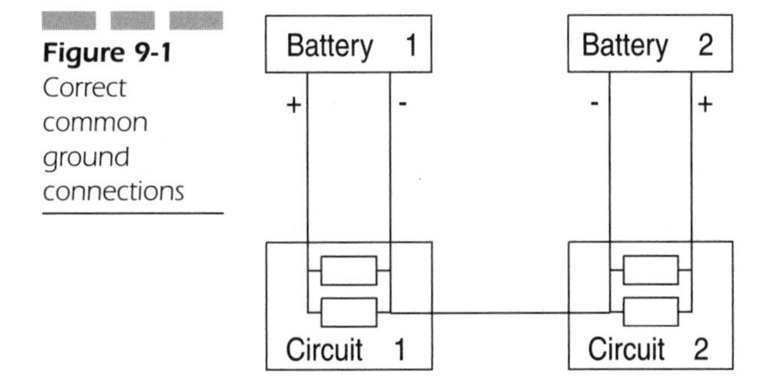

Figure 9-1 Correct common ground connections

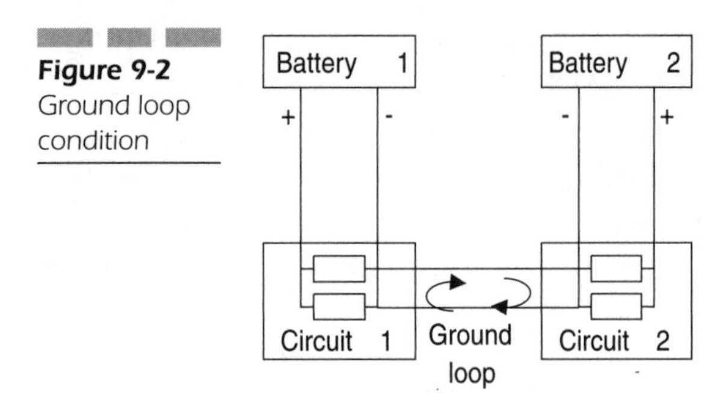

Figure 9-2 Ground loop condition

Using One Battery for Both the Motors and the Controller

For smaller robots, this should be the rule. Small robots don't need or use really high current motors and don't use high-voltage battery packs. We don't have much space in a Mini Sumo robot, for example, and even less weight allowance. In this case, we only want what is necessary onboard. So here are our reasons for using a single battery pack:

- Using less than about 12V for motors, usually in the 6 to 10V range, is well within the effective input voltage range of common and inexpensive voltage regulators.
- The motors pull less than 2 to 3 amps current. We're usually using L298s or units with similar current restraints, which is no problem and it won't cause large voltage sags.
- The space or weight allowance for extra components is limited.

Any motor that pulls 100 ma or more should use *Nickel Cadmium* (NiCd) or *Nickel Metal Hydride* (NiMh) batteries for power. Alkaline batteries have too high of an internal resistance and their voltage will sag badly under the current draw.

A Simple Circuit to Protect the Controller from Motor Brownouts Fortunately, it's not difficult to protect our controller boards from brownouts or voltage sags due to motor current draw. These sags are of a short duration and not a large magnitude if proper batteries are used. This voltage sag protection circuit will do the trick nicely and is inexpensive and simple to implement (see Figure 9-3).

This circuit works by charging C1 to the battery voltage minus the diode voltage drop during normal circuit operation. When motors draw a large current, the voltage will sag somewhat. When this happens, the

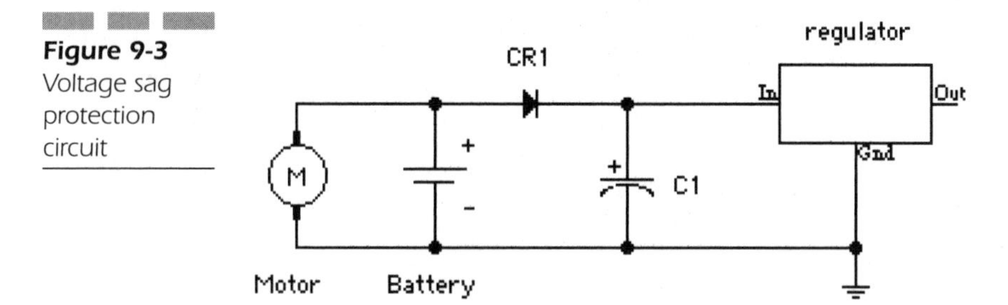

Figure 9-3 Voltage sag protection circuit

voltage on the anode side of the diode CR1 will drop. This reverse biases CR1 and prevents current flow. Meanwhile, on the cathode side of CR1 (the side the arrow points to), the capacitor C1 maintains the correct voltage and the voltage regulator never knows there was a voltage sag.

To minimize the loss of voltage to the power circuitry, you should use a Schottky diode to block reverse voltage conditions. A Schottky diode has a very low forward voltage drop while conducting, usually about 0.35V or less compared to a standard 1N4004 diode voltage drop of about 0.65V or more. We recommend the 1N5817 diode, which has a 1-amp maximum current specification and a fast switching time to reverse bias. If you are using a 6V battery pack, you should be using a *Low Voltage Dropout* (LDO) regulator as well. It is important that C1 be large enough to supply the current needed by the electronics during the sag. Table 9-1 shows some suggested values for C1 in the circuit based on the current draw of the electronics, not the motors. These are very conservative values; capacitors are cheap.

Power Line Conditioning

DC motors as well as even the driver chips and other digital logic generate noise on the power bus. We use bypass capacitors to filter out transient spikes, usually ceramic 0.1 uF capacitors placed liberally on a

Table 9-1 Suggested C1 values

Current draw	C1 value
< 100 ma	100–220 uF
100–300 ma	330–680 uF
300–1,000 ma	1000–3300 uF
> 1 Amp	Get a commercial power supply.

board and always placed close to noisy chips like motor drivers. Also, because motor drivers, especially H-bridge drivers, frequently charge the power bus with shoot-through current, we place larger electrolytic capacitors near them to absorb this excess energy. The electrolytic capacitors can be anything from 47 uF to 470 uF depending upon the manufacturer's recommendations. These capacitors, both bypass and surge suppression, are not optional in your designs. You see them in all of our designs in this book.

Connecting Everything Together

You've got motors, controller boards, driver boards, and a few other things. Great, now you need to connect them together. That means wires and connectors. Here's some suggestions on wire sizes, connector types, and techniques.

Wire: How to Select a Proper Wire Gauge

Connecting wires between low-current controller boards is not the issue here. Using plain old ribbon cable wire works just fine. Our issue in this chapter is dealing with high-current wiring, what size, how to route it, and what connectors to use. In Chapter 7, "Motor Control 101, The Basics," we give this wire gauge table, but it bears repeating here (see Table 9-2). Read Chapter 7 for more discussion on wire gauge selection and current handling estimations.

Connection Strategies

Connecting two wires together or connecting a wire to a device or PC board can be done in a variety of ways. The following are some suggestions. It's a good idea to use some form of connector from the driver board to your motors. This makes the motor easier to remove for servicing or replacement. Make sure to route your wires so that they don't get in your way later. Also, make sure that high current wires are as far from delicate electronics as possible.

Table 9-2
Wire gauge selection

AWG	Amps
10	30
12	20
14	15
16	10
18	5
20	3.3
22	2.1
24	1.3
26	0.8
28	0.5

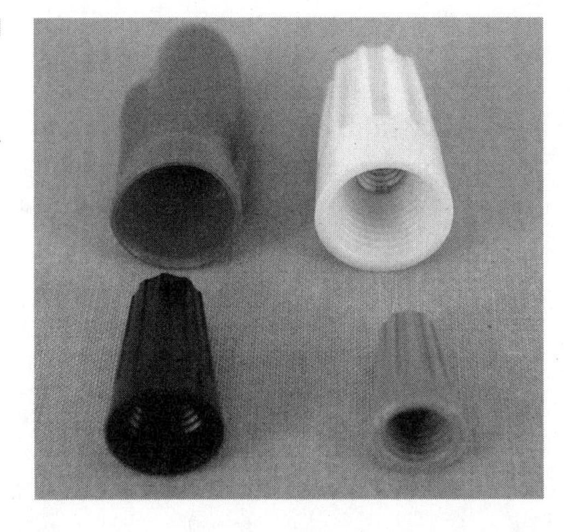

Figure 9-4
Wire nut selection

- **Wire nuts** This is a convenient, nonpermanent way to bind two wires together. They come in a wide variety of sizes, as Figure 9-4 shows. Wire nuts are easy to use; just lightly twist two or three wires together, screw on the wire nut, and you're done. Make sure your wire nut is of the appropriate size for your wires.
- **Barrier strip or terminal block** These enable the efficient and organized routing of wires you may want to change around later. You

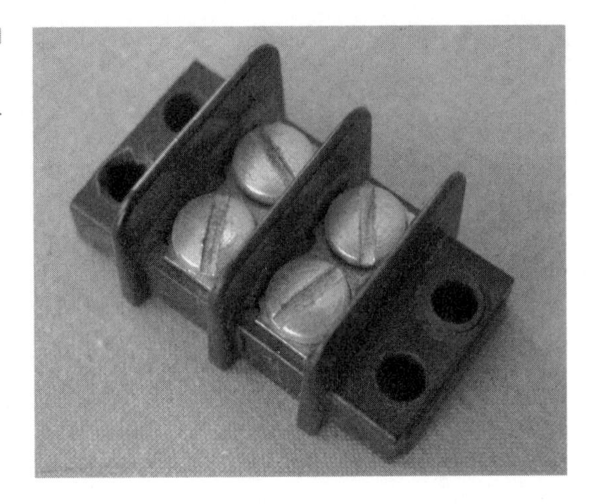

Figure 9-5 Barrier strip

mount the terminal block on a convenient surface and, using your screwdriver, attach wires to their appropriate locations. Figure 9-5 shows a small terminal block (barrier strip) example.

- **Soldering** When your wires don't need to be changed around and/or you're drawing a very high current through them, nothing beats just soldering. You can cover the splice with heat-shrink tubing so shorts don't occur. Heat-shrink tubing is superior to electrical tape because it will never unwind. Figure 9-6 provides an example. Heat-shrink tubing comes in a large variety of sizes, so one of them is perfect for any wire you care to use. Just remember that the tubing

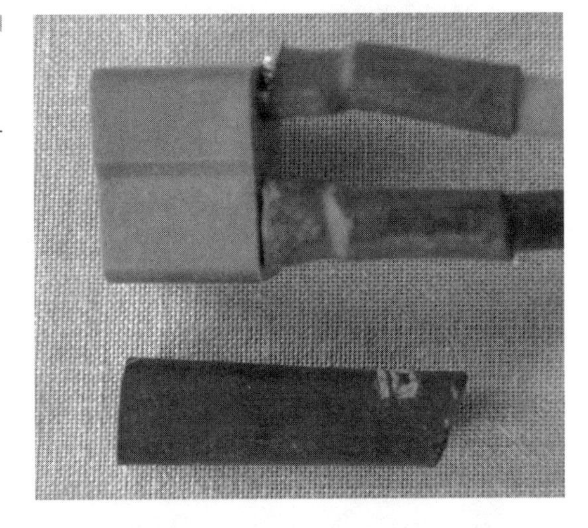

Figure 9-6 Wire soldering and shrinktube

shrinks at least 40 percent when heated. A hair dryer is not hot enough to shrink it; we tend to use a hot soldering iron to shrink it.

- **Connectors** Connectors make your life easier when it's time to repair or replace a subsystem. Keyed connectors will prevent you from plugging your electronics in with a reverse polarity, which is often fatal to them. Low-current (100 ma or less) connectors like those in Figure 9-7 can be used between low-power motors and their drivers, or between the driver boards and the main controller. For high-current motor connections, we recommend connectors designed for 15 amps or more. The Molex connector (Digi-Key part #WM2308) in Figure 9-8, for example, is found on many toy cars and robots for

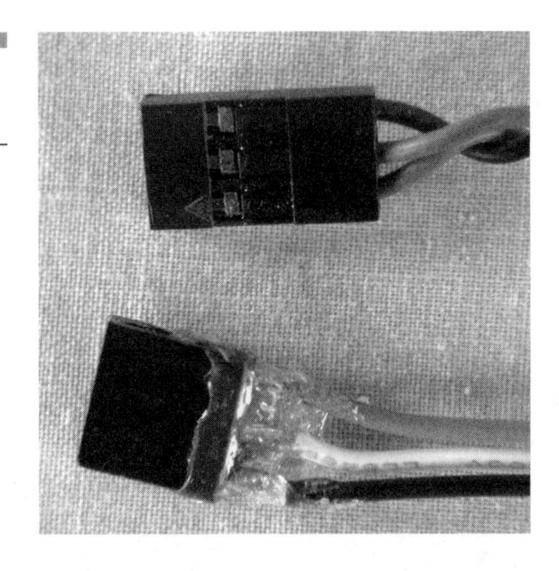

Figure 9-7 Low current connector

Figure 9-8 Tamiya-style plugs

battery connections. It is rated to 15 amps. This connector is often called the Tamiya® connector because that large RC car manufacturer uses it. The connector set in Figure 9-9 is favored by many RC car enthusiasts for its very high current rating (in excess of 50 amps) and robustness. These are called Dean's® Ultra Plugs and can be found at any shop or mail order house that carries radio-control cars.

- **PC board terminal blocks** A good motor driver PC board will have terminal blocks similar to Figure 9-10. A small screwdriver is needed to attach the wires, which makes for a solid connection from the board to the wires and is very neat. As an added bonus, these are stocked by Radio Shack.

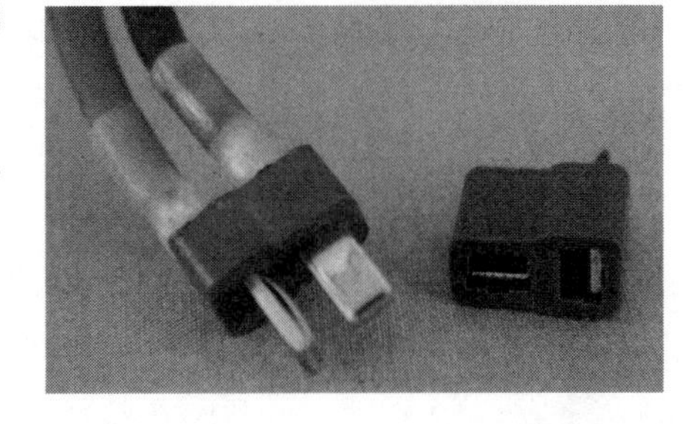

Figure 9-9 Dean's Ultra Plugs

Figure 9-10 PC board terminal block

Interfacing to a Microcontroller or Computer I/O

Okay, you have your motors wired, you have your driver boards wired, and now you need to interface your driver boards to your microcontroller or computer. Three standard methods can be used to go about this: the buffering interface, isolation interface, and serial interface. They all have their advantages and disadvantages.

The Buffering Interface

Buffering simply means that a current-limiting, voltage-matching, or current-amplifying interface exists between two subsystems. Back in Chapter 7, we introduced the 74LS06 hex buffer as the interface between the 5V microcontroller *input/output* (I/O) and the 12V+ transistor drive systems. Here are two more common buffering techniques and why they are used.

Resistor (Passive) Buffering It is common to place 330-ohm resistors between a microcontroller and other interface chips on an experimenter's prototyping boards. A hobbyist may make a mistake during programming and end up with each chip configuring a pin as an output. This isn't instantly lethal to either chip, but if one chip brings its I/O line high while the other brings its I/O line low, you have a potential high current short to ground through both chips. A 330-ohm resistor limits that current to 15 ma maximum, which is well within most microcontroller I/O pin limits. This resistor does not affect signal levels because it's so small. At the end of this chapter, you will see examples that use the buffering resistor interface between subsystems.

IC (Active) Buffering Active buffering is most often used as the interface between your PC and your hobby experiments. The buffer actually increases the power of the signal from your computer (as it's commonly used) because the PC parallel port can supply very little current, it's sensitive to electrical noise, and it's kind of expensive. These are all good reasons to put a barrier in the form of a signal buffer between the parallel port and your project. Figures 9-11 and 9-12 show two common buffers used in this fashion.

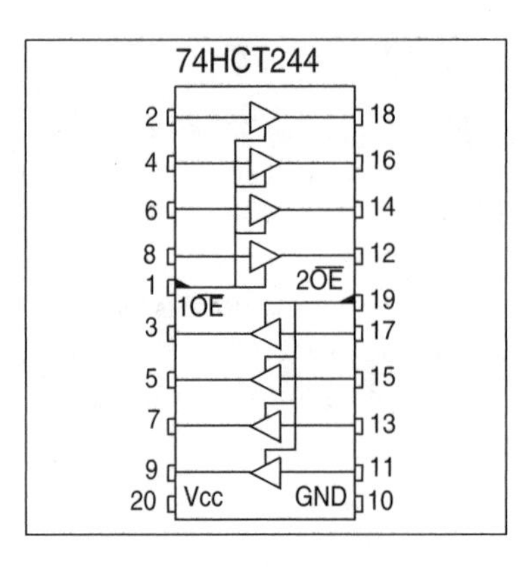

Figure 9-11 74HCT244 octal buffer

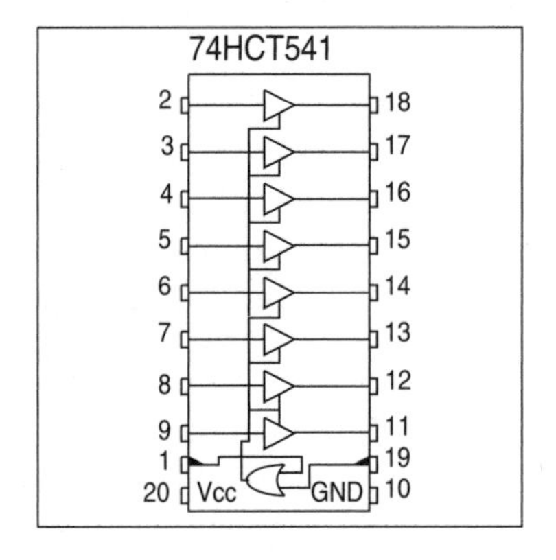

Figure 9-12 74HCT541 octal buffer

This is a very popular buffer. It consists of two independent 4-bit tri-state output buffers, each with its own active-low enable line. Tri-state means that three outputs are possible from the buffer: 0 or logic low, 1 or logic high, and high-Z or high impedance, which means it's not really outputting any signal level at all. Active-low means that a 0 or low enables the buffer to function as intended and pass the logic level through from the inputs to the outputs.

The 75HCT244 buffer is commonly used on "bussed" systems, meaning that multiple chips can write to a common set of wires. It does the same thing as the 74HC244, but all the inputs are on one side of the chip and all the outputs are on the other side. This is convenient for 8-bit systems to use, because it simplifies wiring. For the data to be passed from the inputs to the outputs, both active-low enable lines must be brought low.

Another reason to use an active buffer is to interface between two systems that use different voltages. For instance, in Chapter 7 we developed a discrete four-transistor H-bridge motor driver (refer to Figure 7-6). Our microcontrollers have 5V systems, and the H-bridge uses 12V to power the motor. Our buffer between those two systems is the 74LS06 hex (as in 6) buffer. This buffer can be the interface between a 5V system and a system whose operating voltage can be as high as 30V. The pinout for that chip is given in Chapter 7.

The Isolation Interface

This interface actually electrically isolates the two subsystems from each other and passes information digitally by an optical connection. These interfaces are called opto-isolators, and Figure 9-13 is a representation of one such device. The isolation interface is often used between a controller and a high-current/high-voltage motor driver. The key element to research if you plan on optically isolating your motor driver is the rise time of the opto-isolator with respect to your *Pulse Width Modulation* (PWM) frequency and the required rise-time of your motor driver's PWM input lines. Some opto-isolators are quite slow.

For instance, let's say we want a 20 kHz PWM frequency and we want to make sure that we can get a 10 percent duty cycle through our opto-isolator. Setting up our numbers we get the following:

$$\text{Frequency} = 20 \text{ kHz, Period} = 1/\text{Freq} = 50\mu\text{s}$$

$$10\% \text{ duty cycle} = 50\mu\text{s} \times .1 = 5\mu\text{s}$$

Here is a slow opto-isolator, the PS2702. Its specs of interest are as follows:

- Rise time = 200 μs
- Fall time = 200 μs
- Total delays = 400 μs

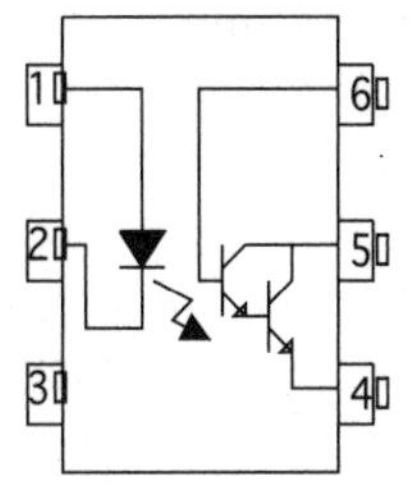

Figure 9-13 The 4N30 opto-isolator

We can see right away we are in trouble. In 5 μs (our 10 percent duty cycle), this opto coupler won't even be halfway switched on.

Here is a fast opto-isolator, the 6N137. Its specs of interest are

- Rise time = 50 ns
- Fall time = 12 ns
- Total delays = 62 ns

This is looking good. To be conservative, and because we like nice round numbers (like 10, for instance), let's say that we want a margin for the switching time that will be 10 times the duty cycle. So, if our duty cycle is 5 μs, we want all switching delays to be 10 times less than that, or 500 ns. At 62 ns, the 6N137 passes our requirements with flying colors.

It pays to read the spec sheets before you buy. Follow their recommended configurations carefully.

When we are using opto-isolators we don't need or want to have our grounds be common. The power systems on either side of the isolator are distinct and isolated from each other, so *don't* connect their grounds together. You are trying to protect the rest of your robot from the hazardous voltages on the other side of that optical wall.

The Serial Interface

The serial interface takes many forms. The three most popular interfaces are the asynchronous serial, the *Inter-Integrated Circuit* (I2C), and the *Serial Peripheral Interface* (SPI). The latter two interfaces are synchronous in that they have both clock and data line(s). We will not discuss all the details of these interfaces; that is a topic deserving of an entire book alone. In relative speeds, asynchronous serial is slowest, the next fastest is I2C, and the fastest is SPI.

Asynchronous serial is a two-wire interface in its typical implementation: one line to send data and one to receive. If your microcontroller is only sending data, you only need one I/O line. The two ends of the "conversation" need to have agreed upon a transmission speed or they can't communicate. This is a common communication interface from a PC to a robot. Many times a single I/O line is used to both transmit and receive data. You will save one data line but will sacrifice being able to send at the same time you are receiving data. We will show two examples of an asynchronous serial interface at the end of this chapter.

Two forms of this interface exist hardware-wise: RS232 and TTL. RS232 defines voltage levels on the transmission lines. The RS232 logic 0 voltage is +3V to +12V, and the logic 1 is −3V to −12V. Any voltage between these levels is undefined. A TTL asynchronous serial is not inverted like RS232. A TTL serial logic 0 is a TTL low, which is 0.8V or lower, and a logic 1 is TTL high, which is 2.4V or higher. This latter specification is often called non-inverted serial.

I2C is also a two-wire interface. One I/O line is the clock signal; the other is data. They are labeled SCL and SDA respectively. I2C communication can be from 2 to 10 times faster than an asynchronous serial. The speed of I2C is completely arbitrary because the master controller sends out a clock stream that defines when data appears on the SDA lines as valid. So, this clock does not even have to be of a consistent frequency usually. Unlike asynchronous serial, each device on the I2C bus has a unique address so each device must be addressed correctly by the master in order to respond.

SPI is a three-wire interface, and the I/O lines are *Serial Clock* (SCK); *Master In, Slave Out* (MISO); and *Master Out, Slave In* (MOSI). The master controls the SCK line, sends data out on the MOSI line, and receives data back on the MISO line. A slave device on the SPI bus sends data out the MISO line and receives instructions and data from its MOSI line. The SPI interface is usually 10 times faster than the I2C bus. Again, because SPI is a synchronous serial interface, the SCK line defines when data is valid, so it does not have to be a consistent frequency usually. The other reason the SPI bus is fast is that the master can be receiving data on its MISO line at the same time it is sending data out on its MOSI line.

Buffered and Serial Interfaces Examples: Stepper and PWM Signal Generation

Interfacing drivers and controllers can be done in too many different ways to cover them all in a single book and still write about anything else. Here we give an example of passive buffering and an example of a serial interface. We include schematics and source code to help explain each step.

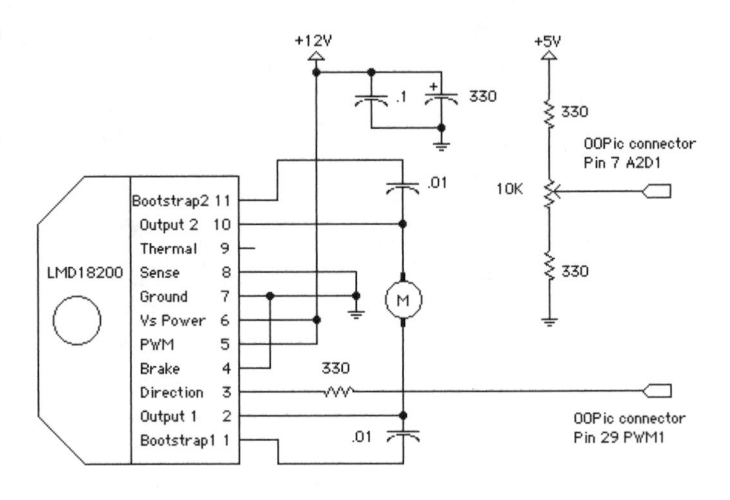

Figure 9-14 OOPic/LMD18200 interface

An OOPic PWM Interface Example

This example shows how to use the OOPic microcontroller to operate an LMD18200 H-bridge motor driver (see Figure 9-14). We are using passive buffering via 330-ohm resistors. You've seen parts of this circuit before, but for completeness' sake, it's all here. Note that 330-ohm resistors on either leg of the potentiometer (pot) connect to the OOPic A2D port. You can consider these to be passive buffers for protection; they prevent the A2D port from being shorted to ground or +5V at the extremes of the pot adjustment. The LMD18200 is being controlled in locked antiphase PWM mode and you change the speed of the motor by adjusting the pot.

Here is the OOPic code for this example:

```
//testPWM.osc (C language syntax)
//OOPIC I version A.1.7, Compiler version 4.0
//
// This is all done in objects.  The A2D object is linked to
// the PWM object to change the PWM duty cycle.  I tried
// 1.2Khz, too noisy, 4.9KHz, still sang, 20KHz, yum, silent
// and smooth.  This project demonstrates locked antiphase
// PWM.  It was tested on an LMD18200 with the PWM line
// pulled high and the PWM from the OOPic connected to the
// Direction line.  This code will compile and run on an
// OOPic I or II.
//
// Copyright Dennis Clark 2002
// Permission is granted to use this code any way you like,
// as long as its origins are disclosed.

oPWM PWMA = New oPWM;
oA2D rate = New oA2D;
oMath connect = New oMath;
```

```
sub void main(void)
{
    OOPic.Node = 2;                    //For debugging

    /* PWM ouput
       PWM1 = I/O 18 = Pin 29
    */

    /* A2D input
       A2D 1 = I/O 1 = Pin 7
    */

    PWMA.IOLine = 0;                   //PWM1
    PWMA.PreScale = 0;                 //div by 1 (5 MHz clk)
    PWMA.Period = 255;                 //20KHz
    PWMA.Operate = 1;                  //Start it.

    rate.IOLine = 1;                   //A2D1
    rate.Operate = 1;                  //Turn it on

    connect.Mode = 7;                  //Just pass input1 on.
    connect.input1.link(rate);         //from A2D
    connect.output.link(PWMA);         //to PWM
    connect.Operate = 1;               //start it

    do
    {
        //Don't have to do anything, its all in the objects.
    }
} //End main()
```

A Stamp II and TTT PWM Controller to 754410 H-bridge Example

The following is an example of serial buffering. The Stamp II sends instructions to a PWM controller chip, which in turn interfaces to the 754410 H-bridge driver (see Figure 9-15). Note that we are also using passive buffering on the serial interface line. This chip is one of my designs. For more information, see my web site, www.techtoystoday.com.

Here is the source code that runs this example:

```
'{$STAMP BS2}
'TTT PWM and servo control for Parallax Stamp II
'PWM chip is on I/O Port 6 of the Stamp II.
'We are driving a 754410 quad half bridge.  Both
'direction bits are used to set the motor direction.
'Control of the TTT PWM controller is via 9600 baud
'8N1 serial protocol and single byte commands.
' 0 = 0% duty cycle, 127 = 100% duty cycle.
'
'Copyright Dennis Clark 2002
'Permission is granted to use this software any way
'you please as long as its origins are disclosed.
```

```
N9600 con 84                    '9600 8N1 to talk to PWM chip
MOT   con 6                     'serial PWM chip I/O port
SPKR  con 4                     'Speaker

'Duty cycle variable for PWM
duty_0 var byte

pause 1000
debug "TTT PWM Example",cr
freqout SPKR,300,1000

'Exercise the serial PWM chip
debug "forward ",cr

'set direction bits for one direction.
serout MOT, N9600, [129]

for duty_0 = 0 to 126 step 2
    serout MOT, N9600, [duty_0]
    pause 200
next
pause 1000

debug "reverse ", cr

'Set direction bits for the other direction.
serout MOT, N9600, [130]
for duty_0 = 126 to 0 step 2
    serout MOT, N9600, [duty_0]
    pause 200
next

'Turn off the motor (direction bits = '0')
serout MOT,N9600,[128]

debug "All done"

end
```

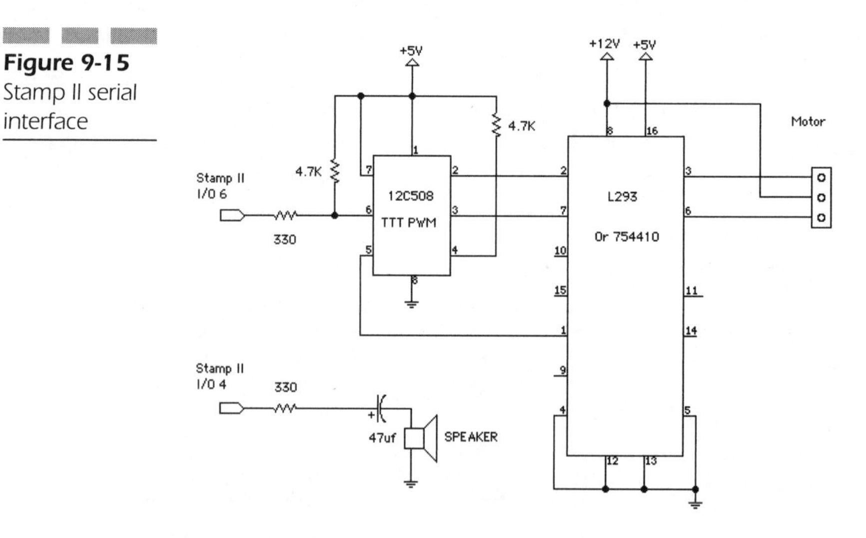

Figure 9-15 Stamp II serial interface

CHAPTER 10

Wheels and Tank Tracks

Okay, so now you have your motors, and you've attached them to your robot. You're almost there, but what kind of wheels or treads should you use and how big should they be? Finally, the $64,000 question is "How do I attach the wheels?" Myriad solutions can answer this quandary, and we are going to show you several. Because, as we stated earlier, we don't expect you to have a fully outfitted machine shop at your disposal, we're going to use off-the-shelf components where possible and show simple modifications of innocent, related hardware when necessary. You can relax again; it won't be that bad or that expensive!

All of the *Do It Yourself* (DIY) projects given here have been tried and found to work. Most of the DIY projects require you to be steady of hand and discerning of eye to get everything to line up. Take your time, use proper safety precautions, and be patient.

Even if you are one of the worst machinists to ever trouble the sleep of a shop instructor, you can still successfully build the DIY projects in this chapter. So be brave and of good cheer; you can do it.

Wheel Diameter, Torque, and Speed

The diameter of the wheel (or drive sprocket in a tank track) has a great influence on the apparent motor torque and speed of your robot. In general, as your wheel diameter increases, so does your speed; however, your torque decreases. There's no such thing as a free lunch.

Way back in Chapter 2, "Motor Types: An Overview," we learned about torque and how it's measured. Torque is the strength of a motor's angular pulling capability at a given distance from the center of the motor hub. Thus, torque falls off linearly as it is measured further from the center of the motor hub. For instance, a motor with a torque of 100 ounce-inches (7200 gram-centimeters) at 1 inch (2.54 centimeters) will have a torque of only 50 ounce-inches (3600 gram-centimeters) at 2 inches (5 centimeters). As you can see, when your wheel diameter increases your torque falls off.

A robot's speed increases, however, as its wheel diameter increases. This is also a linear function. You can calculate your robot's speed if you know the diameter of your wheel and the revolutions per minute (RPM) of the motor. This process is called *wheel rollout* and is very simple (see

Figure 10-1). First, calculate the distance your robot travels with one rotation of the wheel. This is done as follows:

$$D = \pi d$$

Where D is the distance in whatever units your diameter d is measured. Multiply D by the RPM of the motor and you have your speed in units per minute. Do the appropriate multiplies or divides to get miles/kilometers/feet per second/hour. For a tank track, calculate rollout with the diameter of the drive sprocket.

Wheel and Track Selection

Many variables must be evaluated before selecting wheels: robot size and weight, terrain surface, motor power, and, of course, aesthetics (it can't be ugly, right?). Tank tracks have many similar issues and a few that are quite different from wheels.

Weight and Your Wheels: What Type of Wheel to Use

In general, the heavier your robot is, the more robust your wheels or tracks must be. We're not going to dive into materials engineering to be technically perfect in this chapter; we're just going to give some useful

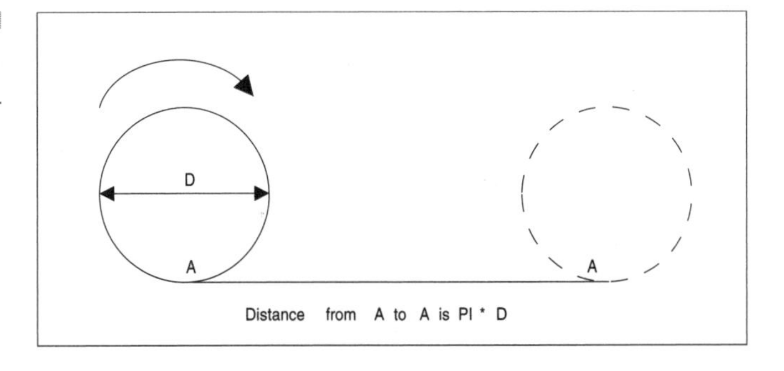

Figure 10-1 Wheel rollout

pointers and when we discuss specific wheels, we'll tell you what they are and are not good for.

Light robots, those under about 2 pounds, can use soft foam wheels. These lighter robots won't deform the soft material very much and will run well with little friction concerns (see Figure 10-2). After 2 pounds, your wheel material needs to be more rigid; the hollow rubber wheels that aren't inflated will work well up to about 6 to 10 pounds. Above 10 pounds, you need to consider solid rubber material or inflated wheels. Lawnmower wheels, both solid and inflatable, will be your choice up to about 40 pounds. After 40 pounds, you start looking at go-cart, minibike, and similar high-performance inflatable tires.

When dealing with tank tracks, your issues are similar. Up to about 2 pounds in weight, those soft, rubber toy tracks work just fine. From 3 to 10 pounds, you need to use heavier rubber that isn't so squishy to avoid stretching the tracks and throwing them from their drive wheels (see Figure 10-3). Over 10 pounds, you need to start looking at reinforced nylon-stranded rubber tracks, similar to automobile *V-belts* and *timing belts*. These heavier robots may even use nylon-stranded belts with metal treads depending on their environment.

These are just general suggestions. Exceptions will occur in specific circumstances when you are willing to accept the higher wear, increased static, and rolling friction to get better traction. In some cases, you will use harder rubber wheels on lighter robots to reduce rolling friction if you have low-powered motors. Another exception would be if you are angling your wheels outward from perfectly vertical. This reduces the tendency to bias your forward direction when you don't have perfectly mounted motors. You will want a more rigid tire in this case because you are driving on the sidewalls of the tire instead of its bottom. Consider

Figure 10-2 Wheel deformation with weight

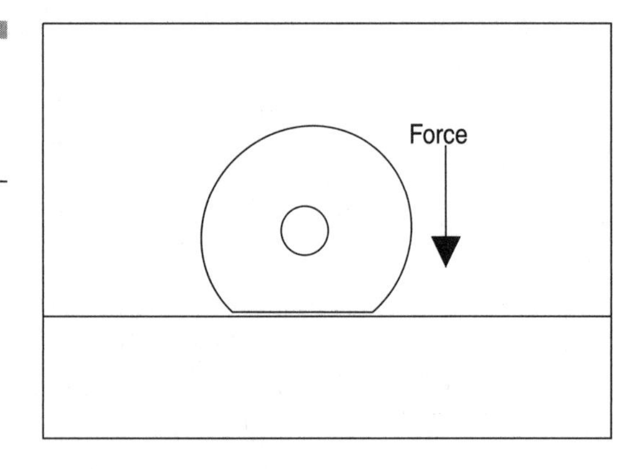

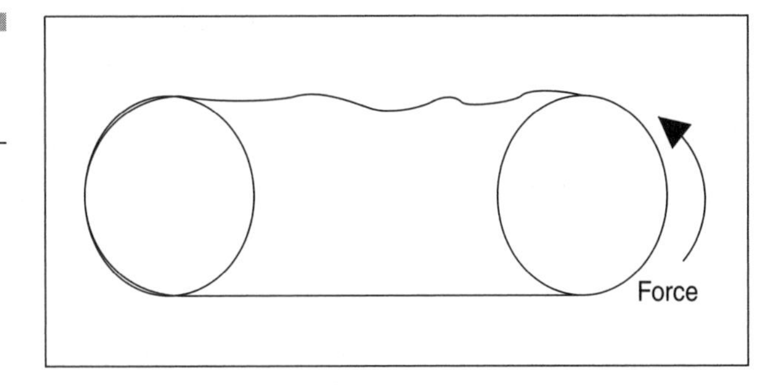

Figure 10-3 Tank track stretch

your tire material and how much it squishes or stretches when your robot is moving.

Terrain Conditions and Tire or Track Material

Let's look at different surfaces and what our cars and bicycles use. On paved roads, we use tires that are highly inflated and don't deform very much. When we are out four-wheeling or mountain biking, our tires are inflated less. In general, if you have a smooth surface, you use more rigid wheels or more highly inflated wheels. If you are off-road, you want softer and squishier wheels to get better traction on an uncertain surface. Use those same considerations with your robot and its expected environment. In a nutshell, on a high-friction surface, use a lower-friction tire; on a low-friction surface, use a higher-friction tire.

If you have an uneven surface, you want more aggressive treads on your tire so that you can increase your friction by getting between elements of the terrain. If your surface is smooth, you want a smooth surface to maximize the frictional contact with the environment. However, if your smooth surface has dust or other small debris on it, you want to have some tread so you can increase the pressure on your driving surface to help reduce the effects of the debris. A wet surface demands more grooves in your tire to help channel water away from your contact area, again, to increase friction or *grip* on the surface.

Some materials are deceptive. For instance, the soft, squishy Dave Brown® Lite Flite wheels don't work well on tile or slick, smooth surfaces, but they work great on carpets or pavement. Sometimes you need to experiment to find the right material for the task. Later on we'll give some specific tire choices and their good and bad points. Sand is a low-traction surface if your cross-sectional area in contact with the surface is narrow; it's a high-traction surface if your cross-sectional area is high, however. So, if you are planning to run in sand, plan accordingly and use wide tires. Narrow tires will just burrow in. However, wider wheels will have a higher coefficient of friction and take more power to move. You'll have to experiment to find your acceptable limits.

Tank treads have some different issues. Because tracks have a *much* larger contact patch (the part of the wheel/track that is in direct contact with the ground at any given moment), the squishy issue is diminished. On rough or uneven terrain, your tracks need to have treads to increase their frictional contact with the driving surface. When your tank tracks are on a smooth surface, you want the tread to be smooth. However, tank tracks have other problems with the surface on which they drive. Although tank tracks would appear to have a much larger wheel diameter (the diameter of the track, right?), in fact their torque is developed at the diameter of the drive sprocket. A tank track just has a higher coefficient of friction than a simple wheel would.

A tank track has three essential components: a drive sprocket, an idler sprocket, and the track. Figure 10-4 shows these components. The drive sprocket is the sprocket that is connected to the drive train of the vehicle. The idler looks exactly the same but just rotates and helps keep the track aligned. Obviously, the track is that which trundles along the ground to move the vehicle.

When constructing a tracked vehicle, you may find that you aren't getting very good traction. If you have chosen your tread material for good friction with your drive surface, then your problem is probably with how

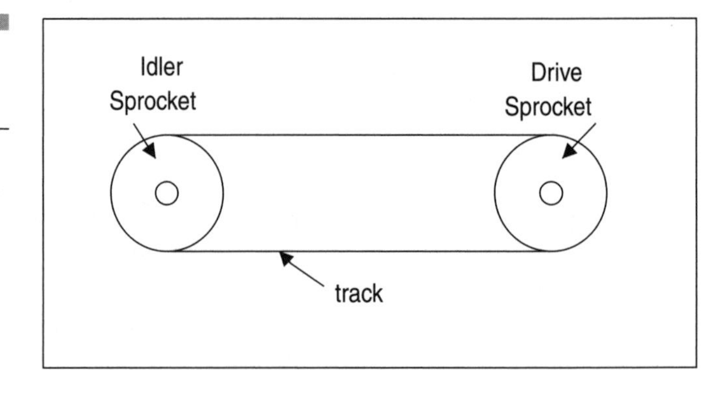

Figure 10-4 Tank track components

your contact patch interacts with your surface. If your track is long with respect to your sprockets, you will have a large portion of the tank track that is unsupported in contact with the ground. Basically, your track flexes over instead of gripping the terrain. The solution is bogies (see Figure 10-5). Bogies will aid in transferring the weight of the vehicle to the track and then to the ground. The best traction solution is to use as many bogies as will fit into the given space, but any number that you use will help the traction considerably.

Tank tracks are intended to be high grip in line with the direction of travel and low grip at right angles to that travel. In other words, a tank may get great traction going forward, but if driving across an incline (as opposed to driving up or down the incline), it may find itself sliding sideways down the hill. A tank-tracked vehicle can't be high friction in its lateral directions or it would not be able to turn. Tank steering is often called *skid steering* because it literally must skid sideways to turn (see Figure 10-6). If you don't believe us, watch a tank drive on TV or look at the tracks made in the dirt by a tracked construction vehicle.

The other major problem with tank tracks is that they tend to throw their tracks (the tracks become derailed or treads break) when a high grip track is used on a high grip surface. This is caused by the skidding action during turns. The skidding places lateral pressure on the treads. On a high-friction surface, this lateral pressure is often enough to derail the treads. On a surface like sand or carpeting, it's pretty easy to throw a track. Plan for this and be aware. Small debris can become lodged into tracks, also causing a thrown track. Pea gravel is death to small tanks!

Figure 10-5 Adding bogies to tracks

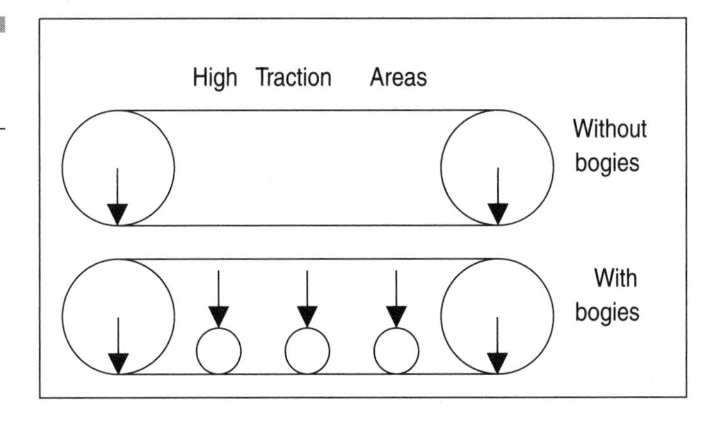

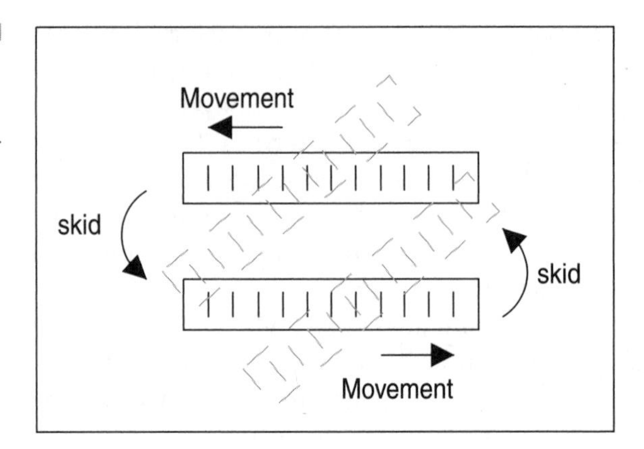

Figure 10-6 Tank track skid steering

Types of Wheels and Where to Get Them

Nearly an infinite number of resources for wheels is available. You can use wheels that were intended to be vehicle wheels or wheels gleaned from devices such as printers that were clearly not intended to be vehicle wheels. We'll show you our list of favorite sources for robot wheels.

Radio Control (RC) Car Wheels

Figure 10-7 shows a variety of sizes and styles of *radio control* (RC) car tires and wheels. RC car tires can be anything from 2 inches (5cm) to 6 inches (15cm) or larger in diameter. They can be as narrow as a poker chip to as wide as your palm. Just about anything you can imagine needing in a small robot can be had in an RC car tire. These wheels are made with a variety of tread materials for a variety of driving surfaces, vehicle weights, and styles. Everything from high-density foam to high-tech rubber is used. A variety of tire diameters and widths are also available. These tires can cost anywhere from $4 a wheel to $10 a wheel, depending on style, material, and where you buy them. To find these types of tires and wheels (also called rims, and always sold separately), check out your local hobby shop or any one of a number of online RC stores.

RC tires have as many exotic rubber compounds and specialized tread patterns as their full-size automobile counterparts. You can get smooth slicks, rain tires, tires for loose dirt, and off-road monster truck treads.

Figure 10-7
RC car tires

Choose the one that will work best for your application. Later on in this chapter we'll show you some ways to attach these tires to various motors.

LEGO® Wheels

LEGO supplies a large variety of tires and wheels in their various Technic™ and Mindstorms™ building kits (see Figure 10-8). LEGO also supplies a special accessory pack that contains a collection of many types of wheels, so you don't have to buy a whole kit to get them. This can be expensive though, so your local thrift shop or toy store may have a better deal on a kit that has the wheels you covet. Wheels are made with a variety of diameters, treads, and material used, so you can choose the one

Figure 10-8
LEGO tires and wheels

that fits with your robot and its terrain. The web site, www.lugnet.com, can point you to online shops and hobbyists that have these items for sale as well.

The LEGO wheel needs to use the LEGO axle to mount properly, and we'll show you how to construct such a mount later in this chapter. Your sources for such wheels are the LEGO Shop-at-Home store, shop.lego.com, and www.pitsco-legodacta.com.

Hobby Airplane Wheels

Dubro® and Dave Brown® make aircraft wheels of many diameters that work very well as robot wheels (see Figure 10-9). These wheels are easy to drill out, glue, or bolt with minimal modification. The Dubro Low Bounce™ wheels are made of soft but reasonably rigid material that gets good grip on smooth surfaces that have dust and debris on them. The Dave Brown Lite Flite wheels are *very* light wheels that get good traction on rough surfaces like carpeting and pavement. These wheels are quite inexpensive and available at all hobby shops that cater to the RC aircraft hobbyist. These wheels are commonly used with RC servos that have been hacked for continuous motion. You only need to get a longer hub screw and bolt them to the round servo horns.

Figure 10-9 RC aircraft wheels

Omnidirectional Wheels

A Carnege Mellon University robotics lab recently published designs for their *Palm Pilot Robot Kit* (PPRK) that includes three omnidirectional roller wheels. These wheels will allow movement in any direction, not just the direction that the wheel faces (see Figure 10-10). Originally intended as rollers for assembly-line tracks, it was discovered that they make really good wheels for a truly holonomic[1] drive system. These wheels are quite expensive and may only be obtained from a very short list of distributors. Expect to pay from $12 to $140 each, depending upon size and construction material. The sources for omnidirectional wheels are www.acroname.com and www.mrrobot.com.

Wheels Made from Creative Sources

Sometimes no one makes exactly what we want in a wheel. In those cases, we look to our ingenuity to find wheels in unusual places. A couple of commonly used types of wheels, which you don't find often sold as wheels, are pulley and O-ring wheels and printer pinch-roller wheels (see Figure 10-11). Both of these wheels are highly rigid and made from high-density rubber that gives them a very low rolling friction. This is highly desirable for robots that don't have a lot of power or need very small wheels to fit into high, constrained places. Obviously, these wheels are quite inexpensive if we've garnered them by "dumpster diving" out behind repair shops. We'll show attachment options later on in this chapter.

Figure 10-10 Omnidirectional wheels

[1]A holonomic robot has no restrictions on its movement direction. It can move in any direction from any point. This is your $20 word for the day.

Figure 10-11 Creative wheels

Larger robots require larger tires. Your local hardware or home-improvement store will have a good selection of lawn mower, wheel barrow, and lawn tractor tires that will certainly fit the bill.

Types of Tank Tracks and Where to Get Them

Our robot tank tracks will be made primarily of some kind of rubber or nylon. Because most of us don't have machine shops, the more exotic materials will not be covered. That said, a variety of sources for functional robot tank tracks is available.

Toys or Models with Tracks

Walking through a toy store, we can see many construction- or military-tracked vehicle toys. If one of these platforms fits the bill, we suggest you just get the toy and hack it to your means. Let's face it; it'll probably be easier than constructing your own. Sometimes these toys only have one motor in them and can't steer. In those cases, you're just going to get the sprockets, bogies, and tracks for your own purposes. Another very good option is a model of a tank that is motorized. You'll get motors, gearboxes, track materials, and a vehicle that is not yet assembled to do with as you

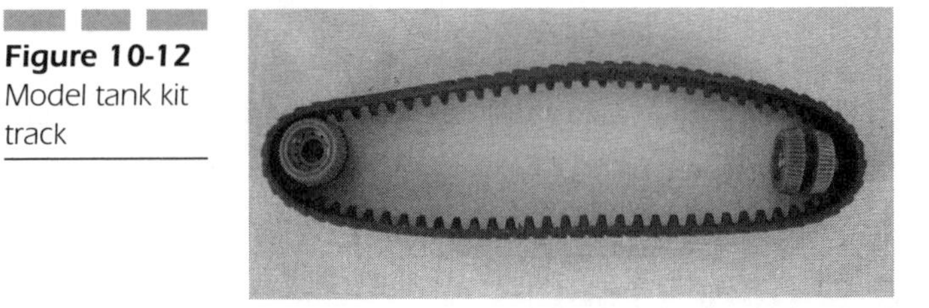

Figure 10-12 Model tank kit track

like. Figure 10-12 shows a tank track from a Tamiya® model tank kit. These types of toy tank tracks are very soft rubber and are only suitable for small robot projects of under 2 pounds usually.

LEGO Tank Tracks

Good old LEGO. We previously mentioned an accessory pack made by LEGO that has a selection of wheels and tires. That accessory kit also has two track assemblies and drive sprockets for them. Figure 10-13 shows these solid rubber tracks that will work well on just about any surface you'd dare run your robot. Several LEGO Technic kits also have tracks in them. The drive and idler sprockets use the same axles as the LEGO wheels do; we'll show you how to attach these axles to motors in the construction section of this chapter. These tracks can be used on robots of the 1 to 5 pound range successfully.

Another type of LEGO track that can be used in some cases is the conveyer belt links. These chain together in single-plate links and mesh with the LEGO gears. The conveyer belt links are not high traction on smooth

Figure 10-13 LEGO tracks

surfaces but work well on soft vinyl flooring and dense carpets (*not* shag!). They are available as accessory packs from the Pittsco LEGO Dacta® educational store (www.pitsco-legodacta.com). This store also carries LEGO wheels, axles, and other LEGO construction supplies.

V-belt and Pulley Tracks

These options are useful for the larger robots of the 10- to 50-pound range, depending upon the width of your V-belt. V-belts are used in automobile engines to connect the various components to the engine's power. They are also used in various farm implementations, machine tools, and fans. You need to be using strong motors and a strong chassis for this option because the components can be heavy. These tracks get very good traction on many surfaces. You can find V-belts and timing belts at an auto parts store, but for smaller, more flexible belts, go to your local hardware store. V-belt pulleys are not easy to find and, again, your hardware store can supply some.

Another option for V-belt pulleys is inexpensive plastic or nylon rope pulleys. Grind off the center pin to remove the pulley from the hanger. Figure 10-14 shows a robot track setup built from 2-inch (5 centimeter) rope pulleys and a 21-inch (53.3 centimeter) V-Belt from an Ace® Hardware store. The idler sprocket spins on its brass bushing; the drive sprocket is affixed to a 1/4-inch steel shaft tightly pressed on, and then with a 4-40 set screw run all the way from the pulley trough to the shaft. The track (V-belt) is only pulled as tight as is required to prevent it from ever jumping out of the pulley trough. The steel rod the drive sprocket is on can either be directly coupled to your motor with a shaft coupler or the drive sprocket shaft can be floating with either a gear or chain link connection to your motor. The former is used if you are using a gearhead motor, the latter if you need a step down from a higher-speed motor.

Figure 10-14
V-belt and pulley tracks

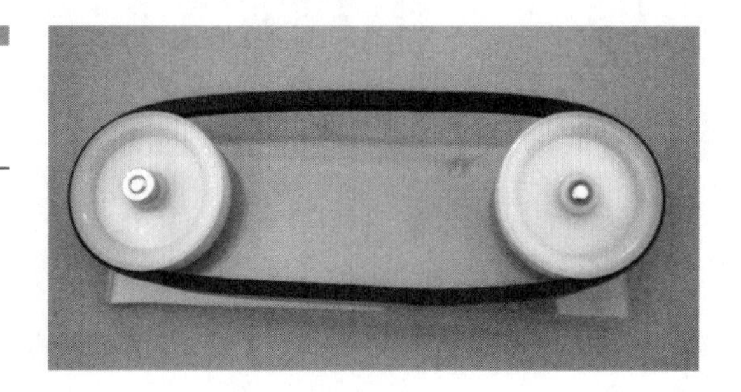

The Eternal Dilemma: Mounting Shafts to Hubs to Wheels

You've been waiting for this part by now. We'll now show you how to find or create axles, hub adapters, shaft couplers, and wheel hubs. Although you don't need a machine shop to do these projects, you *will* need a drill, a good vise, and a very steady hand. A drill press and a hobby drill (such as a Dremel® tool) will come in handy. If you have plenty of funds, you can buy everything we show you how to make here, but we're only going to give you the products that are affordable on a hobbyist's budget or a way to build them inexpensively yourself.

The Easy One: Attaching Wheels and Hubs to RC Servos

Since hobby servos already have control horns that attach to the servos with a screw, and since these control horns are designed to have wires and such attached to them, it's pretty easy to attach a wheel to an RC servo. Because all of us have been clamoring for solutions to our robotic projects, companies have been obliging us with new products; we'll show you some of these new time-savers.

Just about any type of wheel will attach to the control horn of an RC servo if you are careful when you center the wheel. Because you can use a longer screw to hold the wheel to the servo, you can get a firm connection. You can use a 4-40 screw if your servo uses nylon gears; if you have a metal spline gear, you will need to use a 2-millimeter screw for most servos. Then use a hot glue gun or Shoe-Goo® to affix the wheel to the round control horn. An example of this type of wheel attachment is shown in Figure 10-15.

Servo City® and Jameco® (www.jameco.com) sell sprockets and belts that connect directly to the Hitec servo hub (see Figure 10-16). Servo City also sells plain bore sprocket versions that can be press-fitted to shafts. Soon they will be releasing a sprocket series with set screws in the hub. They also sell 32- and 48-pitch gear sets that are plain bore and servo mountable, but building gearboxes is a difficult and exacting process. Link belts and sprockets are more forgiving of inexactitude in spacing. The sprocket and link system can be used to power a floating axle very

Figure 10-15 Wheel glued and screwed to the control horn

Figure 10-16 Servo-mountable sprockets and a link belt

simply when your robot weighs too much to put all of that mass on the output shaft of your hobby servos.

Attaching Hubs and Shaft Couplers to Motors

This is the next most difficult task. Most of the time you will have motors with smooth shafts or hopefully a flatted shaft (a shaft with a flat side ground in it). You will also have wheels that you like and that fit your robot. The question to ask is how to attach the wheel to the shaft. If your motor shaft is long enough, you can attach a hub like Figure 10-17 to your shaft and your wheel to the hub. If you're not so lucky and your wheel can't reach the hub because the motor shaft is too short (not uncommon with stepper motors, for instance), then you'll need to extend the shaft with a shaft coupler. Be careful that you don't extend the shaft of the motor too far or the radial stresses placed on the output bearing of the motor may cause it to wear out more quickly.

Figure 10-17 A sleeve coupler

Another reason to use a shaft coupler is when you have a motor shaft that is not the correct size for any hub. You could also use a shaft adapter to change shaft sizes. W.M. Berg,®Inc. (www.wmberg.com) carries a line of shaft adapters that are perfect for the usual shaft sizes used in hobby robotics. Many distributors carry shaft couplers, also known as sleeve couplers of various kinds. You will find a short list of mechanical component suppliers at the end of this chapter. If you are going to connect directly to another shaft or to your wheel from the motor, then you want a rigid or sleeve coupler.

If your shaft is floating, then you can use Oldham or spider couplings (see Figure 10-18). Both of these shaft couplers are forgiving of imperfect alignments between the motor and the final shaft. The Oldham shaft coupler will forgive parallel shaft misalignment, while the spider coupler will forgive axial angle misalignment. Make sure that the spider coupler that you use is up to the stresses that occur in your drive train. An undersized coupler will self-destruct if the motor it is connected to is very strong and the robot stalls.

Figure 10-18 A spider coupler

Figure 10-19 Sizing up a shaft to a hub

Modifying a Shaft to Fit a Coupler Sometimes all you need to do is a little adaptation from the motor shaft to your hub or wheel. Figure 10-19 shows how to use brass tubing to make a motor shaft a little larger to fit with a hub adapter. You can use successive diameters of brass tubing to increase your shaft size. Each size of brass tubing will fit perfectly into the next size up.

Creating an RC Car-Wheel Shaft Coupler Figure 10-20a shows a DIY aluminum RC car-wheel hub made from an RC aircraft spinner adapter from C.B. Associates® Inc. model 5204 that costs about $2. We then filed the sides down until they fit nicely into the RC wheel hex mount. We chose this adapter to have a screw size on the wheel side whose screw would fit through the wheel and a hole size on the shaft side that could be drilled out to a 1/4-inch shaft size. We used a 4-40 set screw that we drilled and tapped a hole for. Figure 10-20b shows a Delrin™ rod modified to use between a gearhead motor and an RC car wheel. We got this material at a local plastic fabrication shop for $2 for 2 feet, which

Figure 10-20a An aluminum DIY shaft adapter

Figure 10-20b A Delrin DIY shaft adapter

was good as it took a couple of tries to get it right. This plastic is also available at McMaster-Carr and other machine specialty catalog stores. We simply ground six sides into the Delrin using a hobby mototool until it fit the RC wheel. Again, we drilled a setscrew hole and tapped it for a 4-40 screw.

Use a vice to hold your part while you are tapping a hole. A metal tap is very hard and sharp, which means that it's very brittle. You can snap a tap off, even when tapping aluminum, if you aren't careful. Fortunately, taps are not very expensive.

Already-Made Hub Adapters, Just Have the Right Size Shaft Sometimes it all just comes together. You have a 1/4-inch motor shaft, a simple wheel, and then you can use 1/4-inch hub adapter, as shown in Figure 10-21. This hub adapter is carried by Jameco Electronics and Servo City. You can affix it to either a motor shaft or a floating shaft.

Figure 10-21 Hub adapter

Creating a LEGO Axle Shaft Adapter If you are using LEGO wheels, it's easiest if you use the LEGO axle to attach them. You can make an adapter for a motor shaft with the LEGO axle, as shown in Figure 10-22 and Figure 10-23. Figure 10-22 shows two layers of brass tubing, one inside another for strength, and a single 1/4-inch collar holding the adapter to the motor shaft while glue is holding the LEGO axle into the sleeve. At least 1/4 inch (about 1 centimeter or so) of the LEGO shaft is left producing. The collar snugs everything down to the motor shaft very securely. Figure 10-23 shows one collar holding the adapter to the motor shaft and one collar holding the LEGO axle inside the tubing. Since we just hate damaging our LEGO parts in any way (such as with glue), we prefer this latter method.

These shaft collars are available at any hobby shop that caters to the RC aircraft hobby. They come in a variety of sizes, so surely one will work for your project. The brass tubing is also available at your hobby shop or

Figure 10-22 Gluing the LEGO axle adapter

Figure 10-23 Not gluing the LEGO axle adapter

often at arts and crafts or model-railroad-based hobby shops. Each size will fit perfectly into the next size up, which makes this tubing ideal for your purposes of shimming a shaft to a desired size.

Modifying a Brass Sleeve Bushing into a Shaft Coupler This shaft coupler started out life as a bushing in a printer assembly. It has a 1/4-inch inside diameter that makes it ideal for use as a shaft coupler. This sleeve is one inch (2.54 centimeters) long. Drill and tap a 4-40 (or 2 millimeters if you please) hole 1/4 inch or less (about 0.6 centimeters) from each end of the sleeve. Now you can use a plain 4-40 (2 millimeters) screw in the hole, or you can cut off the head of the screw and, using a Dremel tool's cutoff disk, carve a single slot into the raw end of the cut-off screw (see Figure 10-24). We prefer to leave the head of the screw on if possible because the screw is stronger and we can tighten it better. This is a very stout shaft coupler. Many common parts can have these sleeves in them as bushings, pulleys, and printers are just a few examples.

Creating a Flexible Shaft Coupling Besides the spider coupler, Oldham coupler, and bellows coupler that you can buy, you can also make your own flexible shaft coupler. Figure 10-25 shows a flexible coupler made from RC aircraft fuel tubing. This tubing is tough and flexible, and it grips well. It comes in three sizes: small, medium, and large. We've found that it works well in smaller robots of under 2 pounds (1 kilograms). To make your coupler more secure, use more fuel tubing. This is not recommended for sumo robots; the coupler will not take a lot of torque before it starts to twist. How much? We have not quantified that rating, so we'll leave it as an experiment for the student. As mentioned in Chapter 3, hose-clamps can be used (space permitting) to strengthen the tubings grip on the shaft.

Figure 10-24 DIY shaft coupler

Figure 10-25 A fuel-tube flex coupler

Using a Floating Drive Shaft

A floating shaft is a drive shaft that is not directly connected to a motor shaft via a coupler. It may be connected to its drive motor by a gear train, chain link, spider coupler, or another flexible coupler. The weight of the robot is resting on the support system for the floating shaft instead of on the output bearing of the motor.

We highly recommend that if you use a floating shaft that you support the shaft with bearings. A small selection of bearings is shown in Figure 10-26. Shown are two Oilite™ bearings and a ball bearing. An Oilite bearing is a sintered (made from ground-up material with lots of void spaces) that has at least 30 percent of its space taken up by oil. The shaft basically floats on a coating of oil that is "sucked" from the pores of the bearing. A ball bearing is one you are more familiar with and has an inner and outer *race* with balls between them. We recommend that you use *flanged* bearings on your projects. Figure 10-26 shows a flanged bearing in the upper right. This type of bearing will self-center so it is parallel with the surface in which it's mounted. Ball bearings will generally tolerate higher loads than oilite bearings of the same size. Ball bearings are also more expensive; this is another one of those price/payback tradeoffs.

Mounting and Supporting the Floating Shaft

We've already mentioned that you should support your floating shaft with bearings so that the shaft and the material in which it's mounted will last longer and wear better. A floating shaft can be supported in two ways: directly in chassis structures or with pillow blocks.

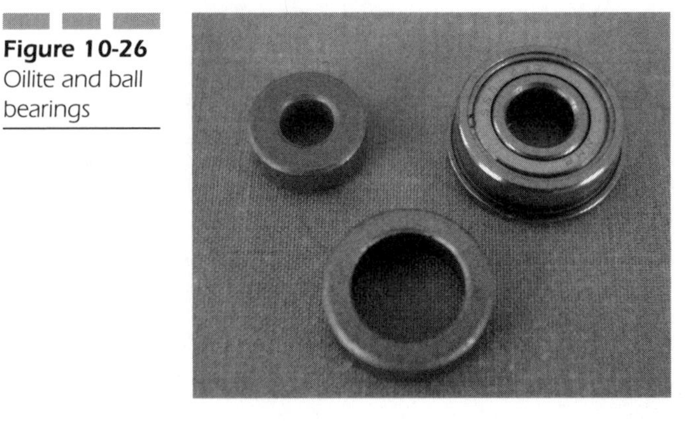

Figure 10-26 Oilite and ball bearings

Mounting the Shaft Through Holes in the Chassis This is the simplest way to mount a floating shaft. However, it is the most finicky because you have to make sure that your panels are all perfectly aligned so that your shaft does not bind. The best way to align your holes is to drill them all at the same time when the panels are all clamped together, as in Figure 10-27. This way you know the pattern of holes will be as identical as possible between panels.

Mounting the Shaft with Pillow Blocks A pillow block is simply a bearing that is mounted in its own surrounding case. A pillow block is useful because it enables you to adjust the mounting of a shaft without redrilling holes in panels or other drastic, one-time-only modifications. Some pillow blocks are self-aligning, while others have built-in shock

Figure 10-27 Drilling panels all at the same time

Figure 10-28 The pillow block

cushioning. An example pillow block is shown in Figure 10-28. The one downside of a pillow block is that it takes up more space because it's not part of the panels or bulkheads of the chassis.

The DIY Pillow Block Making your own pillow block is not difficult. One way of creating one is to get an aluminum or steel L bracket, drill a hole to mount a bearing on one side of the L, and then drill two holes, preferably slots, in the other side of the L. A very simple and useful DIY pillow block is shown in Figure 10-29. When drilling such a large hole, you should do it in stages, creating a slightly larger hole with each successive drill bit. Holes this big are hard to create, so be careful. If you want even greater support, then use square aluminum stock and mount a bearing on one of the parallel sides.

Driving the Floating Shaft

A floating shaft can be powered by three basic means. Each takes up more space than directly mounting wheels to motors, but because the shaft and bearings are supporting the weight of your robot, the output shaft of your motor doesn't have to. The three tools are as follows:

- Shaft coupler, either rigid or flexible
- Spur gears
- Belts or chain-link drives

Connecting to the Floating Shaft with a Shaft Coupler A coupler can be a sleeve coupler, which is rigid, or an Oldham or spider coupler, which enables some leeway in shaft alignments and isolates the motor

Figure 10-29 The DIY pillow block

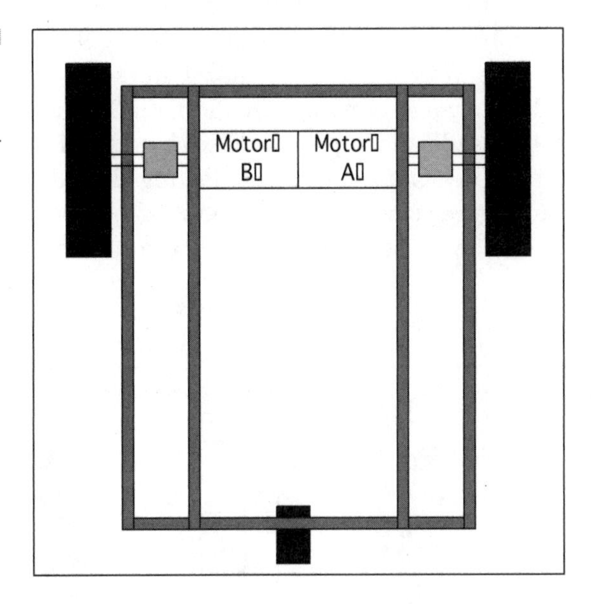

Figure 10-30 Shaft coupler example

somewhat from the output shaft or a flexible coupler like a universal joint, bellows coupler, or even a piece of rubber tubing if your power needs are small. These types of couplers are carried by many distributors. A partial list includes Jameco Electronics, RobotZone, W.M. Berg Inc., Small Parts® Inc., and Sterling Instrument®. Figure 10-30 shows one way to drive a floating shaft with a shaft coupler.

Connecting to the Floating Shaft with Spur Gears Creating a gearbox is nontrivial and we don't recommend it; we suggest instead that you get a gearhead DC motor. These have been engineered to be sound and durable. Several places offer good gearhead motors at a reasonable cost. An incomplete list of our favorites are Techmax at www.techmax.com/small-electric-motors/index.htm, Jameco electronics at www.jameco.com, and All Electronics at www.allelectronics.com.

You can gear up or down if you choose to connect to the floating shaft using the spur gears. The main drawback to gearing is that getting the proper gear mesh and finding the proper gear sizes that fit your motor and your floating shaft are not simple to develop from scratch. We recommend at least a 32-pitch gearing if you are using high-powered motors and a 48-pitch gearing for those robots under about 10 pounds (4 kilograms). A partial list of distributors that supply spur gears are Servo City, W.M. Berg, Inc., Small Parts Inc., and Sterling Instrument, as well as online RC hobbyist shops like Tower Hobbies®. No doubt many more are available, but we've found these distributors easy to deal with. You can also find spur gears of many sizes in both 32 and 48 pitch at any hobby store that sells and supports RC racing cars. Figure 10-31 illustrates one way to design a spur-geared floating shaft drive.

One way to assure yourself that you have a good gear mesh that is not too tight (which can cause excessive wear) or too loose (which can cause

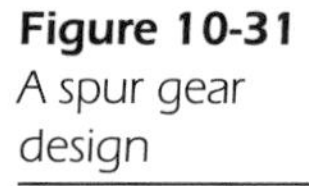

Figure 10-31
A spur gear design

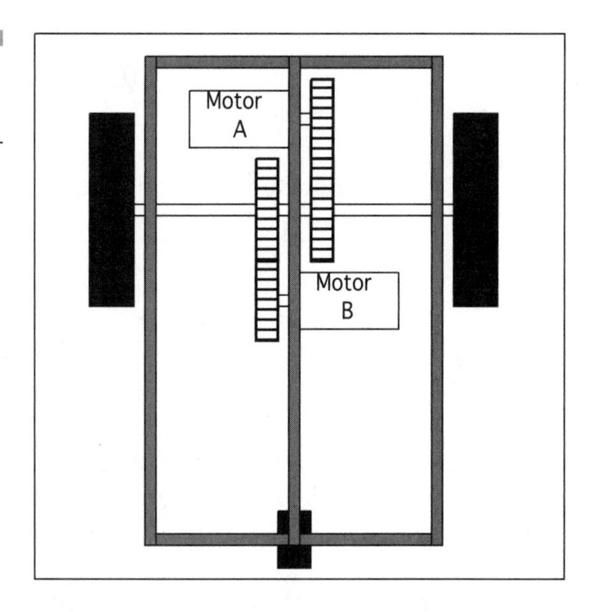

excessive noise and vibration) is to make sure that your motor location can be adjusted. To make your motor mesh adjustable, create an arch-shaped slot for one of your motor mount holes so that the motor can swing a small amount towards and away from the shaft-mounted gear. To get the correct mesh, smash a piece of common notebook paper between the motor pinion gear and the wheel shaft spur gear with your thumb and forefinger. Tighten the motor set screws and then rotate the gears to get the paper out from between the two gears. You now have a good gear mesh.

Connecting to the Floating Shaft with Belts or a Chain You can gear up or down by connecting to the floating shaft with belts or a chain, and a chain link or belt drive is more forgiving than a gear mesh to align. Your design will still work if your alignment is a little off or your sprockets aren't perfectly square. You can often find chain and sprockets at surplus shops such as American Science and Surplus® but not consistently. A new supplier of chain-link drives is Jameco Electronics, which distributes the Servo City microchain link and sprockets for smaller robots. The Servo City's chain drives are unique in that they support direct sprocket connection to Hitec and Futaba® servo hubs. For those larger, heavier robots, W.M Berg, Small Parts, All Electronics®, and Sterling Instrument sell a large variety of chain-link and belt-drive components.

Figure 10-32 shows a chain-drive series distributed by Jameco Electronics. Figure 10-33 illustrates one way to use a chain drive to power your floating shafts. This chain is deceptively strong and capable of pulling up to 7 pounds (3.1 kilograms).

These open hub sprockets from the RobotZone will press fit on very tightly to their assigned hub size. It helps though if your shaft can be

Figure 10-32 The Servo City chain-drive system

Figure 10-33 A chain-drive design

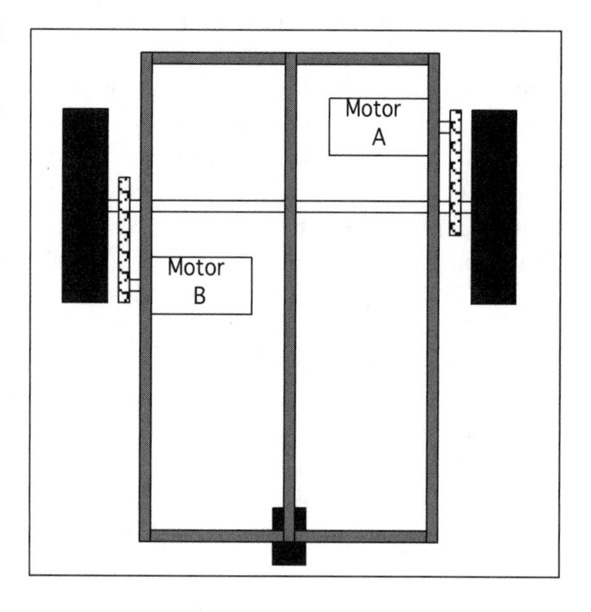

Figure 10-34 A knurled shaft

knurled to give even higher friction to the press on gear. Figure 10-34 shows what we mean by knurled.

Now you've designed your drive system and if you have been using wheels, you may need some dummy wheels to hold up the rest of your robot. The most common "third wheel" to use is the castor. Two very good ways exist for building or adding castors as drag wheels to stabilize your robot. Figure 10-35 shows a picture of them both, a common commercial castor and an RC hobby aircraft tailwheel assembly.

Figure 10-35
Drag wheels

The Final Results

You've seen many kinds of drive train designs, and you've seen how to make some of your own parts as well as where to get parts that you can buy. Now we'll show you two examples of robots that use these designs (see Figures 10-36 and 10-37). Both of these robots are at or under 1 pound (about 500 grams), both use hacked hobby servos for power, and both were made from parts found at your local hobby shop and in your hobby junk box. To build robots, you *must* have a well-stocked junk box.

Figure 10-36
A tracked robot

Figure 10-37 A wheeled robot

Figure 10-38 The Arobot™ drive train

Lest you feel that you have to always follow the "two motor, one drag wheel castor" design rules, here is an example of a robot that does not. This is the drive train for the Arobot™ from Arrick® Robotics (www.robotics.com) in Figure 10-38. This robot uses a single gearhead motor on one rear wheel and two steering front wheels controlled by a hobby servo. This unique design avoids many problems seen with dual-motor designs. For one thing, you can make it go straight without fiddling with differences between two (supposedly) identical motors.

CHAPTER 11

Locomotion for Multipods

Multipods—sounds cool, doesn't it? We could have called them *multipeds,* but the "peds" part of the word implies feet. Our robots tend to have limbs that look more like pseudopods and not feet, so multipod it is. Ever notice how nature doesn't use wheels as a means of transportation? To create more lifelike robots, we try to emulate nature. What nature makes look simple we've found is actually really difficult!

In this chapter, we'll be dealing with creating legged robots using primarily hobby *radio control* (RC) servos. These are easy to mount, easy to interface, and easy to control with a great deal of accuracy. We'll examine four gaits that can be built by the amateur robot creator and we'll show example robots that use variations on the legged robot themes.

Issues Working with Two or More Legs

Walker robots must deal with some unique issues that rolling robots don't have. These issues deal with *gaits* (how a creature moves from point A to point B) and balance. A design aspect of walking robots that is shared with their wheeled brethren is joint stresses, or how to support all that weight on small hobby servo shafts. Both design quandaries have reasonably simple solutions.

Static Versus Dynamic Balancing

When most creatures walk, they shift their center of balance continually so that the process does not involve falling down. This is called *dynamic balancing*, where the robot's controller hardware is constantly getting feedback as to the status of the robot's balance. To compensate for poor changes in balance or orientation, the robot has to alter either its acceleration or its center of gravity. To do this in a robot is complex and often quite expensive. Although the concepts are transportable between robots, every type of robot requires different algorithms and specific fine tuning. Dynamic balancing requires the use of accelerometers in multiple axes, tilt meters, and sophisticated motor control code. It already is a discussion topic for entire books all by itself, so we'll not delve into it here. We'll save that topic for the second edition perhaps.

A statically balanced robot is one that is carefully designed in both physical characteristics and walking style such that it never becomes seriously off balance. To statically balance a robot requires careful planning with foot placement and often really big feet. Each of our robots in this chapter will balance with one of these two statically stable design features:

- A tripod leg arrangement such that at least three legs are always on the ground (see Figure 11-1)
- A really big foot, over which the robot may center its weight and balance (see Figure 11-2)

Figure 11-1 The Lynxmotion® Hexapod using a tripod gait (Source: Lynxmotion, www.lynxmotion.com)

Figure 11-2 The LEGO® Technic™ model with "big feet" (Source: The LEGO Group. www.lego.com)

Servo Shaft Support and Stresses

Whether your hobby servo has bearings or just plastic bushings, it is limited in the amount of *axial* stress it can tolerate. This stress is not rotational, so to get a good feel for it, hold your arm out straight with your palm up and put a book on your palm. That strain you feel in your shoulder and elbow is axial stress (more or less). Another common type of stress felt in our drive motors (servos here) is *axial* stress, which is stress that is pushing the shaft into or pulling the shaft out of its motor. Hobby servos don't mind this as much and are well protected from it, but it creates a great deal of friction that robs us of power. Both of these stresses should be minimized (see Figure 11-3).

To reduce axial stress, we will use the servo control horns with linkages to the joints that they are to move. To reduce axial stress, we'll make sure that no servo directly supports the joint or leg that it is supposed to move. More explicit examples will be shown in sections that deal with our example designs.

Simple Linkages

Several simple linkages are strong enough that our small walking robots can make use of them. Each linkage type has its pros and cons. Even if we are trying to reduce our robot's complexity, usually achieving complex movements can be done in ways that would appear to need many more

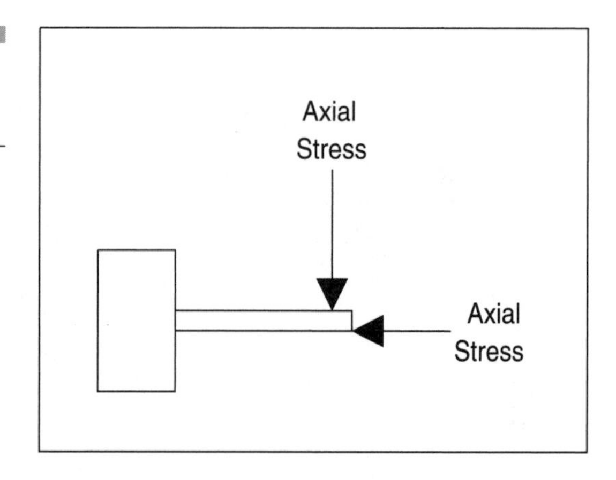

Figure 11-3 Shaft stress types

degrees of freedom (DOF) than are actually needed by our robots. Let's investigate these concepts a bit.

The RC Car Ball and Cup Linkage

The ball and cup linkage (see Figure 11-4) is a common linkage used in RC car-racing kits. This type of linkage is strong and can be used to span reasonable distances without the control linkage bowing or slipping. Its main disadvantages are that it is bulky to implement, it moves the actual pressure point off the center of the control horn and may present a lever arm flex that can allow slop in the system. If we use strong control horn and linkage materials and keep the spanned distance short, then these pitfalls can be avoided.

One of the greatest advantages of the ball and cup linkage comes into play if it is configured with a *turnbuckle*. A turnbuckle is the joint that connects the two cups at the ends of the control arm span together. It is formed and threaded in such a way as to enable the user to rotate the turnbuckle and cause the cups to move away from each other or towards each other. In other words, one side of the turnbuckle is right-hand threaded and the other side is left-hand threaded. This feature enables the linkage to be adjusted without removing any of the parts. Although more expensive than other linkage options, it's probably the most versatile.

Figure 11-4 The ball and cup linkage

Z-bends and Using RC Aircraft Servo Linkage Arms

The least expensive linkage option is the *Z-bend*, so named by the shape of the wire used to connect to the control horns. Figure 11-5 shows what a Z-bend looks like and how it is connected to a control horn to form linkages. You can create a Z-bend with two pairs of needle-nosed pliers, but it is difficult to make a good bend every time. If you are going to be making a lot of Z-bends, your local hobby shop that caters to the RC airplane hobbyist will have a specialized tool to make perfect Z-bends in a variety of wire gauges. Figure 11-6 shows such a useful tool.

A Z-bend linkage is simple to install, but it requires that one side of the linkage have its control horn removed to complete the connection. Because of this permanent link length, a Z-bend linkage must be made correctly the first time and cannot be adjusted. This lack of adjustment is the downside of a Z-bend link. The advantages of a Z-bend link are that it is very low profile, so it can fit into tight spaces and work well. The

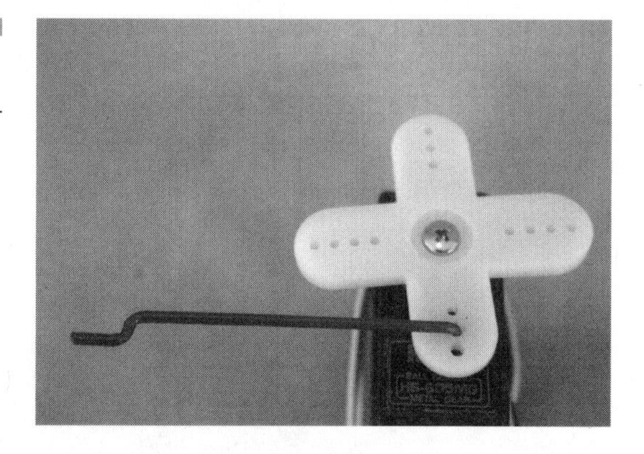

Figure 11-5
Z-bend linkage

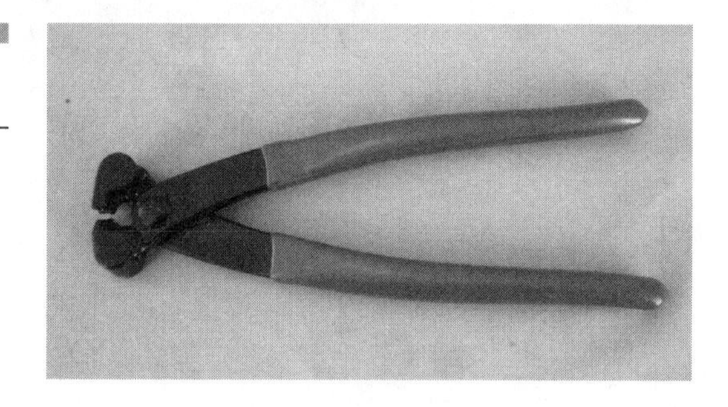

Figure 11-6
The Z-bender

pressure point on the linkage is also directly in line with the force generated by the control horn. This causes less flex in the control horns and makes for a more repeatable motion.

The RC Aircraft Clevis Linkage

Halfway between the Z-bend and the ball and cup linkage is the clevis linkage. A clevis is a removable linkage terminator that clamps onto the control horn in such a way that the transfer of force is directly along the linkage wire from the control horn like a Z-bend. Because the clevis can be screwed onto the end of a linkage wire, it is adjustable. Often a clevis will be installed on one end of a link and a Z-bend on the other, making the linkage more adjustable and just as strong as a Z-bend. Figure 11-7 shows an installed clevis link. In high vibration environments, it is common to clamp the clevis so that it can't open up. The accepted manner in which this is done (in the RC aircraft field) is to use a piece of clear fuel tubing forced up onto the clevis, as shown in Figure 11-8.

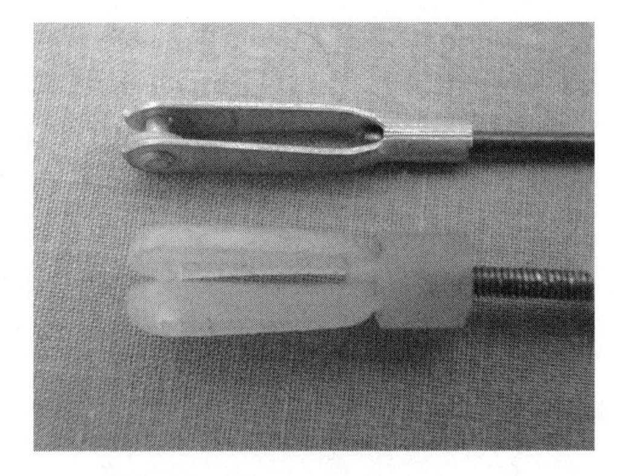

Figure 11-7 The clevis

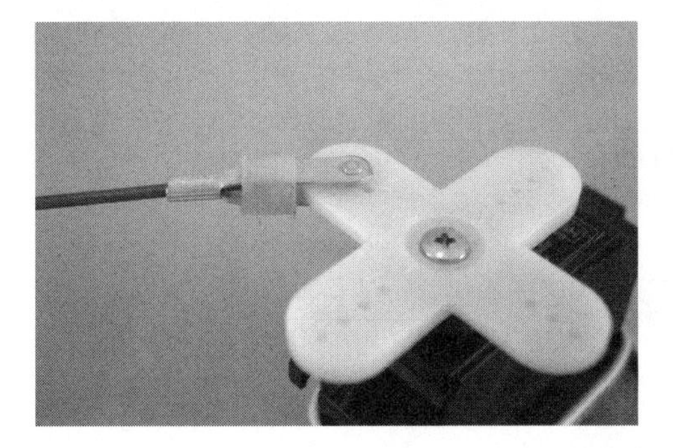

Figure 11-8 Clamping a clevis link

Power Versus Speed: Choosing Linkage Locations

A trade-off exists between how far we can move a limb and how much power is available while we're moving it. Way back in Chapter 2, "Motor Types: An Overview," we discussed the inverse relationship between power and the length of a moment arm. Remember that torque is defined as the force that is produced at a given distance from the center of a rotating shaft. The product of that distance from the center and the force delivered at that point is the torque. What we lose in torque when we move further away from the center of the motor shaft we gain in total movement and speed. Because the outer edge of a lever attached to a shaft must travel farther than a point farther in towards that shaft, and it must transition in the same time as the inner location, the outer location of the control horn is moving faster than any point inward from it. Figure 11-9 clarifies these concepts.

This suggests that if you want more movement, and incidentally more speed, you connect your linkage to an outer option hole in the control horn. If you want more power, connect to an option hole farther in towards the motor shaft. Inversely, placing the linkage farther out on the controlled surface being moved will deliver more torque but move the limb or joint less distance. Figure 11-10 shows what this means.

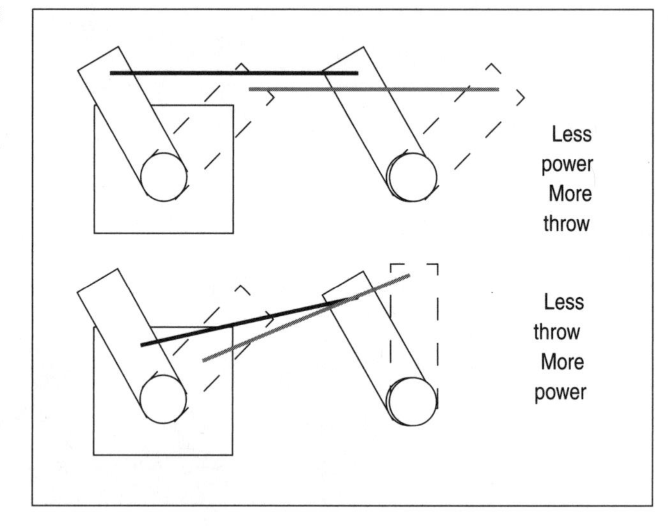

Figure 11-9 Power, speed, and distance at the source

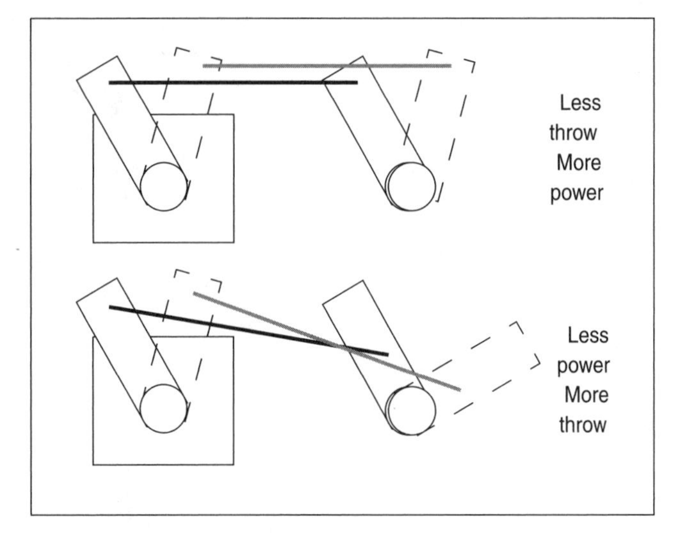

Figure 11-10 Power, speed, and distance at the target

Common Linkage Arrangements

You've seen how to make a linkage, now we are going to show you how to make the legs that you will use to move your walking robot around.

Piano Wire: The Common Linkage Connection

Piano wire is a strong, stiff wire that is bendable but will retain its position after being bent. One can assume that it was once used inside a piano to play music. Regardless of its origins, it can be found in any hobby shop in many diameters. The usual length of a piece of piano wire is about 2 to 3 feet (about a half-meter to almost a meter).

Diameter of Piano Wire to Use The diameter of piano wire to use depends strongly on how you are planning to use it. A linkage that needs to move a reasonable mass, both pushing and pulling, will require heavier wire, and one that only pulls or moves a very light mass needs only a light wire. Also, the longer the run, the heavier the wire needs to be so that it remains rigid when in use and doesn't flex. Figure 11-11 shows a

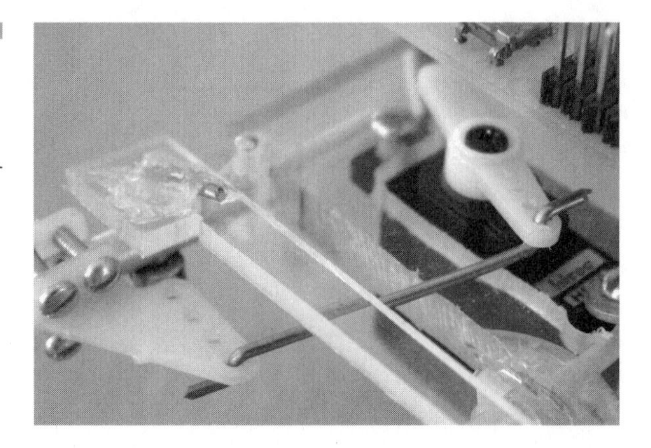

Figure 11-11 A good piano wire connection

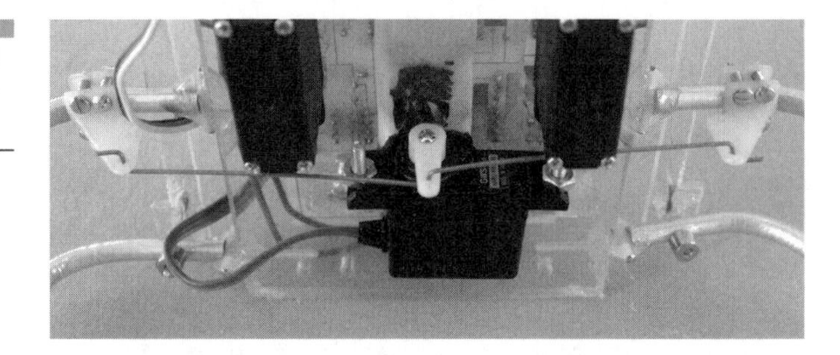

Figure 11-12 A light use for piano wire

good Z-bend piano wire connection: short, heavy, and straight between a servo and the control horn. Figure 11-12 shows a piano wire connection made with very light wire; all the force of lifting will be done by pulling on the wire. Finer piano wire makes routing the wire through all the control horns easier.

Figure 11-13 shows a variety of piano wire diameters arrayed side by side for comparison. Your choice of piano wire will be dictated by how much axial stress the wire will need to tolerate, how much weight you will need to move, and, our next discussion topic, how the wire will be routed.

Routing the Piano Wire You will usually want to keep your piano wire as short as possible to reduce the flex of the coupling while pushing. However, sometimes you will need to bend the piano wire to go around an obstruction or to make a corner that is part of your design. If you need to bend the wire to make a connection, make sure you use even larger wire so that it won't flex when pressure is applied. Also, keep your

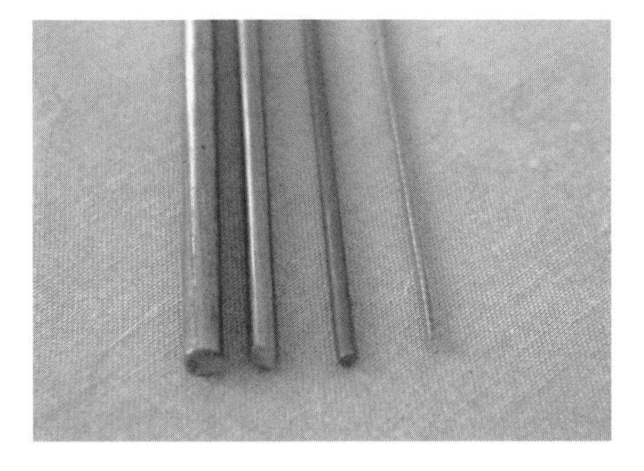

Figure 11-13 Piano wire sizes

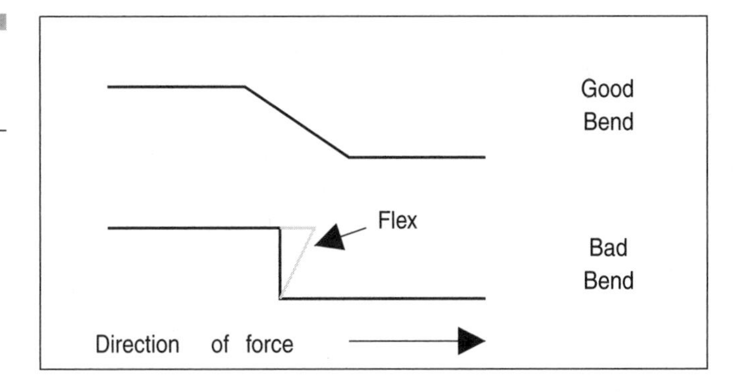

Figure 11-14 Wire route bends

bends to 45° or less so that the force path is still nearly parallel with the force transmitter (the wire). A 90° bend is the worst kind of bend; the wire will always flex at the point where the line of force is perpendicular to the line of the wire. See Figure 11-14 for more explanation of the force vector issue.

If you have very long wire paths or you have bends in your piano wire that you are unsure about, then use wire supports in one or more places along the wire path. Supports will give the piano wire greater rigidity by supporting the wire and reducing flex during compression strokes (pushing the wire). Good supports can be made by simply bending a small piece of metal or forming plastic to support the wire. You can either support the bare wire with a formed support, as in Figure 11-15, or you can sleeve the entire wire in plastic before clamping it in the support. The former method is needed if your wire has bends in its path, while the latter works well for very long but straight wire paths.

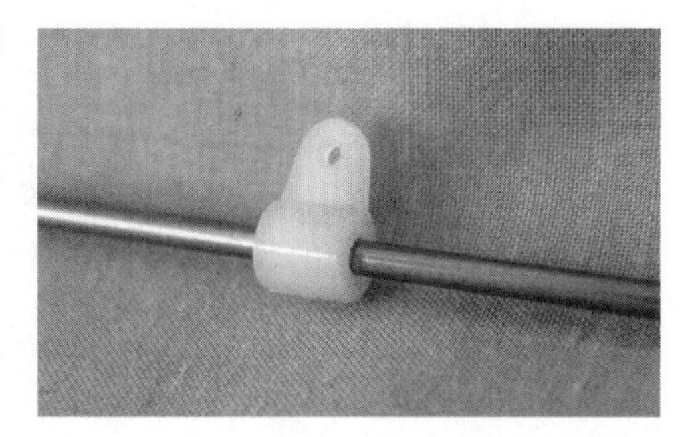

Figure 11-15 Supporting a wire linkage

Chain Links

Sometimes you need to get a lot of movement in a joint, and you want it to be proportionally consistent over the whole range of motion. A piano wire push rod will move the greatest distance, while the control horns at the servo and the target joint are parallel to each other and the force is perpendicular to the control horn motion. But when the control horns are not perpendicular to the motion, then you get less actual movement for the same distance moved on the control arms. Figure 11-16 shows a graphical explanation of this discussion. Basically, your motion will be

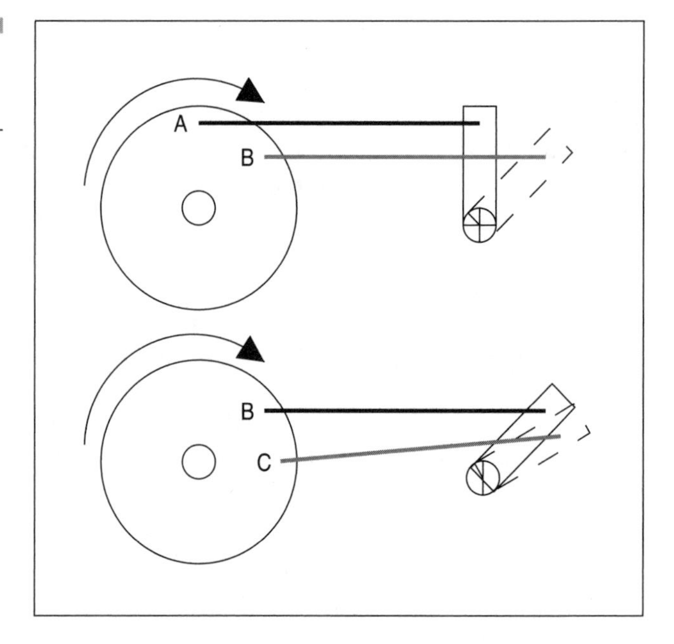

Figure 11-16 Nonlinear motion

Figure 11-17 Chain link joint actuator

slower and less pronounced at the extremes of the movement and more pronounced during the middle part of the movement. This nonlinearity is often not desirable.

A solution exists for this dilemma. Use a chain and sprocket linkage between the servo and the moving limb. Not only is the motion exactly proportional between the servo and the moving limb, but you can set the proportion of movement by selecting sprocket sizes that are different for the servo and the limb joint. With a chain and sprocket drive, if you make the sprocket on the servo smaller than on the limb joint, you can boost the torque at the joint while lowering its speed. Conversely, you can make the sprocket larger at the servo than at the joint and speed up the motion at the joint (at the sacrifice of power). Figure 11-17 provides an example of such a chain and sprocket limb joint. The drive motor and limb actuator are always perfectly synchronized and the movement is linear.

Multiple DOF Servo Mounting

First off, DOF means degrees of freedom, as stated earlier, and describes the number of joints or places that a limb may bend. Most animals have six or more DOF per limb. It is uncommon for a robot to have more than three. To get multiple DOF per limb, we must have a way to connect, mount, and orient our hobby servos. This section outlines some ways to make robot limbs. We'll make these example designs simple enough to create in your own home. More sophisticated products can be found out there, but without expensive machinery, we can't make them at home.

The Basic Leg We use hobby servos to simulate the articulated leg in our robot designs. The major reasons for this choice are that the hobby servo is very strong for its size and weight, it is a fully self-contained motor controller and positional sensor, and it's simple to interface to. This simplicity and completeness is a major boon when you consider that any walker will use from 3 to 18 joints.

This leg design can be used in any multilegged robot of four or more legs with minor modifications. Its design is such that it can be built using the simplest of tools. You will need a way to cut metal or plastic, a drill, a tap, two pairs of pliers, and an assortment of screwdrivers.

Figure 11-18 shows the construction of the upper leg swivel plate. It is a simple aluminum or plastic plate to which a servo is attached as shown. The servo is held in place using common hobby servo mounting

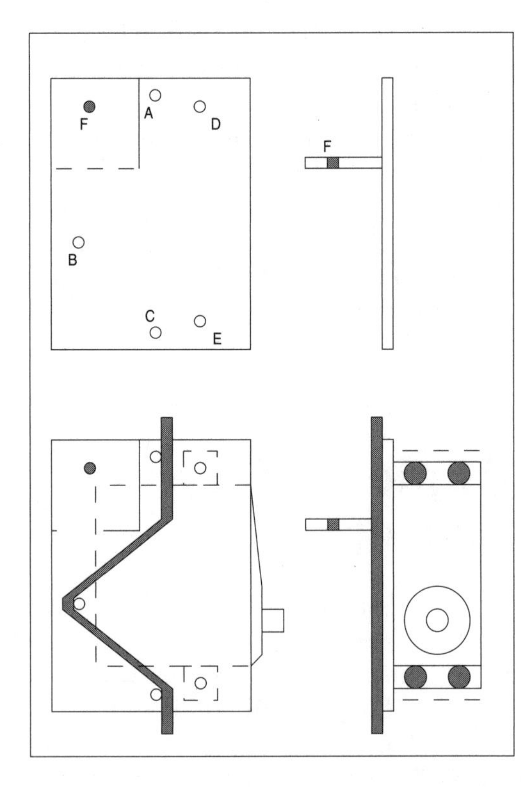

Figure 11-18 Servo pivot plate

blocks, which are available at your local hobby shops or at www.servocity.com, for instance. The holes A, B, and C are tapped to use screws to hold the piano wire shown in gray. A bend in the piano wire prevents the wire from rotating under stress and coming loose. The actual bends that you use are not important as long as the whole assembly comes out flat and the upper and lower protrusions are in line with each other. These protrusions are the pivot points that enable the plate to swivel like a hip. The holes D and E are to hold the servo mounting block to the plate. The hole marked F is for the servo linkage to rotate the plate.

Figure 11-19 shows the construction of the upper leg. This leg is attached to the control horn (we prefer the disk) of the servo that is attached to the upper leg swivel plate. The leg holds in it the servo for the lower leg. Since this part of the leg does not need as much strength as the upper leg servo and the swivel servo, it can be a smaller servo.

Figure 11-20 shows a full, three DOF leg assembled. This arrangement of the servos and leg segments enables an insect robot, for instance, to elevate or drop its body by changing the geometry of the leg segments.

Figure 11-21 shows a simpler, two DOF leg made from this design. This leg is not as versatile as the three DOF leg but is simpler to make and works well.

Figure 11-22 shows a section of a robot using the two DOF legs and the placement of the servo that will control the upper leg swivel mount. A is the linkage rod between the swivel servo and the swivel mount. B is

Figure 11-19 Upper leg plate

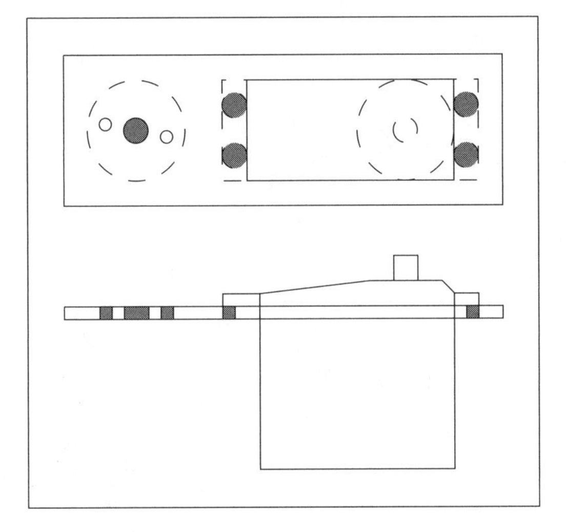

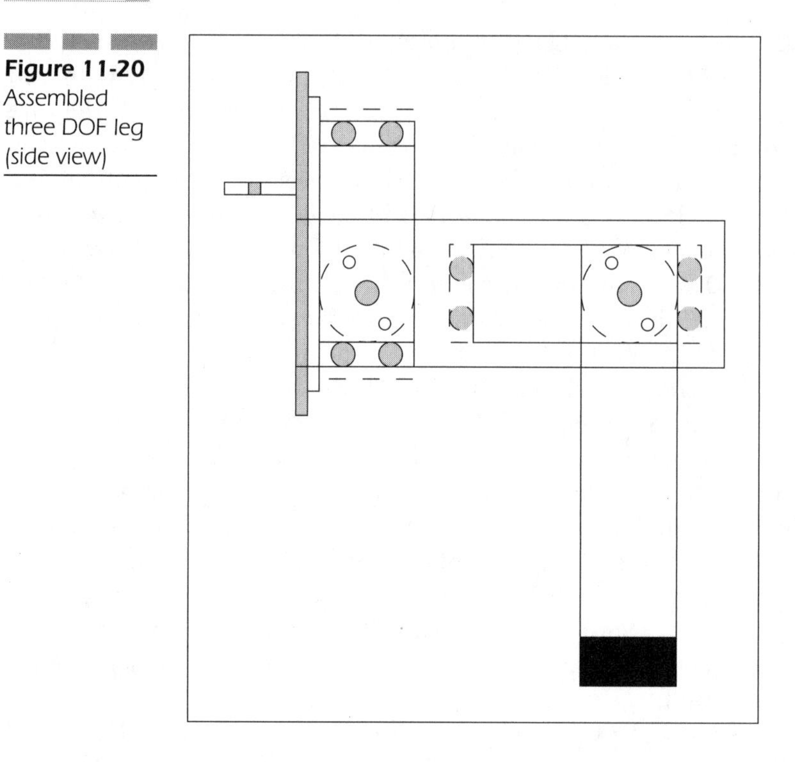

Figure 11-20 Assembled three DOF leg (side view)

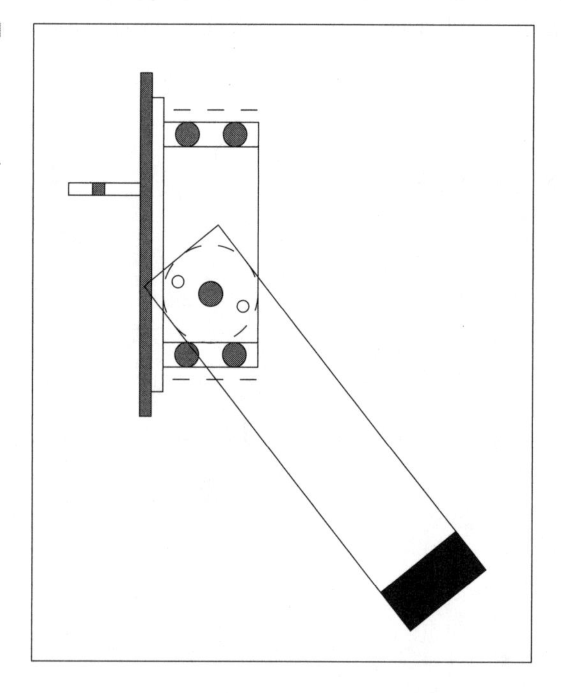

Figure 11-21 Assembled two DOF leg (side view)

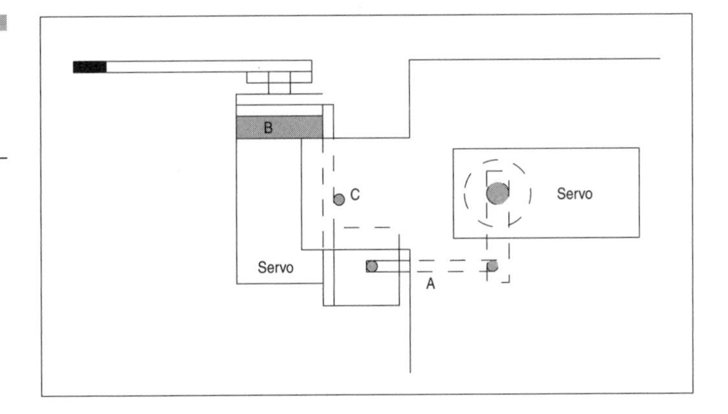

Figure 11-22 Leg mount assembly (top view)

a servo mounting block (one of two needed to hold the servo snugly to the swivel plate) and C is the swivel joint. This must match the hole location on the bottom of the robot chassis as well as the bottom swivel joint on the swivel plate.

The Bug Legs Arrangement Arthropods have a distinctive leg arrangement. They stick out the sides perpendicular to the body's travel path (except for the crab, which likes to go sideways). Mount your swivel plates as shown in Figure 11-23 to simulate insects and spiders. Figure 11-23 shows the mounting of a couple of the two DOF legs.

The Mammal Legs Arrangement To arrange the legs to move like a mammal would move, simply swivel the plate 90°. A mammal's legs move in parallel with the animal's direction of motion. Use the three DOF legs as shown in Figure 11-20 and rotate the swivel plate to mount the legs as shown in Figure 11-24. This will give rudimentary hips as well as the thigh and calf parts of a mammal leg. Arranging the position of the upper and lower legs will be covered in the next section, "Gaits."

A Special Leg for Bipeds This is a design for a biped with the least possible (and still functional) DOFs that we have seen. All the joints have as stress relief for the servo shaft a second pivot point at the back of the servo that is aligned with the servo output shaft. These pivot joints enable the servo to move freely with the least possible axial stress on the shaft.

Figure 11-25 shows the upper leg servo arrangement and pivot point location. We show a suggested servo mount bent into an L shape to

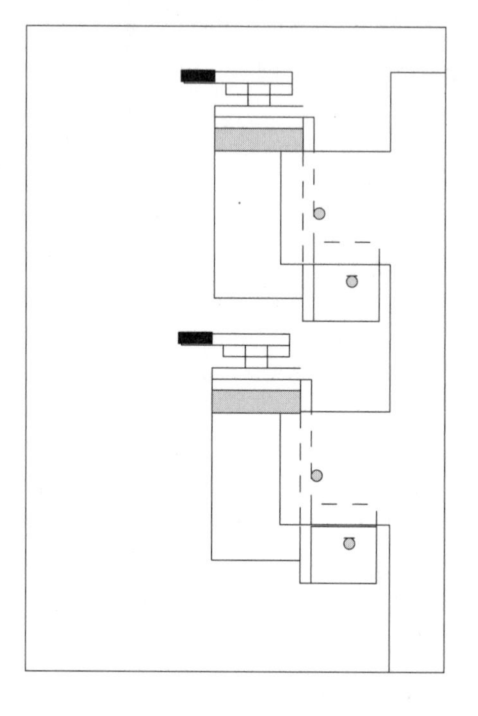

Figure 11-23 Insect leg arrangement (top view)

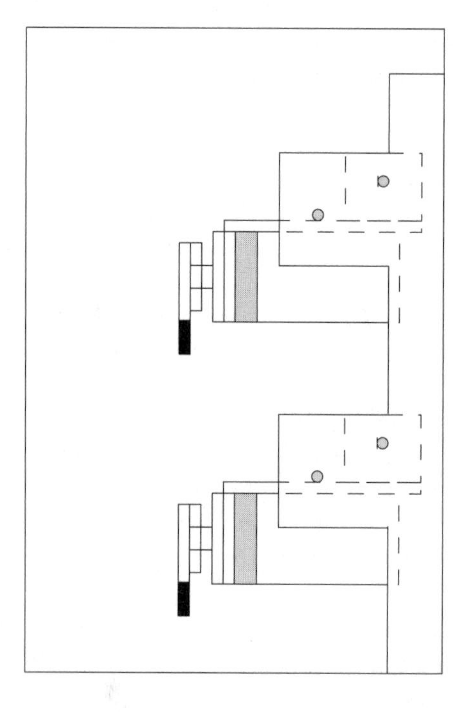

Figure 11-24 Mammal leg arrangement (top view)

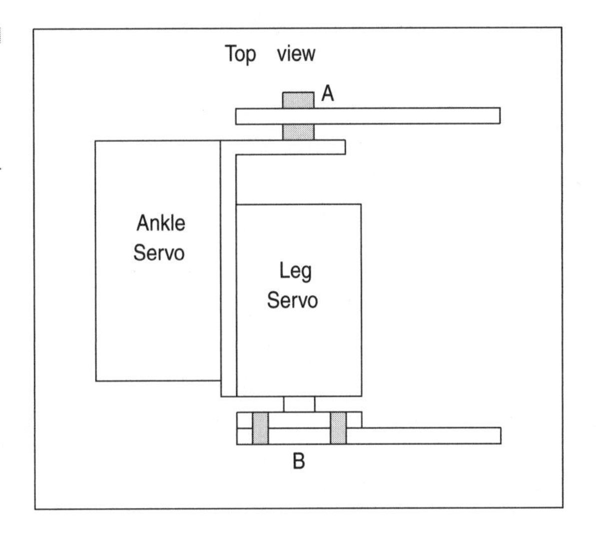

Figure 11-25 Upper leg arrangement (top view)

enable the positioning of the rear pivot point. We recommend a nylon bushing and nylon bolts to minimize friction and make the connection as simple as possible. We also recommend using aluminum as the servo mount material because it is light, tough, and easy to work with. Point A shows the pivot joint between the servo mounting plate and the robot chassis. Point B shows where the servo's control horn (here we use the disk) mounts to the main chassis of the robot.

Figure 11-26 shows the ankle construction, which is the most complex part of the leg. This servo must rotate the foot of the robot's leg. When a leg is lifted, this means that the weight of the entire robot is on that one leg, without any other support, and all that weight is an axial stress on the hobby servo. To improve the life expectancy of these ankle servos, precautions are taken.

Figure 11-26 shows how to create a servo mount such that *no* axial stress takes place on the servo output shaft at all. The shaded area, denoted A in Figure 11-26, is the nylon bushing and bolt that enable the foot to pivot at that point. The shaded area B shows where the control horn (here we're using a standard control horn) attaches to the wooden block C, which is in turn bolted to the foot. The B attachment point is also a pivot, enabling the control arm to swivel the foot around on bushing A. D is round so that the wooden block C can be carved out, enabling it to freely move around the perimeter when the foot is rotated.

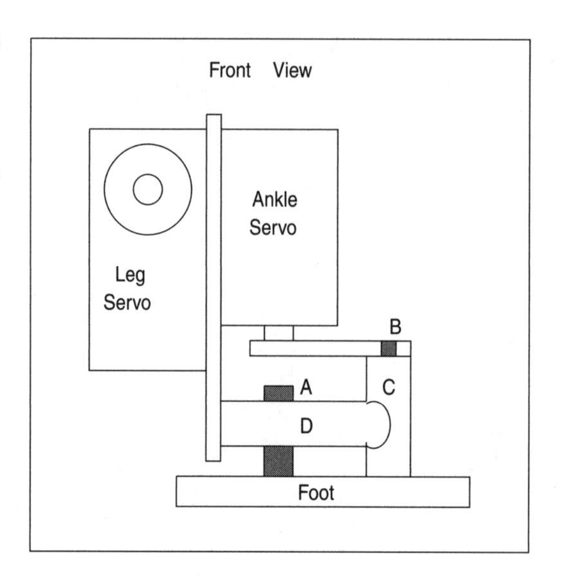

Figure 11-26 Ankle arrangement (front view)

Gaits

Now that we have our walkers built, how do they walk? Read on and we'll get you started down the road to teaching your robot how to walk.

A Two-Legged Shuffle

A clever person can create some unique ways to solve a complex problem. For instance, let's discuss a bipedal robot on the cheap. Mark Whitney of Acroname® (www.acroname.com) in his second attempt at a bipedal walker created Mrs. Stampy. We don't really know if the name was a nod to the Parallax® processor that controlled Mrs. Stampy or her mode of locomotion. Mrs. Stampy has two legs, each with two DOFs powered by hobby servos and a single head servo. The legs are the complex part and were detailed in Figures 11-25 and 11-26. Figure 11-27 shows Mrs. Stampy all dressed up for the Trinity Fire Fighting competition. The figure shows the result of our construction efforts.

To understand how Mrs. Stampy gets around, stand up and imagine walking in such a way that you may not swing your legs nor move your hips. All motion comes from lifting your legs up to the side and rotating your ankles. Here is the movement sequence Mrs. Stampy uses:

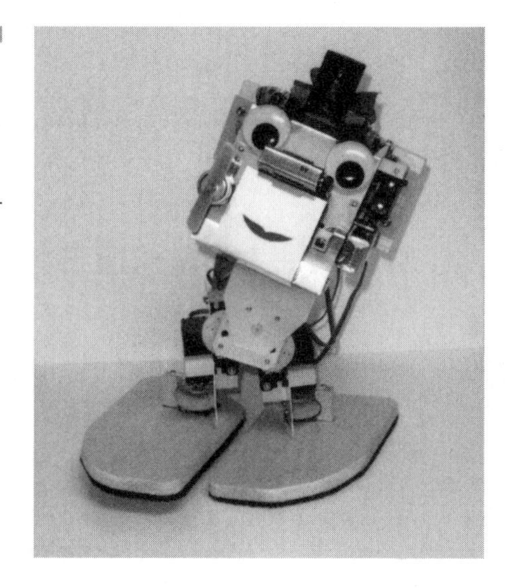

Figure 11-27 Mrs. Stampy (Source: Mark Whitney and Acroname)

1. Swing the head so that it is centered over the foot that is staying on the ground. This puts almost all the weight on this large, flat foot for stability.
2. Rotate each leg servo to lift the other foot into the air and have it remain parallel with the ground. This is why you need to use both leg servos. If you only rotate one leg servo, then the foot won't be parallel with the ground. The edge might actually be *on* the ground.
3. Rotate the ankle servo for the foot on the ground to move the elevated foot forward (the whole body will rotate) while rotating the elevated foot to be parallel with the static foot to go straight forward. The big feet have a cutout on the inside back edge that enables the foot to be rotated a full 90° without hitting the other foot. By adjusting the angle the feet are placed at, you adjust how far or even if Mrs. Stampy is turning. Figure 9-27 clearly shows what this will look like when this step is completed.
4. Rotate the two leg servos such that the elevated foot is placed back on the ground.
5. Repeat step 1, exchanging the roles between the two feet.

Making a bipedal walker can be done in other ways, but this one is the simplest that we have seen successfully used. Mark has video clips of Mrs. Stampy walking on the Acroname web site. Robots of Mrs. Stampy's ilk are limited to flat surfaces with minimal obstacles to avoid because she certainly cannot step over them.

A Four-Legged Walker, Mammal Style

Take a look at a dog sometime or a cat, or even a human crawling on his hands and knees. Note how the limbs are jointed. The knee joints point forward and the elbow joints point backwards (see Figure 11-28). This suggests that the majority of power to thrust forward is in the back legs, while the front legs are designed to reach or pull forward. Keep this observation in mind when coordinating a gait for a quadruped. *Many* quadruped gaits exist: the creep, walk, trot, pace, canter, gallop, and so on.

We're going to show how to trot as a quadruped. When trotting, two legs are off the ground and two are on the ground. These legs are paired diagonally; that is, when the front left leg is off the ground, so is the right rear leg. At the same time, the front right and left rear legs are on the ground. But what about the upper and lower legs? How do they move in relation to each other? Dan Michaels (www.oricomtech.com) has been doing research on mammalian perambulation and has suggested the following pattern as a good trot. Since we are not dynamically balancing our robots, we must either not lift the legs very far and suffer a bit of a wobble on the diagonals, or we must use the big feet static balance methodology in our robots.

Figure 11-28 Quadruped limbs

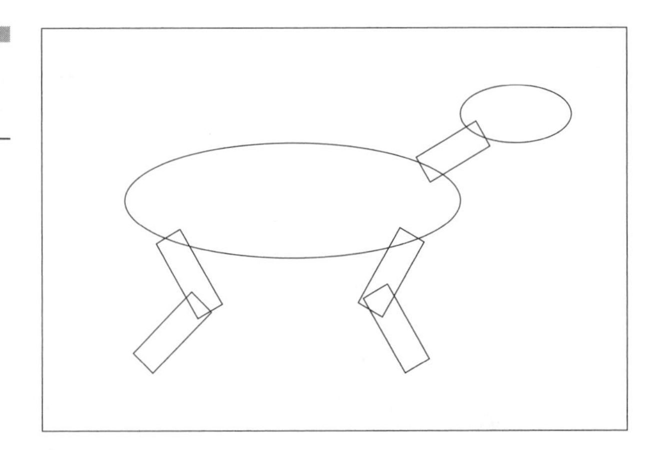

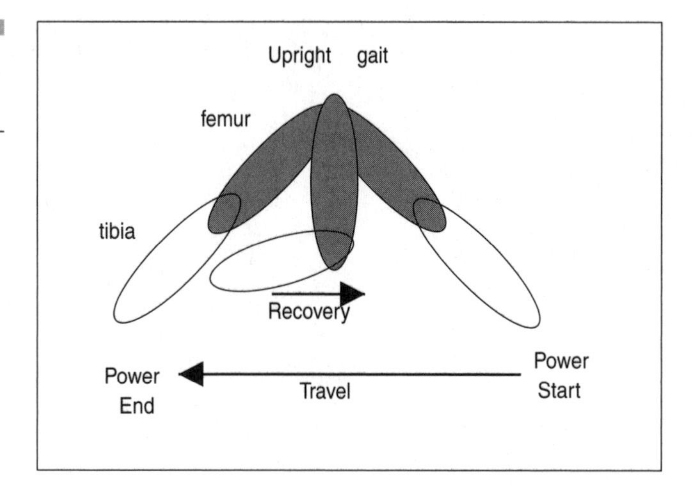

Figure 11-29 Upright walking

Each leg during its full movement phase using both the upper leg (femur) and lower leg (tibia) generally has a movement pattern as shown in Figure 11-29, which we'll call *upright* walking. When a leg is being used to propel the animal, it starts its step at the Power Start position and moves to Power End. The Recovery phase is used to bring the leg back to the Power Start position to begin another step.

The tough part of getting this walk to work is in the timing of the motions. If the body of the robot (or animal for that matter) is short, then the front leg must not complete its power phase before the rear leg finishes its recovery stroke on the same side or a leg collision will occur. This is yet another timing issue that needs to be addressed if your quadruped is to look graceful when in motion. In general terms, here is the gait pattern for a quadruped robot trot, assuming that we are starting with one diagonal set of legs already set to move:

1. Move one set of diagonal legs back at the speed you want them to move and the angle of the tibia and femur you want to use. These legs must stay in synchronization with each other.
2. At the same time, angle the tibia of the other diagonal pair of legs such that they clear the ground and move them forward in preparation to begin their next power phase.
3. When the legs in step 1 have reached the end of their power phase, extend the tibia of the legs referenced in step 2.
4. Go to step 1 and start again, this time reversing the diagonal pairs of legs being used to power and recover.

Figure 11-30 Nico the quadraped (Source: Dan Michaels and www.oricomtech.com)

Clearly, some adjustments on when to bend the tibia and how fast to move the legs will need to be taken. By changing the angle of the tibia in relationship to the femur, you can change the apparent *mood* of your robot. A greater angle will have the appearance of a stealthy or cautious movement, and a more upright angling may look more cheerful. Yes, we're talking body language here, but isn't that the point? To have our robots look and act more lifelike?

You can use this gait as the springboard to creating other useful or interesting gaits in your robot. Be creative; experiment with various leg orientations and synchronization techniques. Figure 11-30 shows Nico, Dan Michaels' experiment in quadruped motion. Yes, those are #2 pencils for legs.

With long enough legs, this quadruped is quite capable of handling imperfect terrain, inclines, and minor obstacles in its path (stepping over them).

A Six-Legged Insect

No discussion of legged robots is complete without a discussion of bugs. The arthropod is perhaps the most successful beast on the planet, so it must have something going for it. Most of the commercial walking robots for sale either completed or as kits are bugs (see Figure 11-31.) Perhaps this popularity is because the insect is a very stable platform for statically balanced robots and the gaits for an insect are the easiest to implement of any discussed so far.

Figure 11-31 A Lynxmotion® hexapod III kit

Figure 11-32 Lynxmotion Hexapod IIIS kit

Hexapods, the commonly used term in robotics to describe an insect-type body, may be implemented in a variety of fashions. The most complex use three DOF limbs as described in this chapter's leg construction section. More common variants like MIT's Genghis and the Lynxmotion® Hexapod IIIS kit (see Figure 11-32) use only two DOF legs successfully. Finally, for the definitive word in minimization, the three servo hexapod kits can be used where each leg has one DOF, but the front and back legs on either side are locked together and the middle two legs are locked but are reversed from each other as shown in Dennis' cricket robot in Figure 11-33.

All these robots use a form of walking called the *tripod* gait. The tripod gait gains static balance by making sure that three legs are always in contract with the ground at any given time. The pattern formed by the *grounded* limbs is a triangle with the robot suspended within. The three-servo hexapod has no choice in the matter; it will either have all its

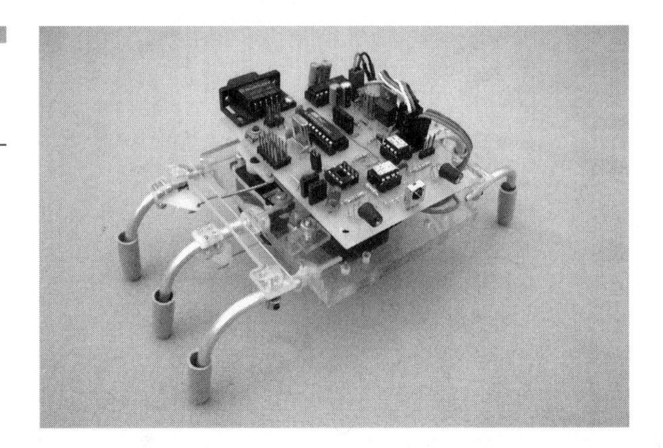

Figure 11-33 Cricket DIY hexapod

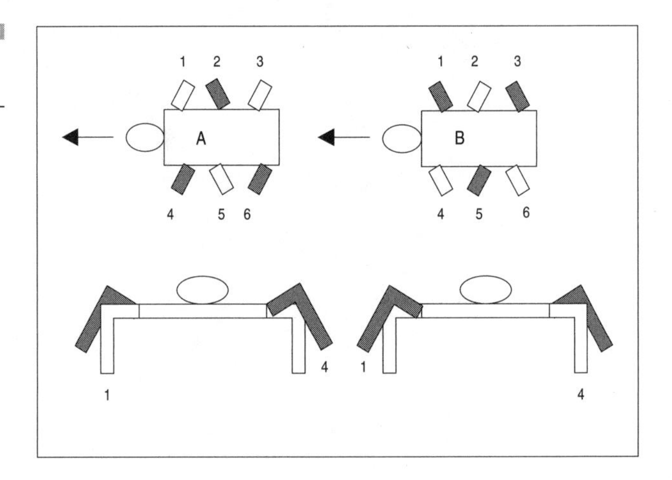

Figure 11-34 The tripod gait

legs on the ground or three legs. The other hexapods' programming dictates the form and pattern of the leg motion, but the most stable is the tripod gait. Let's use Figure 11-34 to describe the basic pattern for a tripod gait robot.

This is the gait timing for a basic tripod gait (refer to Figure 11-34):

1. Lift legs 2, 4, and 6 and move them forward, as shown in Figure 11-34(A).
2. While legs 2, 4, and 6 are moving forward, legs 1, 3, and 5 are on the ground. Move these legs towards the back of the robot.
3. Lift legs 1, 3, and 5 and move them forward, as shown in Figure 11-34(B).

4. While legs 1, 3, and 5 are moving forward, legs 2, 4, and 6 are on the ground. Move these legs towards the back of the robot.
5. Go to step 1.

This timing procedure will have your robot moving forward. To go in reverse, reverse the procedure. Instead of lifting the legs and moving them forward, lift and move them towards the rear. To turn, have one side go forward and the other in reverse (this will pivot the robot). To turn a more gradual rate, have one side's legs move less distance forward than the other. If you have three DOFs in each leg, you can do even more fancy steps and pose rearing up or squatting down, for instance. A two DOF or three DOF legged hexapod has the capability to deal with terrain that is not perfectly flat. It can raise its legs high enough to go over many obstacles. With enough sensors, it is quite capable of climbing inclines as well.

If you have built a three-servo hexapod, your choices are *much* more limited. Figure 11-35 shows the gait that you can use to move your robot. This gait is much the same as Figure 11-34, but the main difference is that the front and rear legs can only move forward and backward; the center legs can only move up and down. Your hexapod will tend to waddle because you are pivoting on the center leg that is on the ground when the front and rear legs on the other side are moving the robot forward (on that side).

The three-servo hexapod can do one other gait; we call it the crawl. We leave the discovery of this gait as an exercise for the student. The three-servo hexapod is limited to flat surfaces and would be highly challenged if it had to get over obstacles.

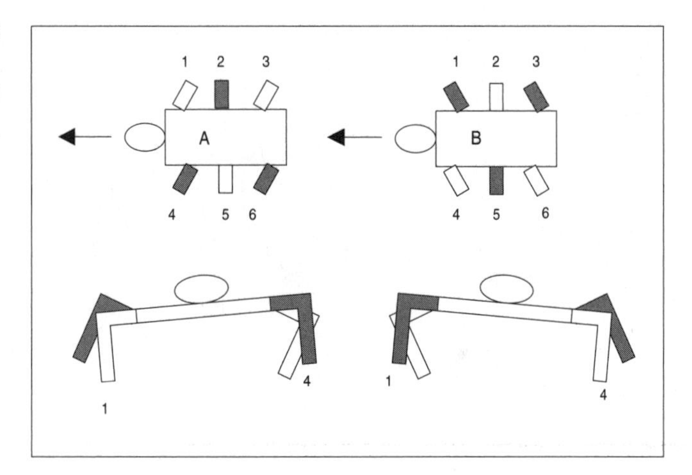

Figure 11-35 The three-servo hexapod gait

An Eight-Legged Spider with Synchronized Cams

Up until this point, all our walkers had legs constructed using hobby servos as the drive motors at each joint. This eight-legged walker uses two motors, one on each side of the robot to drive the four legs on each side. This walker uses a modified form of the tripod gait, one that has four legs on the ground at all times, two on each side. The basic static balancing proposition is exactly the same, however. We have seen several robots built upon this principle. The main advantage of this type of walking system is its simplicity; mechanically, it is very elegant. The main drawback is that, like the three servo hexapod, it is limited to flat or nearly flat environments.

Figure 11-36 shows a great toy made by WowWee® (www.wowwee.com) called the Cyber Spider™ that uses this type of walking design. We have modified this Cyber Spider to be fully autonomous. The original model was remote controlled via an *infrared* (IR) transmitter pendant. These models don't contain a lot of space to house more electronics, but you can always piggyback a controller board if you want or use it as a template to make your own walker using a similar methodology. You could also hack out the motor box of the Cyber Spider and build another robot entirely from just this and perhaps the legs. This is a cool toy. Get one and you won't be disappointed with it.

This two-motor drive design walks by timing the lifting and pushing of the legs with respect to each other on that side. No special effort is

Figure 11-36 Cyber Spider (Source: WowWee, R&D division of Hasbro Robotics)

made to synchronize the legs on the left side with the legs on the right side. The end result is that two legs, spaced every other leg, are on the ground on each side at any given moment in time. The legs are double jointed to create a form of universal joint near their point of contact with the leg cams, as you can see from Figure 11-36. This enables each leg to move up, down, forward, and back as the cam wheel rotates. Figure 11-37 shows how the timing is achieved and how the legs move.

Force/Tactile Feedback

Now we have our robots walking. To us, it seems fascinating and perhaps even spooky to see our walking robots move about. They seem more real. After about five minutes, even with bumper antennae out in front, we discover some very real problems with legged robots. They'll fall off the edge of the table or their legs will get snagged in things like table legs and shoes. Drat. Let's look at some simple ways to detect these annoying events.

The Generalized Contact Switch Circuit

In general, we're going to put simple contact switches in place to determine if a leg has touched down or if a leg has bumped into an obstacle

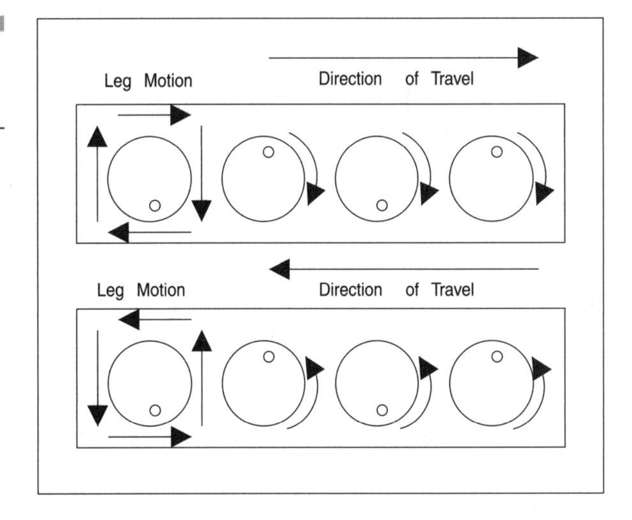

Figure 11-37 Cam-driven legs

that is restricting its movement. This type of a sensor can be used to great effect on any embedded processor you choose; it does not need an analog-to-digital converter to read its value. At the end of the chapter, we'll show you a way to use an analog sensor called a *bend sensor* to read how much a leg is being blocked.

Here in Figure 11-38 we have the basic contact switch circuit, workable with any type of microcontroller using any language compiler or assembler. The 330-ohm resistor is used as a current-limiting protective buffer in case of accidents. The switch can be made from just about anything from two pieces of bent wire to a micro switch.

How to Know When the Leg Has Touched Down

The basic question "Has my foot touched the ground?" can be answered in a variety of ways, with the simplest being a simple pushbutton. There are thousands of variations on this theme, but you really want a *Single Pole, Single Throw* (SPST) momentary contact switch. SPST means two wires are attached to the button. The button can be either *Normally Open* (NO) or *Normally Closed* (NC). Figure 11-39 shows a micro switch glued to the end of a robot leg. The wires are running up the inside of the tube and connect to the microcontroller's *input/output* (I/O) pin from there. You can place these buttons on every leg, but in reality the front legs are a must and the rear legs are quite useful in case our robot is backing off the end of a table.

All the momentary contact buttons we have used require significant force to depress them. This is both good and bad for us. You don't want the button pressed if your robot is on a surface that is giving way. On the

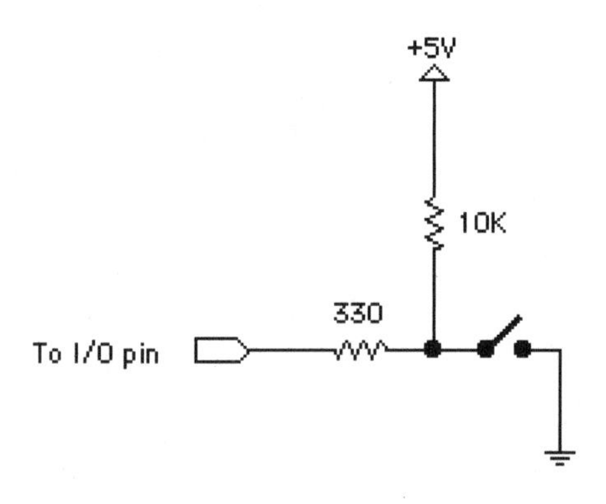

Figure 11-38 Contact switch circuit

Figure 11-39 Foot-down button mounting

other hand, you don't want a button that is so hard to press that your robot doesn't weigh enough to push it down. You'll have to test and see what works for your robot based on its weight and the force it exerts on the legs where the foot buttons will be used. These buttons can be slick, which is not a good idea on a walker that may be having traction issues already. We recommend that you glue a rubber "boot" on the bottom of the button, which is to say, on the actual button itself.

How to Know If a Leg Is Blocked

One of the main irritations of a legged robot in general, and a hexapod in specific, is how often the front legs get tangled up in a table leg or other hard-to-avoid obstacles. One of the ways to know that you are getting into trouble is to put sensors on the legs to see when this happens.

A simple way to do this useful task involves a micro switch and piano wire. Figure 11-40 is self-explanatory about this process. Our preferred glue for holding the piano wire to the micro switch is ShoeGoo™. Look for it at your local hobby shop or Google[1] for an online shop that carries it. You

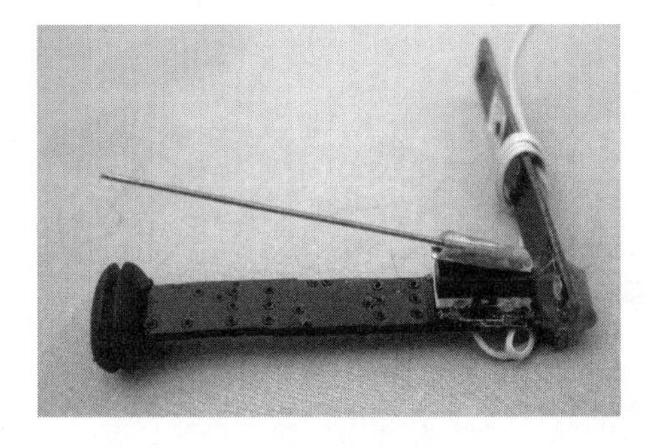

Figure 11-40 A leg-blocked sensor

[1]The joys of a living language! This means simply to use a search engine online.

can hold the switch to the leg with wire-ties or more ShoeGoo. Gosh, you can even bolt it on (as it was intended to be used) if you have the space.

Detecting a blocked leg can also be done in other perhaps more elegant ways. These include a current sensor, similar to those detailed in Chapter 7, "Motor Control 101, The Basics," on the power lines of the servos that control the motion to the legs in jeopardy, or even a limit switch that detects when a servo does not reach its expected end-of-throw position.

One final word must be made on this subject, because we've been dying to use the bend sensor *somewhere*. You could use a bend sensor arced in such a way that the amount of pressure could be read from it to determine the severity of the blockage. To use a bend sensor, your controller needs to be able to read an analog voltage. Use the circuit shown in Figure 11-41 to read the sensor, which will look something like Figure 11-42. You can get bend sensors from www.jameco.com or from eBay® in the form of the ancient Nintendo® Powerglove™, from which you can glean up to four bend sensors.

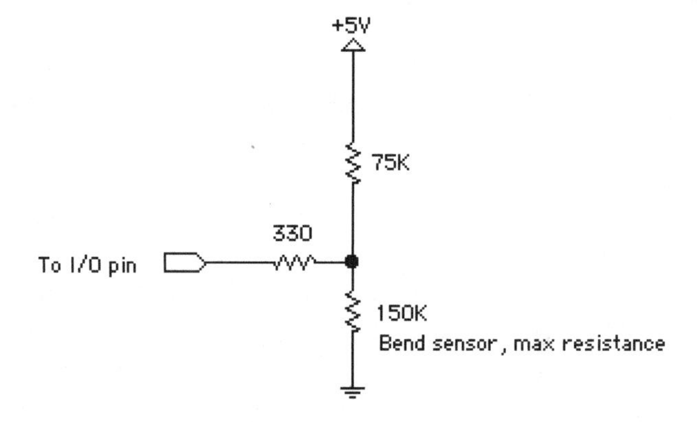

Figure 11-41
Bend sensor circuit

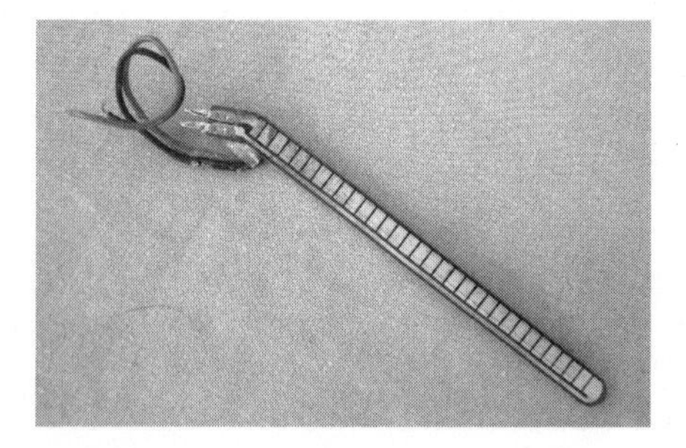

Figure 11-42
A bend sensor

APPENDIX 

Glossary

AC Alternating current. Current that periodically reverses polarity. A household current is AC.

amp, ampere A unit of measure for electrical current flow, defined as coulombs per second.

armature The spinning central component of the motor that holds the windings that generate the electromagnetic field. It also contains the commutator. See Chapters 2 and 3.

balance, dynamic The act of balancing that depends on either feedback or motion or both. See Chapter 11.

balance, static The act of balancing that does not depend upon feedback or motion. See Chapter 11.

bearing A device that reduces the friction of a shaft mounted in some fashion. See Chapters 3 and 10.

bipolar Relating to stepper motors, this is a stepper motor whose windings require current to flow in both directions for operation. See Chapter 5.

braking, dynamic A method of slowing or stopping a DC motor by removing the power and shorting the back EMF produced by the motor through a resistance. *See also* braking, regenerative.

braking, mechanical A method of slowing or stopping a motor using mechanical techniques. The brakes in your car are a good example of mechanical brakes.

braking, regenerative A method of slowing or stopping a DC motor by removing the power and routing the back EMF produced by the motor back to the battery. *See also* braking, dynamic.

brownout In electronics, a condition where the voltage of a circuit drops low enough to cause a circuit to malfunction. See Chapters 7 and 9.

buffer An electronic device that reduces or eliminates the electrical interaction between two distinct circuits. See Chapter 9.

bypass Electronically, this is the act of removing or reducing voltage spikes on the power line. See Chapters 5, 7, 8, and 9.

canstack Relating to stepper motors, a canstack stepper motor has two sets of windings that appear to be in two separate rings or cans that are stacked one on top of the other. See Chapter 5.

capacitor An electronic device that resists the change in voltage in a circuit. See Chapters 7 and 8.

CEMF Counter Electromotive Force. The current generated within an inductor when the applied current is removed and the electromagnetic field collapses back towards the windings. See Chapters 3, 5, 7, and 9.

clamp diode The diodes used to reduce voltage spikes across an inductor that are caused by CEMF. See Chapters 5 and 7.

closed loop A control circuit that relies upon feedback from a sensor to correct the results of its control. See Chapter 8.

CMOS Complementary Metal Oxide Semiconductor. A type of digital circuit characterized by a wider range of acceptable power lines, anywhere from 3.3V to 12V, whose logic low is very close to 0V and logic high is close to its power line voltage.

coefficient of friction The multiplier used with other force calculations to describe the force required to overcome the skidding or rolling friction a surface experiences when in contact with another surface. This is always in reference to two surface materials. See Chapter 2.

commutation The switching of current to the opposite direction in a motor so that the Lorenz Force keeps the motor spinning. See Chapters 1, 2, and 3.

commutator The part of the DC motor armature that contacts the brushes and switches the direction of current flow. *See also* commutation. See Chapter 3.

coupler A mechanical device that connects one shaft to another. Electronically, it is a means of connecting a signal from one circuit to another. See Chapters 3, 10, and 11.

DC Direct current. The current generated by batteries. DC current flows in only one direction.

DC brush motor This common DC motor uses brushes and a commutator to deliver current to the windings, causing the motor to rotate. See Chapters 2 and 3.

DIY Do it yourself.

duty cycle The percentage of "on" time in a single time period of a signal. On is defined as the nonzero voltage for the signal. See Chapters 2 and 7.

DVM Digital voltmeter. Electronic test device that measures voltage, current, resistance, and sometimes many other useful quantities.

dynamic torque The torque a motor delivers while it is in motion. See Chapter 5.

EMI Electromagnetic interference. This is the noise generated by electronics and motors that can be transmitted, without contact, to other electronic devices.

feedback The act of measuring the output of a circuit and making the result available to the input controls of a circuit. See Chapter 8.

frequency The number of cycles of current change that occur within a second. Frequency = 1/period.

gearhead motor A motor that has a gear reduction integrated with it. See Chapters 2 and 3.

Gray code A form of binary encoding where only one bit changes at a time as the encoder moves from state to state.

Hall Effect In electronics, a device that responds to changes in a magnetic field.

hysteresis A form of electronic memory that requires a greater change in voltage to change the state back than it did to change the state initially.

inductor An electronic device that resists a change in current. See Chapters 3 and 7.

left-hand rule With your left hand, point your fingers forward and your thumb up. Your fingers represent the direction of current flow, and your thumb the direction of magnetic flux (north to south), while your palm faces in the direction of the Lorenz Force. The Left-Hand Rule is also sometimes termed the Lorenz Force Law. See Chapters 1 and 3.

locked antiphase A form of PWM where both the magnitude and direction of a motor's rotation is encoded in the duty cycle of a waveform. See Chapter 7.

Lorenz Force The electromagnetic force exerted perpendicular to current flow in a magnetic field.

open loop A control circuit that does not benefit from or rely upon feedback to see what the results of its control is. See Chapter 8.

period The amount of time it takes for a cycle to complete once. Period = 1/frequency.

permanent magnet DC motor A motor using permanent (as opposed to electro-) magnets as stators. All nonstepper motors we work with in this book are PM motors.

PFM Pulse Frequency Modulation. A form of PWM where the frequency of a fixed size pulse is varied to create a specific current in a motor. It is not commonly used to control DC motors. See Chapter 7.

PID Proportional Integral Differential. In the context of this book, the act of using feedback to precisely control a motor's speed. See Chapter 8.

pillow block A simple device for mounting a shaft. Essentially a way to mount a shaft in an adjustable manner. See Chapter 10.

pitch Relating to gears, the number of teeth per unit measure. In the United States, pitch is measured in teeth per inch. See Chapter 3.

pole A magnetic orientation, either north or south. It can be either permanent magnet or electromagnet. See Chapters 2, 3, and 5.

power A unit of work. In the context of this book, it is the work generated by a motor. See Chapter 2. *See also* watt.

PPM Pulse Position Modulation. Similar to PWM, PPM has the information encoded on a signal by the width of the pulse, not the duty cycle. See Chapter 7. *See also* PWM and PFM.

PWM Pulse Width Modulation. The practice of changing the duty cycle of a waveform for the purposes of altering the current in a DC motor (in the context of this book). See Chapter 7. *See also* PFM and PPM.

quadrature encoding A means of encoding shaft speed and direction using a dual pulse train where each waveform is 90 degrees out of phase with respect to the other. See Chapter 8.

recirculating diode Diodes that provide a CEMF current with a path to ground through an H-Bridge circuit. See Chapter 7. *See also* clamp diode.

relay An electrically controlled switch.

resistor An electronic device that resists current flow. See Chapters 7 and 8.

robot You can supply your own definition. We're not going to restrain your imagination!

rotor In a stepper motor, this is the spinning central component that holds the permanent magnets that are attracted to the stator coils in the motor can. See Chapter 5.

RPM revolutions per minute. A unit of measure for the rotational velocity of any object. See Chapters 2, 4, and 5.

sag (voltage) The occurrence of a sudden but temporary drop in voltage on the power line.

Schmitt trigger An electronic circuit that uses hysteresis to reduce the affects of electrical noise in a circuit. See Chapter 8.

sign-magnitude A form of PWM where only the speed of the motor is encoded in the duty cycle of a waveform. See Chapter 7.

stator In a stepper motor, these are the windings in the can of the motor that are energized, causing the motor to advance in steps. See Chapter 5. In a PM DC motor, the stators are permanent magnets that provide the magnetic flux in which the armature spins.

tachometer A device that indicates the speed of a rotating object (such as a motor output shaft). Rotation speed may be indicated by the frequency of an output pulse train or by a varying voltage.

torque This is the angular force exerted at a given distance from the center of the motor shaft. See Chapter 2.

TTL Transistor-Transistor Logic. A type of digital logic characterized by its requirement of 5V power and a logic low equal to 0.7V and logic high equal to 2.4V.

unipolar Relating to stepper motors, this is a stepper motor whose windings require current to flow in only one direction for operation. See Chapter 5.

volt A unit of measure for electrical force.

watt A unit of measure for power. $P = VI$. See Chapter 2.

winding The name given to the coils that are wound on either the armature of a DC brushed motor or the stators of a stepper motor. See Chapters 3 and 5.

APPENDIX B

Tables, Formulae, and Constants

This is a selection of useful tables, formulae, and conversion constants that are consolidated in one place. Not all tables, formulae, and constants or conversions from the book are listed here.

Tables

Table B-1

Selection of AWG wire sizes and metric equivalents (solid wire)

AWG	Amps	Metric (mm)
10	30	2.59
12	20	2.06
14	15	1.63
16	10	1.3
18	5	1.02
20	3.3	0.813
22	2.1	0.635
24	1.3	0.508
26	0.8	0.406
28	0.5	0.330

Stepper Patterns for Unipolar Steppers

Table B-2

Wave (one phase at a time) pattern

	A	B	C	D
1	On			
2		On		
3			On	
4				On

Table B-3

Two-phase pattern

	A	B	C	D
1	On	On		
2		On	On	
3			On	On
4	On			On

Table B-4

Half-step pattern

	A	B	C	D
1	On			
2	On	On		
3		On		
4		On	On	
5			On	
6			On	On
7				On
8	On			On

Stepper Patterns for Bipolar Steppers

Table B-5

Wave (one phase at a time) pattern

	A_1	A_2	B_1	B_2
1	–	+		
2			+	–
3	+	–		
4			–	+

Table B-6 Two-phase pattern

	A_1	A_2	B_1	B_2
1	−	+	−	+
2	−	+	+	−
3	+	−	+	−
4	+	−	−	+

Table B-7 Half-step pattern

	A_1	A_2	B_1	B_2
1	−	+	−	+
2			−	+
3	+	−	−	+
4	+	−		
5	+	−	+	−
6			+	−
7	−	+	+	−
8	−	+		

Formulae

Ohm's Law and Derivatives

$$V = IR \text{ volts} = \text{Amps} \times \text{ohms}$$

$$I = \frac{V}{R} \text{ amps} = \text{Volts/ohms}$$

$$R = \frac{V}{I} \text{ ohms} = \text{Volts/amps}$$

$$P = VI \text{ watts} = \text{Volts} \times \text{amps}$$

General Electronics

$$Frequency = \frac{1}{Period}$$

$$Period = \frac{1}{Frequency}$$

Motor and Power Physics

$P_m = T\omega$ Mechanical power = Torque × angular velocity in watts.

$P_{max} = 1/4\ T_{max}\omega_{max}$ The formula for the maximum power a motor will deliver.

$F_w = mg\sin\theta$ The formula for weight or the force due to gravity.

$F_f = \mu mg\cos\theta$ The formula for friction, be it skidding, dynamic or rolling.

$$\mu_s = \frac{F_r}{F_n}$$ Use this to find your coefficient of

friction of your material. See Chapter 2.

$C = \pi D$ The circumference of a wheel.

$$v = \frac{d}{t}$$ Velocity is distance per unit time.

Constants and Conversions

Radians per second = RPM × 0.10472

Newton-meters (Nm) = Gram-centimeters (g-cm) × 0.0000981

Ounce-inches (oz-in) = Gram-centimeters (g-cm) × 0.0138874

Pounds (lbs) = Kilograms (kg) × 2.2046

Grams (g) = Ounces (oz) × 28.3494

Feet (ft) = Centimeters (cm) × 0.0328

Radians (rad) = Degrees (°) × 0.0147453

CHAPTER C

Resources

Here we list many books, web pages, articles, companies, and gurus that helped us write this book. These resources were invaluable to us and may be of great service to you as well.

Books and Magazines

These resources represent what we had in our library, not what you can buy right now. In many cases, newer book revisions are available.

Computational Principles of Mobile Robotics by Gregory Dudek and Michael Jenkin. Published by Cambridge University Press, 2000.

Contemporary College Physics by Edwin R. Jones and Richard L. Childers. Published by Addison-Wesley (2nd Edition), 1993.

Electrical Fundamentals by J.J. DeFrance. Published by Prentice Hall, Inc., Englewood Cliffs, NJ, 1969.

Mobile Robots: Inspiration to Implementation by Joseph L. Jones and Anita M. Flynn. Published by A.K. Peters, Wellesley, Massachusetts, 1993.

Nuts & Volts. Published by T&L Publications, Inc., Corona, CA. www.nutsvolts.com.

The Robot Builder's Bonanza by Gordon McComb. Published by McGraw-Hill, 2001. www.books.mcgraw-hill.com.

TTL Cookbook by Don Lancaster. Published by Howard W. Sams & Company, Indianapolis, Indiana, 1987.

. . . and, because Dennis *always* harps on people to read this book when they start into robotics . . .

Vehicles: Experiments in Synthetic Psychology by Valentino Braitenberg. Published by MIT Press, Cambridge, Massachusetts, 1984.

Web Sites and Groups

A lot of good information can be found out on the Internet. Here are many of the places where we found inspiration or information. Although this list is not exhaustive, it is a good place to start.

www.4qd.co.uk This is a manufacturer site, but they have lots of papers on H-bridges as well as good design tips and explanations.

www.allegro.com A massive collection of information about controlling and driving DC motors of all kinds.

www.ams2000.com Advanced Micro Systems have many white papers on stepper motor control.

www.cs.uiowa.edu/~jones/step A good place to find answers to questions about stepper motors.

www.dprg.com This club publishes some good articles on robot hardware and construction.

www.findchips.com Your first stop to finding parts and where to buy them in the United States.

www.google.com Know your tools. This should be the first place you look to find a datasheet or a definition.

www.joinme.net/robotwise A place to go for answers and sharing experiences with legged robots.

www.national.com Read their application notes; that's all I have to say.

www.oriocomtech.com This site has some good research on legged robots and biomimetics in general. There is good work done here.

www.physlink.com This is a free site that tries to answer your tough questions online.

www.school-for-champions.com On occasion, you find a great educational site. This is one of them — for the physics of motion.

www.seattlerobotics.org A club whose newsletter *The Encoder* is a must read for the amateur robotics engineer.

www.st.com SGS Thompson, another maker of H-bridge drivers, whose pages contain a wealth of information.

www.swampgas.com/robotics/rover.html Michael's web-operated roving robot — you can give him suggestions for the errata "in person".

www.techtoystoday.com The web site of your's truly Dennis Clark, which specializes in robotics source code, instructional programming hints, sensor projects you can do yourself, and a variety of hobby kits for robotics.

www.trcy.com The home web site for the Robotics Club of Yahoo. This site and its links to YahooGroups® can get you in touch with a lot of enthusiasts.

Specialty Robotic Component Suppliers

Although we don't claim that our list of suppliers is exhaustive, we think you'll find that it's reasonably complete. One word of caution, however: Some of these suppliers, particularly a few of the surplus electronics retailers, also carry mechanical parts and vice versa. We have noted this in the descriptions of the supplier, but in the interest of avoiding redundancy, we have chosen to avoid placing suppliers in more than one category.

Acroname, Inc. Supplier of both robot kits and a wide variety of sensors, motor controllers, and other accessories. Acroname sells many hard-to-find robotics supplies.

4894 Sterling Dr.
Boulder, CO 80301-2350
www.acroname.com

Arrick Robotics Supplier of robotics kits and accessories for both hobbyists and research. Lots of good information is available on this site.

P.O. Box 1574
Hurst, TX 76053
817-571-4528
www.robotics.com

BotParts This BattleBot-oriented business sells a variety of mechanical and electronic parts and assemblies.

Attn: Brock Schippers
139 Williamsburg West Ct.
Nashville, TN 37221
www.botparts.com

Cruel Robots As might be gleaned from the name, Cruel Robots specializes in BattleBot parts and accessories, with a particular emphasis on wheels and drive train parts.

32547 Shawn Drive
Warren, MI 48088
www.cruelrobots.com

Diverse Electronic Services This outfit sells high-power motor controllers as well as a variety of *radio control* (RC) devices and accessories.

1202 Gemini St.
Nanticoke, PA 18634-3306
570-735-5053
http://divelec.tripod.com/index.html

Future-Bot Future-Bot carries parts ranging from motors to *infrared* (IR) emitter-detector pairs to acrylic domes. They also carry a Motorola 68HC11-based microcontroller module, the P-Brain.

Future-Bot Components
203 Pennock Lane
Jupiter, FL 33458
561-575-1487
www.futurebots.com

HVW Technologies This Canadian company carries a complete line of sensors, microcontroller parts and kits, motor controllers, and much more. Very reasonable prices too.

HVW Technologies Inc.
3907—3A St. NE Unit 218
Calgary, Alberta T2E 6S7
403-730-8603
www.hvwtech.com

IFI Robotics IFI specializes in high-current motor control electronics and radio control systems.

9701 Wesley St.
Suite 203
Greenville, TX 75402
903-454-1978
www.ifirobotics.com

Kadtronix Purveyors of the prebuilt Workman Robotic Platform as well as a variety of drive systems and drive train assemblies.

info@kadtronix.com
321-757-9280
www.kadtronix.com

LEGO® Shop at Home We don't think we need to tell you what you can find here.

Shop at Home
555 Taylor Road
P.O. Box 1310
Enfield, CT 06083-1310
800-453-4652
www.legoshop.com

Lynxmotion Although primarily known for their robot kits, including various wheeled platforms and a variety of walker configurations, Lynxmotion also sells a wide variety of sensors as well as electronic and mechanical assemblies.

P.O. Box 818
Pekin, IL 61555-0818
866-512-1024
www.lynxmotion.com

Mondotronics (The Robot Store) Mondotronics sells a variety of robot kits, toys, books, parts, and assemblies—quite a stock.

4286 Redwood Hwy. PMB-N
San Rafael, CA 94903
800-374-5764
www.robotstore.com

Mr. Robot Mr. Robot distributes components and kits ranging from microcontroller boards to sensor assemblies to more exotic items, such as wireless cameras. They also sell complete kits, including the Talrik™ line of robot platforms.

4286 Redwood Hwy. PMB-N
San Rafael, CA 94903
800-374-5764
www.mrrobot.com

Pitsco LEGO Dacta Vendor of LEGO Technic™ components, motors, gears—you name it. Lots of other small robotics components, kits, and robot arms too.

P.O Box 1707
Pittsburg, KS 66762-1707
800-362-4308
www.pitsco-legodacta.com

RobotParts.org RobotParts.org provides custom machining services for the amateur robotics enthusiast. This company specializes in small, one-off part runs.

37 W. Wheelock St.
Suite D
Hanover, NH 03755
www.robotparts.org

Robot Power Robot Power sells the Open Source Motor Controller. This high-current motor controller is available both fully assembled and as a kit.

31808 8th Ave. S.
Roy, WA 98580
253-843-2504
www.robot-power.com

Robotic Power Solutions This company specializes in high-capacity battery packs and chargers, along with related accessories.

305 9th St.
Carrollton, KY 41008
502-639-0319
www.battlepack.com

SozBots SozBots provides wheels, drive train parts, and motor controller hardware for smaller robots. They also sell a controller board in the elusive 5-amp range.

340 N. Frederic St.
Burbank, CA 91505
www.sozbots.com

Zagros Robotics Zagros's ever-expanding product line includes motor controllers, MCU boards, sensors, and more. They specialize in a complete line of robotic platform kits, the Max™ Series.

P.O. Box 460342
St. Louis, MO 63146-7342
314-768-1328
www.zagrosrobotics.com

Mechanical: Motors, Hardware, and Materials

C and H Sales Company Offers a tremendous selection of surplus DC and AC motors. They are also a fine source of mechanical assemblies, such as bearings, shaft couplers, and sprocket and chains. C and H also does a brisk business in electronics surplus.

2176 E. Colorado Blvd.
Pasadena, CA 91107
626-796-2628
www.aaaim.com/CandH/index.htm

Grainger A large outlet that emphasizes machine tools and supplies. Although the emphasis is on industrial supply, Grainger also sells a wide variety of hardware and metal stock. They also maintain a substantial brick-and-mortar retail presence with several locations in North America.

www.grainger.com

Herbach and Rademan Another good source of electromechanical surplus, H&R carries a wide selection of small motors and mechanical hardware. They also stock a great deal of electronic surplus components, test materials, and books. H&R is a particularly good source for surplus batteries.

353 Crider Avenue
Moorestown, NJ 08057
404-346-7000
www.herbach.com

McMaster-Carr If you can name it, they stock it. A premier source for a variety of drive-train-related hardware, McMaster is also a great source for materials such as plastic and metal stock.

P.O. Box 740100
Atlanta, GA 30374-0100
404-346-7000
www.mcmaster-carr.com

MECI Stocks an array of surplus items, ranging from office supplies to electronic parts, and they can always be counted on to offer a reasonable selection of small DC motors.

340 E. First St.
Dayton, OH 45402
800-344-4465
www.meci.com

MetalMart.com Serving as the Internet retail outlet of Metal Express, Metal Mart sells metal stock cut to order.

Metal Express Headquarters
W229 N2464 Joseph Road
Waukesha, WI 53186
262-547-3606
www.metalmart.com

MSC Industrial Supply Co. Carries a wide variety of mechanical tools but also stocks lots of mechanical hardware, from nuts and bolts to belts and pulleys. Their catalog claims over half a million items.

75 Maxess Road
Melville, New York 11747-3151
800-645-7270
www.mscdirect.com

Servo City Very cool RC servo drive train components and supplies. Sells Futaba and Hitec servos and radio equipment.

20 Industrial Blvd.
Winfield, KS 67156
877-221-7071
www.servocity.com

Small Parts Inc. As the name implies, Small Parts stocks lots of small mechanical parts and assemblies. Small Parts also carries a wide assortment of tools and materials, such as metal and plastic stock.

13980 NW 58th Court
P.O. Box 4650
Miami Lakes, FL 33014-0650
1-800-220-4242
www.smallparts.com

Solarbotics Ltd. A good source for small robot motors and BEAM robot kits and supplies.

179 Harvest Glen Way NE
Calgary, Alberta T3K 4J4
403-232-6268
North America Toll-Free: 1-866-B-ROBOTS (1-866-276-2687)
www.solarbotics.com

Techmax Sells many small and medium-sized DC motors for reasonable prices. A unique source for motors and many other surplus items of delight, this is an online-only shop.

Bob@techmax.com
www.techmax.com

W.M. Berg, Inc. Anything mechanical you can think of, this company has.

499 Ocean Ave.
East Rockaway, NY 11518
800-232-BERG
www.wmberg.com

Electronics Supplies

All Electronics Sells surplus electronics components. You will also find a selection of small DC motors.

P.O. Box 567
Van Nuys, CA 91408-0567
1-888-826-5432
www.allelectronics.com

Alltronics Sells surplus electronics components. Their inventory usually includes a few DC motors suitable for robotics use.

P.O. Box 730
Morgan Hill, CA 95038-0730
408-847-0033
www.alltronics.com

American Science and Surplus This company offers lots of novelty items and toys, but some useful stuff too. They have recently added a robotics section to their online catalog worth having a look at.

P.O. Box 1030
Skokie, IL 60076
847-982-0870
www.sciplus.com

B.G. Micro Surplus shop for motor drivers and electronics.

555 N. 5th St., Suite 125
Garland, TX 75040
800-276-2206
www.bgmicro.com

Circuit Specialists Sells electronic components, kits, assemblies, tools, and test equipment. This outfit is a particularly good source for printed circuit board fabrication supplies.

220 South Country Club Drive #2
Mesa, AZ 85210
800-528-1417
www.web-tronics.com

DC Electronics Sells a variety of RF kits and components as well as a few microcontrollers and related accessories.

P.O. Box 3203
Scottsdale, AZ 85271-3203
800-467-7736
www.dckits.com

Digi-Key A major mail-order retailer of new electronic parts and equipment. If you need it, they probably stock it.

701 Brooks Avenue South
Thief River Falls, MN 56701
800-344-4539
www.digikey.com

Electronic Goldmine Sells all manner of surplus electronics parts and assemblies.

P.O. Box 5408
Scottsdale, AZ 85261
800-445-0697
www.goldmine-elec.com

FerretTronics Control Circuits

P.O. Box 89304
Tucson, AZ 85752-9304
FAX: 240-526-8985
www.ferrettronics.com

Hoffman Industries Sells surplus and overstocked components.

630-898-4029
www.hoffind.com

Jameco Electronics It was hard to determine where to place this one. You can buy electronics, motors, steppers, drivers, and robotic specialty hardware, all here.

Jameco Electronics
1355 Shoreway Road
Belmont, CA 94002-4100
800-831-4242
www.jameco.com

Marlin P. Jones and Associates (MPJA) Carries a substantial inventory of components, assemblies, and equipment, including a selection of DC motors.

P.O. Box 12685
Lake Park, Florida 33403
800-652-6733
www.mpja.com

Mouser Electronics Deals in new electronic parts and equipment, and carries an inventory in excess of 150,000 products.

1000 North Main Street
Mansfield, TX 76063-1514
800-346-6873
www.mouser.com

Newark Electronics A major retailer of new electronic components and equipment. Newark's stock is, to say the least, extensive—you'll need a friend to help you lift the catalog. Here is where you'll find just about any Allegro Micro, Inc. part.

4801 N. Ravenswood
Chicago, Illinois 60640
800-639-2758
www.newark.com

Surplus Sales of Nebraska Carries a variety of components, with an emphasis on military surplus and obsolete parts.

1502 Jones Street
Omaha, NE 68102
402-346-4750
www.surplussales.com

Hobby Supplies

Balsa Products
22 Jansen Avenue
Iselin, NJ 08830
Phone: 732-634-6131
FAX: 732-634-2777
www.balsapr.com

Dave Brown Products Inc., International Hobbycraft
3340 Dundee Road
Northbrook, Illinois 60062-2311
Phone: (847) 564-9945

Tower Hobbies A major web retailer of components for RC enthusiasts. It is a great source for servos, wheels, and so on.

P.O. Box 9078
Champaign, IL 61826-9078
800-637-4989
www.towerhobbies.com

Miscellaneous

US Digital Manufactures and sells precision rotary encoders. They are willing to sell small quantities to the general public.

11100 NE 34th Circle
Vancouver, WA 98682
800-736-0194
www.usdigital.com

INDEX

Symbols

A

B

C

E

F

G

H

I

J–L

N

O

P

Q–R

S

U

V

W–Z